Studies in Systems, Decision and Control

Volume 637

The series "Studies in Systems, Decision and Control" (SSDC) covers both new developments and advances, as well as the state of the art, in the various areas of broadly perceived systems, decision making and control–quickly, up to date and with a high quality. The intent is to cover the theory, applications, and perspectives on the state of the art and future developments relevant to systems, decision making, control, complex processes and related areas, as embedded in the fields of engineering, computer science, physics, economics, social and life sciences, as well as the paradigms and methodologies behind them. The series contains monographs, textbooks, lecture notes and edited volumes in systems, decision making and control spanning the areas of Cyber-Physical Systems, Autonomous Systems, Sensor Networks, Control Systems, Energy Systems, Automotive Systems, Biological Systems, Vehicular Networking and Connected Vehicles, Aerospace Systems, Automation, Manufacturing, Smart Grids, Nonlinear Systems, Power Systems, Robotics, Social Systems, Economic Systems and other. Of particular value to both the contributors and the readership are the short publication timeframe and the world-wide distribution and exposure which enable both a wide and rapid dissemination of research output.

Indexed by SCOPUS, DBLP, WTI Frankfurt eG, zbMATH, SCImago.

All books published in the series are submitted for consideration in Web of Science.

Arun Lal Srivastav · Ashutosh Kumar Dubey ·
Abhishek Kumar · Mariya Ouaissa
Editors

Climate Resilience: Impact of Quantum Computing and Artificial Intelligence on Urban Planning

Springer

Editors
Arun Lal Srivastav
Department of Applied Sciences
School of Engineering and Technology
Chitkara University
Solan, Himachal Pradesh, India

Ashutosh Kumar Dubey
Department of Computer Science
and Engineering
School of Engineering and Technology
Chitkara University
Solan, Himachal Pradesh, India

Abhishek Kumar
Chandigarh University
Sahibzada Ajit Singh Nagar, Chandigarh,
India

Mariya Ouaissa
Networks and Cybersecurity
Cadi Ayyad University
Marrakesh, Morocco

ISSN 2198-4182 ISSN 2198-4190 (electronic)
Studies in Systems, Decision and Control
ISBN 978-3-032-06790-6 ISBN 978-3-032-06791-3 (eBook)
https://doi.org/10.1007/978-3-032-06791-3

This Springer imprint is published by the registered company Springer Nature Switzerland AG
The registered company address is: Gewerbestrasse 11, 6330 Cham, Switzerland

Contents

Climate Science Education: A Tool for Better Future

Tran Thi Viet Ha, Nguyen Thuy Duong, Nguyen Minh Viet, and Do Trung Hieu

Abstract Recently, climate change has threatened to undermine the progress that all humans have focused on to achieve the Millennium Development Goals (MDGs). Climate change also poses one of the most serious challenges to reducing global poverty in the world. Climate science education can act as a powerful tool to equip individuals and communities with the suitable knowledge and necessary skills for mitigating and adapting to climate change impacts. This chapter is going to explore the role of climate science education in fostering informed global citizenship, enhancing adaptive capacities, and inspiring action. It provides an overview of current global efforts, successful models from Europe, the USA, and Asia, and innovative educational approaches such as eco-social systems thinking, art-based learning, and phenomenon-based collaborative teaching. Although there are various challenges in climate science education, such as scientific complexity, misinformation, and limited educator training, climate science education still remains important in building a resilient and sustainable future. The chapter calls for transformative pedagogical strategies and stronger collaboration between scientists, educators, and learners to empower future generations to act effectively on climate issues.

Keywords Education · Active learning · Empowerment-based education · Future · Climate resilience

---

T. Thi Viet Ha · N. Thuy Duong
Faculty of Advanced Technology and Engineering, VNU Vietnam Japan University, Hanoi, Vietnam

N. Minh Viet (✉)
VNU Key Laboratory of Advanced Material for Green Growth, VNU University of Science, Hanoi, Vietnam
e-mail: nminhviet86@gmail.com

D. Trung Hieu
Faculty of Geography, VNU, VNU University of Science, Hanoi, Vietnam

A. L. Srivastav et al. (eds.), *Climate Resilience: Impact of Quantum Computing and Artificial Intelligence on Urban Planning*, Studies in Systems, Decision and Control 637, https://doi.org/10.1007/978-3-032-06791-3_1

1 Power of Climate Science Education

- ***Why do we need climate science education?***

The latest Intergovernmental Panel on Climate Change (IPCC) report once again confirms that greenhouse gas emissions from human activities are the leading cause of global warming [28]. The atmospheric temperature has reached 1.5 °C above pre-industrial levels in 2024 [13, 38]. This is the limit on temperature increase that countries have agreed to pursue efforts to keep by the 2015 Paris Agreement [47]. Global warming has dramatically changed global climate conditions, increasing the intensity, frequency, and extent of extreme weather events such as heat waves, storms, floods, droughts, and forest fires [28]. Severe heat waves were recorded and warned about worldwide in the third week of August 2023, including in Europe, where the climate is temperate. The World Meteorological Organization reported that 2024 was the hottest year, with unprecedented heat waves affecting countries such as Japan, Australia, and Iran [38].

Although countries, especially the European Union, have implemented emission reduction policies, these efforts are not enough to slow down the rate of global temperature increase. All solutions mentioned in the Paris Agreement 2015, including increasing climate change education, should be implemented by all countries. We must accept the reality and find ways to adapt to the impacts caused by climate change. Less understanding of the impacts of climate change within current urbanization trends could lead to severe consequences for future human society and development [41]. To prepare future generations with the understanding and skills to adapt to the impacts of climate change, climate science education (CSE) is essential.

- ***Climate science education and climate change mitigation***

Collected data and calculations on total global CO_2 emissions show that global CO_2 emissions have been increasing continuously at a rapid rate since 1950 (Fig. 1) [8]. The average greenhouse gas emission per capita in 2023 is about 6.59 tons (Fig. 1). Climate models have shown that the global average temperature will increase by more than 4 °C above the pre-industrial level if the rate of CO_2 emission remains at the current one [28]. To have the best chance of avoiding a 2 °C rise in global temperatures, the average global carbon footprint per year must drop to below 3 tons by 2050 if the population can be reduced to 7 billion. To limit global warming to 1.5 °C, the world needs to reach net-zero emissions by around 2050. This requires countries to implement deep and rapid emissions reductions across all sectors [21]. The results of monitoring CO_2 concentrations in the atmosphere show that the COVID-19 pandemic has affected CO_2 concentrations in the atmosphere due to restrictions on people's travel, proving that many different solutions need to be implemented to achieve this target, but individual contributions also significantly contribute to the effort [29].

Climate science education is the basis for understanding climate change's causes and impacts, enabling informed decisions, and moving people from awareness to action by fostering concern and perceived efficacy [34, 49]. Public engagement with

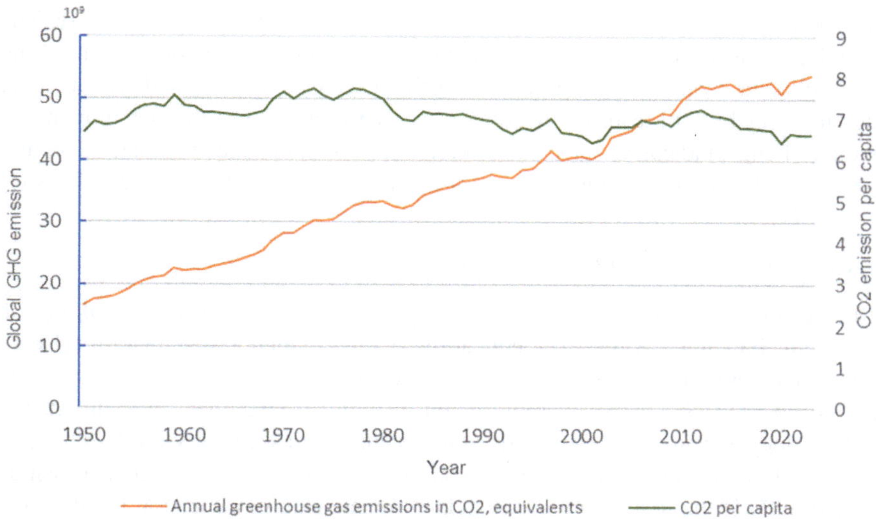

Fig. 1 Global Greenhouse Gas Emissions and CO_2 Emissions Per Capita (1950–2023). *Data source* OurWorldinData

climate change is closely linked to levels of understanding and perceived relevance. Although each individual's contribution to reducing global CO_2 emissions is insignificant, billions of individuals globally will reduce a considerable amount of CO_2 emissions into the atmosphere. Therefore, CSE and raising public awareness about changing lifestyles will significantly reduce global CO_2 emissions.

- ***Climate science education and climate change adaptation***

CSE is a critical component of adaptive capacity [2]. UNESCO analyses the considerable steps required in virtually all education systems in the world (both developing and developed education systems) to prepare young people for the difficult decades ahead (). Education empowers individuals, including workers, youth, children, and women, to create resilient solutions to climate change [5]. Learning about climate change can also provide opportunities for students to develop their knowledge and a range of inquiry, problem-solving, and critical-thinking skills, which is very important to face climate issues [1, 21].

2 Climate Science Education Status

Although sustainable development goals (SDG 13 and SDG 14) focus on awareness and capacity for climate change mitigation, adaptation, and impact reduction, and an approach to education that ensures that all learners acquire the knowledge and skills needed to promote sustainable development, concerns about CSE have been raised in many areas [1, 49, 51]. However, according to [50], only half of the national curricula in the world have a reference to climate change, with low priority. The organization also mentioned that less than 40% of teachers feel confident teaching about climate change, with many lacking the necessary training and resources. A global survey by [48] revealed that 80% of people want schools to teach more about climate change, indicating substantial public support for enhanced climate education [48]. CSE showed more highlights in European countries and the United States [1, 27, 36, 44].

- ***Climate science education in Europe***

Climate education across Europe is evolving, with notable advancements alongside persistent challenges. As of 2021, education for environmental sustainability (EES), which includes CSE, is well represented in education strategies and policy documents across the EU27 but not well represented in national strategies. In addition, the EES strategy appears most frequently at the primary and secondary levels. The EES practice at the university level is enhanced by EU-funded projects such as Erasmus + [10]. Italy is the first country to mandate climate change education in all public schools, dedicating nearly one hour per week to the subject in 2019. Climate change is taught under the courses of environmental, biological, and historical sciences, and is also a part of civic education programs in secondary schools in Italy [12, 20]. According to the European Commission (2024), the topic of sustainability was introduced in all curricula in all EU education systems. However, the inclusion of sustainability competence is unequal among the EU education systems. The subjects are included in the curricula as separate ones, often optional, in only 6 EU countries (Cyprus, Hungary, Romania, Slovenia, Spain, and Sweden). In the other curriculum, the knowledge is included in an interdisciplinary way.

- ***Climate science education in the USA***

The Next Generation Science Standards have improved climate education in the USA, but implementation varies by state, and some regions still lag behind. In addition, Earth Science is given lower priority than other subjects in public high schools [39]. Mandatory CSE is rarely found in the curriculum at the college level [18, 52]. There is a lack of integration of climate change issues in teaching about topics in engineering, leading to misconceptions about CSE persisting across middle school, high school, and higher education [32].

By conducting surveys from graduate students in different training programs from universities in different areas of the USA, Kuster and Fox point out that the introduction of CSE has shown in earth sciences programs or sustainability, or environmental policy/management programs, and some states in the Southeast [27]. The introduction of climate science education has been shown to have lower priority than the others. The other survey showed that about 51% (46 of 90) of public health programs in the USA did not offer any climate change-related course [3].

- ***Climate science education in Asia***

CSE in Asia is growing in recognition and scope, but it still faces equity, resources, and implementation gaps. Some countries have begun integrating climate science education (CSE) into national curricula [14, 43, 55]. However, in lower-income or conflict-affected nations (e.g., Afghanistan, Myanmar), integration is still minimal or non-existent due to limited resources and competing educational priorities. Over 60% of countries in the Asia–Pacific region had no national policy mandating climate education in schools [5, 50]. A significant issue across many Asian countries is the lack of teacher preparedness. Many educators are not formally trained in climate science, leading to low confidence in delivering the content. NGOs such as UNESCO, USAID's LEAF, and ADB offer teacher training and provide teaching materials and open-source curricula [5, 37, 38].

In Vietnam, with the support of the British Council from 2012 to 2014, climate science education was taught in the primary and secondary school curriculum. With support from NGOs such as the Australian Aid and the British Council, the Vietnam Ministry of Education and Training (MOET) developed teaching materials and manuals and organized national training workshops for climate science education [19, 53]. In addition, climate science education was also included in the 2017 General Education Curriculum, especially in subjects like Natural Science, Geography, and Civic Education. Climate science-related courses have also been added to the curriculum of training programs of many universities. For example, Earth Science was introduced as an elective ordinary course for all programs in all the member universities of the Vietnam National University.

3 Methods Used in Climate Science Education

This section will introduce practical methods and applications in the classroom and community for climate science education, including (i) an eco-social approach to education, (ii) art-based learning, and (iii) phenomena-based collaborative learning. The impacts of different climate education methods are analyzed as we review and summarize the practical implementation and the diverse strategies used recently. The success of climate change education is recognized in the ability to transfer the theoretical scientific knowledge into real, practical engagement and empower learners to contribute to a sustainable future. Thus, moving outside the textbooks and engaging in meaningful activities is necessary to empower future generations. Climate science education can range from laboratory experiments to community-based projects that foster a more profound understanding and inspire actual action.

- ***Traditional Science Class Approach***

Recently, climate science education has been most commonly delivered in traditional classrooms, presenting information and theory to learners. The traditional classroom often relies heavily on passive learning, where students absorb information through lectures, textbooks, and standardized assessments. This approach is the most efficient way to provide basic and core knowledge in the field of climate science, such as the greenhouse effect, carbon cycles, and climate modeling. However, its limitations are recognized over time in fostering deep engagement, critical thinking, and action-oriented behavior, which is increasingly apparent. Besides, there is a lack of information on multiple climate change dimensions, including social, economic, and political dimensions, in these traditional science classes. Furthermore, the complexity of scientific data and future prediction can make the learners feel distant and irrelevant to their immediate lives, thus hindering the development of a sense of urgency and personal responsibility.

Research done by Monroe et al. reviewed 959 unique citation records addressing climate change education [33]. Among them, only 49 studies met the criteria for inclusion, including 28 primary or secondary schools and 11 colleges/universities from different countries and regions (the United States, India, Germany, South Asia, Turkey, Greece, Canada, Sweden, Switzerland, England, and Malaysia). The authors also mentioned several teaching methods implemented with diverse and engaging activities recently, however, the complexity and politicization of climate change present significant educational challenges, which were also highlighted. Because the research subjects of the article are diverse in age and level, many students have limited climate science knowledge, and misconceptions are widespread. Simply delivering factual information (traditional lecturing) was less effective than approaches that promote critical thinking and problem-solving skills. Based on the analysis and discussion, the authors call for a shift from purely informational teaching toward more interactive, context-based, and action-oriented learning experiences.

The weaknesses of the traditional science class approach become even more apparent when climate education is delivered to non-major students. For example, a Computer science or Language major student might find the knowledge on atmospheric chemistry or the complexities of climate modeling unrelated and disconnected from their primary major and interests. They might still feel that the foundational knowledge remains valuable, but the delivery method of traditional classes often fails to resonate with their learning styles and motivations. A theory-heavy format and assessments focused on scientific and technological facts can make the students feel like an imposition rather than an opportunity for their future development. This can lead to a superficial understanding of the material, a lack of sustained interest, and ultimately, a failure to internalize the broader implications of climate change for their future lives and chosen fields/majors. To address this problem, it is proposed that educators clearly define the goals of climate science education for different types of learners. The goal of education is not to turn every learner into a scientist with a deep understanding of climate science but to foster a generation of climate-informed citizens across all aspects of their lives who recognize that addressing this complex challenge requires diverse perspectives and collaborative action by many stakeholders.

- ***The Ecological-Social Linked System Approach***

The basis of this method is the deep and inseparable connections between ecological and social systems in the context of climate change. Typically, when educating people about climate science, they will immediately think of purely scientific and technological understanding. However, with the ecological-social linked system approach, the social, economic, political, and cultural factors that promote and are affected by the climate crisis will also be analyzed and clarified. Based on that, this approach aims to empower learners to become the change factors for a just society and a sustainable future. In modern societies with flexible, creative, adaptable, well-informed, and inventive communities, the ultimate goal of human growth is fully developed humanity, which highlights altruism, dedication, and ties to other people and society. As a result, the ecological-social linked system approach has recently been practical in climate science education.

It is well known that the term "sustainability" means fulfilling current generations' needs without compromising future generations' needs while ensuring a balance between economic growth, environmental care, and social well-being. If natural resources, ecosystem services, and human beings are the only instruments for market growth, our debt to future generations will grow and grow. So, the balance between environmental, social, and economic sectors is always the solution to our debt to the next generation. In today's educational science, teachers try to guide students toward autonomy/initiative with the educational philosophy mostly related to learner-centered conceptions, valuing their own opinions with little regard for the ecological realities undergirding them. However, a clarification is needed here, as autonomy can sometimes lead to the misunderstanding that autonomy means prioritizing humans and viewing nature primarily as a resource for humans to exploit. Pulkki et al. recognized that this anthropocentrism had led humans to see themselves as superior to nature and separate [22]. Eco-social autonomy is an attempt to redress the balance by contextualizing autonomy. Hence, it incorporates the idea of self-control while considering humankind's impact on our surrounding ecosystems. The research also formulated eco-social autonomy as an extension of relational autonomy, based mainly on ecological, eco-social, and ecofeminist ideas. Whereas the individualist idea of autonomy suggests that a human being owes nothing to society, eco-social autonomy acknowledges the need to cultivate aspects of self-sufficiency that combine reason, emotional maturity, and will. In another research study by the same author, an eco-social philosophy of education based on the idea of an ecological self was also developed, concentrating on cultivating an epistemic humility grounded in our ecological embeddedness in the world [23].

- ***Art-Based Learning Approach***

Art has had an important role in human societal and cultural change. Art can work as a powerful tool for expressing human hopes, challenging injustice, inspiring movements for freedom and equality, and also, somehow, drive social transformation. The art-based learning approach in climate education is a potential strategy to let learners more deeply face the effects of climate change and handle the difficulties in climate science education. Unlike scientific knowledge, art is about subjective knowing, which is more emotional and human. The important concepts of climate science education, such as sustainable development or a sustainable future, can be re-embedded into the world and the practice of living by art-based learning [35]. Arts offer unique forms of meaning-making that traditional scientific methods often overlook [15].

A study by Davis et al. emphasized that integrating the arts into environmental education fosters deeper ecological literacy by connecting intellectual understanding with emotional experience. Learners engaging with art-based projects show higher retention of scientific concepts, more excellent critical thinking skills, and increased motivation for sustainable action [11]. Art allows learners to process complex climate data through facts and figures, personal reflection, storytelling, and creative expression. The application of the art-based learning approach is not limited to classrooms and lecture halls but can also fulfill its role for the general public. For example, the Cape Farewell Project highlights humanity's destructive impact on the earth and inspires people to act towards a sustainable cultural shift by creating art founded on scientific facts (Capefarewell) [7]. This art project collected and published the stories that move us, if we are to be inspired to make social and behavioral changes in our lives. Recent activities of Cape Farewell include a world tour (2019–2020) of the Cape Farewell-commissioned Pollution Pods, an immersive installation by artist Michael Pinsky made up of five geodesic domes emulating polluted environments in cities globally. This was experienced by over 30,000 visitors (including Greta Thunberg) in London, WHO-Geneva, Bremerhaven, Vancouver, Manchester, Dorset, Melbourne, UN-New York, and COP25 in Madrid. Another example of an art-based learning approach is visual arts, performance, and public installation activities in the National Sustainability Festival—Australia's largest and longest-running sustainability event, inspiring solutions to the sustainability and climate emergency (Climarte) [9]. These activities can foster public engagement with sustainability issues, and it also can show the public how creative expression can inspire environmental responsibility. The art-based learning approach offers a powerful complement to traditional education methods. It revitalizes climate science education by embedding sustainability into learners' lived experiences, cultivating informed citizens and empathetic and creative changemakers ready to address environmental challenges.

- ***Phenomenon-Based, Collaborative Learning Approach***

According to the definition by Times Higher Education, in phenomenon-based learning, learning is framed around and stimulated by events or occurrences (phenomena) [46]. Ever present in the learning experience, the selected phenomenon is the basis for students as they progress through a cyclical inquiry process of engaging, exploring, expanding, and executing. This approach reflects the interconnected nature of the real world, making learning more meaningful and engaging for students [25]. Simply speaking, phenomenon-based learning in science and engineering is an educational approach that uses real-world phenomena or complex problems as the basis for teaching and learning. The basis of this kind of education is that the students can learn about science by looking at the events around them daily. Phenomenon-based collaborative learning can empower the students' attitude because they can ask questions, conduct experiments, and propose solutions together by themselves. This approach aligns with the main goals of climate education, which can foster interdisciplinary thinking, problem-solving skills, and the necessary qualities to address the complexity of the climate crisis.

Phenomenon-based learning is a profoundly collaborative learning approach involving students working together across disciplines, drawing on diverse perspectives and knowledge to explore and understand real-world phenomena. In the context of climate science, collaborative projects like group research on local climate impacts, co-designing sustainability initiatives, or simulations of international climate negotiations have proven especially effective. For example, Sterman et al. present an interactive role-play simulation to educate participants about climate change and international negotiations [45]. In this study, participants in the classes acted as delegates from different countries and negotiated carbon emissions targets to limit global warming. The results of the study pointed out that the simulation activities can significantly improve participants' understanding of climate change issues, the difficulties in achieving international agreements, and the urgency of decisive policy action. Data showed that participants were more likely to support aggressive mitigation policies after the simulation and believe that effective action is possible. The authors conclude that role-play simulations like World Climate enhance climate literacy and motivate engagement. In other research, Elkalaf et al. emphasized the impact of student engagement activities on future climate change adaptation [16]. Based on the results of the research, it was shown that student-led initiatives, such as advocacy campaigns and environmental clubs, foster leadership skills and a sense of responsibility among young people. These activities can lead to real actions and raise awareness about critical climate problems. More importantly, it can inspire students to influence the communities in which they live and contribute to broader efforts toward sustainability. Engaging students in innovative climate-focused approaches, such as simulations, Model UN conferences, and collaborative projects, equips them with essential skills like critical thinking, negotiation, and leadership. These experiences empower students to amplify their voices and engage with policymakers, ensuring that youth perspectives are integrated into climate discussions and decision-making processes.

4 Challenges, Perspectives, and Future Directions

> Our world possesses the knowledge and the resources to achieve the MDGs… Our challenge today is to agree on an action agenda to achieve the MDGs.
>
> –U.N. Secretary-General Ban Ki-moon-

There is no doubt that education is key in empowering peoples to make the right choices and develop behaviors that reorient society toward enhancing resilience and sustainable practices. However, climate science education faces several significant challenges.

The first challenge is about the complexity of climate science. There is no doubt that the science of climate change is highly complex. It deals with a broad range of scientific disciplines, including not only natural science, such as physics, chemistry, biology, environmental science, and geography, but also the economy, policies, art, global trade, cultural impacts, ethical responsibilities, etc. Besides, natural cycles and anthropogenic activities can bring unpredictable changes every second. Understanding topics requires basic scientific knowledge and the ability to connect the concepts, thus making it challenging to create a clear, unified picture. This complexity can be overwhelming for both educators and learners. Without a strong interdisciplinary teaching approach that connects the different fields and relates to daily life experiences, climate science education can become distant, technical, or irrelevant to learners, and the learners may disengage or even distrust the information if they cannot see its coherence or significance in their own lives.

The second challenge that should be mentioned here is the current major misconceptions about climate science. When information in approved textbooks is usually dated, students typically look to educators and scientists as the most trusted sources of information [31]. If these sources are inadequate, they often turn to media reports on the Internet through social network threads and AI platforms like ChatGPT, Gemini, Deepseek, etc., leading to significant misconceptions in climate science. These students also contribute to the views and values within the households and community in which they live; such misconceptions can affect the delivery or add to confusion among the general public about issues such as climate change. Several studies have specifically targeted the challenge of overcoming misconceptions about climate change. One of the most common misconceptions highlighted is the confusion between climate change and the ozone hole [4, 26].

Scientists and educators face several challenges when they deliver knowledge and communicate with learners about climate science. Many educators/teachers may not have received specialized training in climate science, leading to gaps in understanding or confidence when addressing complex topics [6, 31, 40]. In general, teachers tended to maintain the integrity of their subjects rather than be involved in extensive interdisciplinary teaching [17], which is very important in climate science education. Additionally, the politicization of climate change in some regions creates a sensitive environment where educators might fear backlash from parents, school boards, or communities. Teachers must navigate these sensitive challenges in a careful manner, finding ways to present accurate scientific information while encouraging open-minded discussion. It is a matter of fact that not all universities are fully prepared to address climate change in their curricula [54]. The lack of updated teaching materials and necessary support from functional deparment can hinder their ability to integrate climate education meaningfully into the broader curriculum [6, 54].

On the other hand, difficulties from the learners' side also need to be considered. Students may struggle with the nature of climate science, particularly when scientific concepts are not directly linked to their daily lives and their trained majors [42, 54]. Moreover, the scale and seriousness of climate change can provoke feelings of fear, helplessness, or denial, making it emotionally challenging for learners to engage fully [24, 30]. Exposure to conflicting information or misinformation through media and social networks can further lower the understanding and then create doubts. Without supportive learning environments that validate students' emotions and provide clear pathways for action, sustaining their interest and motivation to learn about and respond to climate change can be challenging.

To effectively fight the urgent challenges of climate change, it is crucial to strengthen the knowledge base, information sharing, and networking in climate science education. Climate science education must be rooted in a strong scientific understanding of socio-ecological systems while addressing human behavior's ethical dimensions. Improving the communication process through stronger collaboration between scientists and educators is also very important to reach the goal. Scientists should be encouraged to engage more proactively with lay audiences and students, presenting information that aligns with how people process and understand new knowledge. At the same time, educators must take responsibility for staying current with credible scientific information and teaching students how to critically evaluate sources, especially in an age where misinformation about climate change is widespread online. One important method is teaching students to question the credibility of the information they encounter. These modern and active teaching and learning approaches foster knowledge acquisition, critical inquiry, and reflective judgment. Beyond simply transferring knowledge, climate science education must aim to empower young people. Education for empowerment involves helping students develop a holistic understanding of climate issues, including scientific, ethical, political, and social dimensions. Importantly, the hopeful visions for the future need to be inspired, and reinforce the idea that students themselves have the power to influence outcomes. Empowerment-based education encourages students to articulate their concerns, discuss potential solutions, and ultimately take action on specific climate problems. This approach need to move beyond passive learning to active engagement, fostering an informed and motivated generation to shape a sustainable future. By improving the flow of accurate information and building dynamic networks between scientists, educators, and students, climate education can be more resilient, responsive, and effective in preparing future generations for informed climate action.

5 Conclusions

Climate science education is no longer an option; it is a necessity. As the climate crisis intensifies, education must evolve to deliver scientific facts, inspire engagement, cultivate critical thinking, and foster action. This chapter has shown that effective climate education goes beyond traditional classroom models, embracing interdisciplinary, emotionally resonant, and student-centered approaches. However, various challenges—ranging from scientific complexity to institutional inertia—require systemic changes in educational policies, teacher training, and curriculum development. Empowering educators and learners through active learning methods, cross-sector collaboration, and inclusive educational environments is key to shaping climate-literate, resilient, and proactive societies. A well-informed and engaged youth is our best hope for a just and sustainable future facing global climate uncertainty.

Acknowledgements The authors would like to acknowledge the support of VNU Vietnam Japan University for providing resources and an enabling environment that made this work possible.

References

1. Academies, A.E.: A snapshot of climate change education initiatives in Europe: some initial findings and implications for future climate change education research. In: Cliona Murphy, G.M.S., Aksela, M., Bergman, G., Jones, M., Léna, P., Wilgenbus, D. (eds.) (2020)
2. Anderson, A.: Climate change education for mitigation and adaptation. J. Educ. Sustain. Dev. **6** (2012)
3. Arora, M., Comrie, A.C., Ernst, K.E.: Assessing climate and health curriculum in graduate public health education in the United States. Front. Public Health **11**, 1124379 (2023)
4. Baker, J., Loxton, J., Sherren, K.: Using art elicitation to deliver and evaluate a grade 4 climate change instructional module. Appl. Environ. Educ. Commun. **12**, 130–142 (2013)
5. Bank, A.D.: Climate change and Education Playbook: Investing in Education and Skills for Climate Resilience in Asia and the Pacific. Asian Development Bank, Manila, Philippines (2024)
6. Beach, R.: Addressing the challenges of preparing teachers to teach about the climate crisis. Teacher Educ. **58** (2023)
7. Capefarewell. Available: https://www.capefarewell.com/. Accessed
8. Global Carbon Budget.: P. B. O. V. S. W. M. P. B. O. W. I. D. 2024. CO_2 and Greenhouse Gas Emissions (2024)
9. Climarte. Available: https://sustainabilityfestival.au/organisation/climarte/. Accessed
10. Commission, E.: Education for environmental sustainability: policies and approaches in European Union Member States: Final Report, LU, Publications Office (2021)
11. Davis, S.: The engagement tree: arts-based pedagogies for environmental learning. Int. J. Educ. Arts **19** (2018)
12. De Pascale, F.: Geographical education and climate change perception in secondary school: a case study in Southern Italy. Sustainability **15**, 3255 (2023)
13. (ECMWF), E. C. F. M.-R. W. F.: The 2024 Annual Climate Summary—Global Climate Highlights 2024 (2025)
14. Education, R. O. K. M. O. 2015. National Curriculum Framework of Early Childhood Education,. *In:* SEJONG CITY, R. O. K. (ed.).
15. Eisner, E.W.: The Arts and The Creation of Mind. Yale University Press (2002)
16. Elkalaf, B.M.: The impact of student engagement activities on future climate change adaptation: the case of student simulation models. Future J. Soc. Sci. **3** (2024)
17. Gayford, C.: Controversial environmental issues: a case study for the professional development of science teachers. Int. J. Sci. Educ. **24**, 1191–1200 (2002)
18. Hess, D.C., Brandi, M.: Climate change and higher education: assessing factors that affect curriculum requirements. J. Clean. Prod. **170**, 1451–1458 (2018)
19. Hiển, N.T.: Giáo dục biến đổi khí hậu trong Trường phổ thông qua kinh nghiệm Nhật Bản, Trung Quốc và Singapore. Tạp chí Giáo dục **3**, 216–221 (2019)
20. Indelicato, A., Rocco, R.M.P.D., Martín, J.C., Marinello, V.: Bridging attitudes and evidence: climate change perspectives in Italy. Challenges **16**, 2 (2024)
21. IPCC.: IPCC Sixth Assessment Report Impacts, Adaptation and Vulnerability (2022)
22. Jani Pulkki, S.K.: Ecosocial autonomy as an educational ideal. Relations Beyond Anthropocentrism **10** (2022)
23. Jani Pulkki, J.V., Mullen, J.: Ecosocial philosophy of education: ecologizing the opinionated self. Stud. Philos. Educ. **40**, 347–364 (2021)

24. Jones, A., Buntting, C., Hipkins, R., Mckim, A., Conner, L., Saunders, K.: Developing students' futures thinking in science education. Res. Sci. Educ. **42**, 687–708 (2012)
25. Jongyung, T.: Approaches and techniques of phenomenon-based learning: PheBL. J. Educ. Learn. Rev. **2** (2025)
26. Karpudewan, M., Roth, W.-M., Chandrakesan, K.: Remediating misconception on climate change among secondary school students in Malaysia. Environ. Educ. Res. **21**, 631–648 (2015)
27. Kuster, E.F., Garey, A.: Current state of climate education in natural and social sciences in the USA. Clim. Change **141**, 613–626 (2017)
28. Lee, H., Romero, J.: Climate change 2023 synthesis report. Intergovernmental Panel on Climate Change (2023)
29. Liu, Z., Ciais, P., Deng, Z., Lei, R., Davis, S.J., Feng, S., Zheng, B., Cui, D., Dou, X., Zhu, B., Guo, R., Ke, P., Sun, T., Lu, C., He, P., Wang, Y., Yue, X., Wang, Y., Lei, Y., Zhou, H., Cai, Z., Wu, Y., Guo, R., Han, T., Xue, J., Boucher, O., Boucher, E., Chevallier, F., Tanaka, K., Wei, Y., Zhong, H., Kang, C., Zhang, N., Chen, B., Xi, F., Liu, M., Bréon, F.-M., Lu, Y., Zhang, Q., Guan, D., Gong, P., Kammen, D.M., He, K., Schellnhuber, H.J.: Near-real-time monitoring of global CO_2 emissions reveals the effects of the COVID-19 pandemic. Nat. Commun. **11**, 5172 (2020)
30. Mary, G.V., Taylor, J., Therrien, W., Hand, B.: Science education for students with special needs. Stud. Sci. Educ. **48** (2012)
31. Mcbean, G., Hengeveld, H.G.: Communicating the science of climate change: a mutual challenge for scientists and educators. Can. J. Environ. Educ. **5** (2000)
32. Milovanovic, J., Shealy, T., Godwin, A.: Senior engineering students in the USA carry misconceptions about climate change: implications for engineering education. J. Clean. Prod. **345**, 131129 (2022)
33. Monroe, M.C., Plate, R.R., Oxarart, A., Bowers, A., Chaves, W.A.: Identifying effective climate change education strategies: a systematic review of the research. Environ. Educ. Res. **25**, 791–812 (2019)
34. Moser, S.C., Dilling, L., Kates, R.W., Bostrom, A., Lashof, D., Leiserowitz, A., Ungar, S., Dunwoody, S., Pratt, L.G., Rabkin, S., Agyeman, J., Doppelt, B., Lynn, K., Hatic, H., Mcneeley, S., Huntington, O., Bingham, S., Warner, L., Cole, N., Watrous, S., Mccright, A.M., Regan, K., Chess, C., Johnson, B.B., Tribbia, J., Michaelis, L., Grotzer, T., Lincoln, R., Bateson, M.C., Gershon, D., James, K., Smith, A., Arroyo, V., Preston, B., Atcheson, J., Farhar, B., Young, A., Fraley, N., Tennis, A., Duvair, P., Sadler, S., Usibelli, A., Anderson, S., Meyer, D.S., Averill, M., Jamieson, D., Harriss, R.: Creating a Climate for Change: Communicating Climate Change and Facilitating Social Change. Cambridge University Press (2007)
35. Natalia Eernstman, A.E.J.W.: Locative meaning-making: an arts based approach to learning for sustainable development. Sustainability **5**, 1645–1660 (2013)
36. Nepraš, K., Strejčková, T., Kroufek, R.: Climate change education in primary and lower secondary education: systematic review results. Sustainability **14**, 14913 (2022)
37. Nizam, M.S.M.: Climate change education for Asia Pacific Small Island Developing States, UNESCO (2016)
38. Organization, W.M.: State of the Global Climate 2024 (2025)
39. Plutzer, E., Branch, G., Townley, A.L.: Climate change education in U.S. middle schools: changes over five pivotal years. J. Microbiol. Biol. Educ. **25**, e00015-24 (2024)
40. Pruneau, D., Gravel, H., Bourque, W., Langis, J.: Experimentation with a socio-constructivist process for climate change education. Environ. Educ. Res. **9** (2003)
41. Senevirathne, M., Amaratunga, D., Haigh, R., Kumer, D., Kaklauskas, A.: A common framework for MOOC curricular development in climate change education—findings and adaptations under the BECK project for higher education institutions in Europe and Asia. Progress Disaster Sci. **14**, 100222 (2022)
42. Sharma, A.: Global climate change: what has science education got to do with it? Sci. Educ. **21**, 33–53 (2012)
43. Shashidhara, L.S.: Integrating climate change education across the curriculum. Resonance **24**, 1025–1028 (2019)

44. Sorensen, C., Magalhães, D., Hamacher, N., Sullivan, J.K., Weinstein, H.N.W., Pinho-Gomes, A.-C., Biberman, D., Donaldson, H., Gómez-Duarte, I., Middleton, J., Magaña, L., Urbina, M., Kaseje, M., Cascante-Flores, N., Surenthirakumaran, R., Ivers, R., Sáenz, R., Chen, T.T.-W., Lopez, W., Romanello, M., Zhang, Y.: Climate and health education in public health schools worldwide during 2023–24: a survey. Lancet Planet. Health **8**, e1010–e1019 (2024)
45. Sterman, J., Franck, T., Fiddaman, T., Jones, A., Mccauley, S., Rice, P., Sawin, E., Siegel, L., Rooney-Varga, J.N.: World climate: a role-play simulation of climate negotiations. Simul. Gaming **46** (2014)
46. Sue, L., Cuthbert, K.: Phenomenon-based learning: what, why and how [Online] (2023). Available: https://www.timeshighereducation.com/campus/phenomenonbased-learning-what-why-and-how. Accessed
47. Nations, U.: Paris Agreement [Online] (2015). Available: https://unfccc.int/sites/default/files/english_paris_agreement.pdf. Accessed
48. UNDP.: People' climate vote 2024 result (2024)
49. UNESCO.: Country progress on climate change education, training and public awareness: an analysis of country submissions under the United Nations Framework Convention on Climate Change (2019)
50. UNESCO.: Getting every school climate-ready: how countries are integrating climate change issues in education. UNESCO (2021)
51. UNESCO.: Education for People and the Planet; Creating Sustainable Futures for All. Global Education Monitoring Report (2016)
52. Vallée, M.: How and why US universities fail to impart environmental literacy to all students. Int. J. Sustain. High. Educ. **25**, 60–77 (2024)
53. Vietnam Ministry Of Education, Training, L.A.L., Australian Agency For International Development (Ausaid).: Tài liệu hướng dẫn dạy và học về ứng phó với biến đổi khí hâu [Online] (2012). Available: https://livelearn.org/assets/media/docs/resources/ClimateChange_Edu_teachers_viet.pdf. Accessed
54. Walter, L.F., Sima, M., Sharifi, A., Luetz, J.M., Salvia, A.L., Mifsud, M., Olooto, F.M., Djekic, I., Anholon, R., Rampasso, I., Donkor, F.K., Dinis, M.A.P., Klavins, M., Finnveden, G., Chari, M.M., Molthan-Hill, P., Mifsud, A., Sen, S.K., Lokupitiya, E.: Handling climate change education at universities: an overview. Environ. Sci. Europe **33** (2021)
55. Wilderspin, I.F., Nguyen, D.C., Be, H.Y.: Final Report: Climate Resilient Education Landscape Study in Vietnam. Save the Children International, Hanoi (2024)

Climate Change and Resilience Using a Sustainable Development Perspective

Wasswa Shafik

Abstract The problem of climate change cannot be underestimated, as it is one of the pressing concerns of our lifetime. This chapter demonstrates how Climate change has attracted considerable interest at both the global and national levels in recent years. The observed presence of climatic change has prompted researchers to employ models, primarily econometric approaches, to forecast the impact of climatic change on economic growth, particularly in developing countries. Despite numerous empirical studies that estimate the impacts of climate change on economic growth, knowledge gaps remain regarding the characteristics of these estimated impacts. Critically, most estimations depend heavily on the characteristics of the data and the specific regions in which they are conducted. These studies, in relation to country-specific levels, mainly focus on macroeconomic measures, while microeconomic estimates connected to both climate change damages and economic resilience are in short supply. Despite the problems associated with estimating the economic impacts of climate change, several factors contribute to or influence these impacts. In an ideal world, normal economic systems are efficient, environmentally protective, resilient, and can fully support sustainable development outcomes, regardless of their formal states. Resilience is a crucial concept in the study of various natural and socio-economic systems. To economists, resilience means the contribution of economics and economic systems to the conservation and preservation of human, built, and natural capital.

Keywords Biodiversity conservation · Carbon footprint · Climate adaptation · Community resilience · Disaster risk reduction

---

W. Shafik (✉)
Dig Connectivity Research Laboratory (DCRLab), Kampala, Uganda
e-mail: wasswashafik@ieee.org

School of Digital Science, Universiti Brunei Darussalam, Gadong, Brunei

A. L. Srivastav et al. (eds.), *Climate Resilience: Impact of Quantum Computing and Artificial Intelligence on Urban Planning*, Studies in Systems, Decision and Control 637, https://doi.org/10.1007/978-3-032-06791-3_2

1 Introduction

Climate change indicates a lasting transformation of climate patterns. Human activity over the past half-century has likely altered the atmospheric concentration of these greenhouse gases. It presents numerous hazardous risks, including an increase in global mean temperature, a rise in sea levels due to the thermal expansion of water and the melting of ice, and an increase in severe weather events [35]. Over the last two decades, we have struggled with economic development, which has resulted in increased production and consumption aimed at meeting basic requirements and generating profits for shareholders. However, production generates waste, and concentrated waste contributes to global environmental problems, which in turn exacerbate climate change issues [10]. The term "sustainable development" emerged when the environment, climate change, and profit were discussed in conjunction. The concept is formally defined as follows: "Sustainable development meets the needs of the present without compromising the ability of future generations to meet their own needs" [42].

In recent years, the scientific literature has increasingly focused on climate change as a major strategic threat and a competitive disadvantage for companies that do not respond by increasing their resilience efficiently. Therefore, we investigate climate change from the perspective of corporate sustainability. In summary, it is crucial to investigate the potential impacts and responses, in terms of strategies and policies addressing the risks, on climate change for sustainable organizations. This paper analyzes climate change resilience, focusing on the impact on European SMEs—a magnification of the perceptions of the European business community [4]. Resilience is analyzed using a framework including four dimensions: social, economic, institutional, and infrastructural. Moreover, the paper includes an in-depth analysis of the preferred insurance solutions for climate risk among the SME community in Europe. The conclusions are based on responses from European Economic Area SMEs [27]. The paper, therefore, provides evidence on the current state of SMEs regarding climate change and resilience.

Climate change, defined as a statistically significant variation in either the mean state of the climate or in its variability, persisting for an extended period (typically decades or longer), has given rise to various ideas and strategies. Researchers have used the terms adaptive capacity, vulnerability, and sustainability interchangeably with climate resilience [23]. We will, however, use a different terminology. For us, the idea of resilience will mostly correspond to some sustainability. The problem is that today, we have a very wide range of interpretations of what sustainability is. For example, ecologists have presented a simple idea according to which natural systems have a good degree of resilience because harmony with nature poses fewer challenges to the accumulation of problems than does the constant struggle against an inherently resistant system [11]. A healthy market economy can generally be extremely efficient and stable as long as it remains within expected and known parameters. In addition to delivering stability and expected socio-economic benefits, a market economy also provides the necessary flexibility and resilience to adjust and recover in the face of major shocks [29]. Resilient systems have been widely studied across various disciplines due to their ability to quickly adapt and recover after being subjected to an economic or ecological shock. In the context of climate change, enhancing the resilience of socioeconomic systems becomes increasingly important for policymakers in preparing for and responding to extreme events. The concept of the resilience economy turns focus onto how economic systems function when they are functioning as they should. Resilience can, however, also be a weakness of a system [36, 38].

This type of argument inevitably leads to the conclusion that sustainability is about accepting the existing situation, regardless of how unpleasant it may be. While valid for this chapter, given our challenge of limited progress, the essay acknowledges that such interpretations are not in line with the principles governing international sustainable development efforts. Several textually different definitions of sustainability, produced by the principal governing processes, are presented in Annex II. Despite the somewhat circular nature of those definitions, in particular, in the case of the concept of intergenerational links, some elements recur [1]. In particular, it is almost always stated that sustainability must be viewed as a policy goal that is independent of a fixed stock of specific assets or resources. It should be the explicit objective of policy to maintain the flow of valued goods and services to and from all peoples over time. The goal of sustainable development is the maintenance of the provision of benefits of different types of assets, such as natural resources and physical and human capital. To facilitate the understanding of our later discussion, we present selected examples of these definitions that show some of the underlying principles of sustainable development relevant to reducing the future of the currently poor and increasing the benefits of the yet undiscovered technologies [2].

1.1 Climate Change Impacts on Sustainable Development Goals

The 2030 Agenda for Sustainable Development and its Sustainable Development Goals (SDGs) provide a global policy framework for addressing the challenges faced by present and future generations with the imperative of “leaving no one behind.” The 17 SDGs encompass economic, social, and environmental development objectives, offering a comprehensive and transformative vision of sustainable development. However, such a comprehensive agenda does not fully reflect the urgency of acting promptly against the impacts of climate change [26]. The UN Global Climate Challenge presents this challenge in stark figures. This publication highlights that greenhouse gas emissions continue to rise, extreme weather events are increasing, and millions of people rely on global action to motivate and support efforts to mitigate and adapt to these adverse changes, ultimately making modern societies more resilient. However, the 17 SDGs are closely connected to climate change and its impacts. Various chapters in this book explore these interlinked relationships. Here,

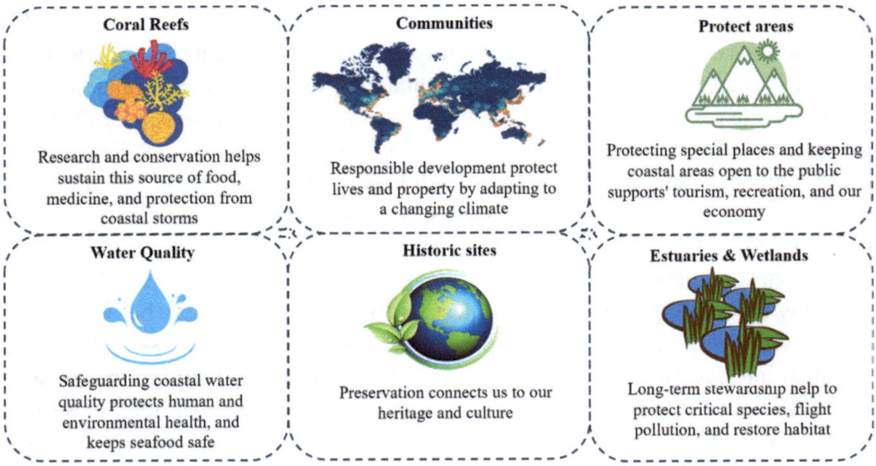

Fig. 1 Significant areas for ensuring climate protection

we summarize those connections as they are understood today and provide comprehensive resources for more detailed analysis in the hope of inspiring more action at the local level, raising ambition, and improving data collection for sustainability indicators on the path to securing future generations [25, 33]. We recognize from the outset those trade-offs within the SDGs and in climate change sanctuaries. Climate change is a preeminent threat to sustainable development, and understanding the various ways in which climate change and human pressures impact sustainable development is crucial for developing effective policies to address these challenges [12]. This understanding should also address equity issues at various levels, including income inequality, climate injustice, and the marginalization of individuals and communities due to inadequate support. However, achieving substantial progress on these critical dimensions within such a comprehensive and universally recognized agenda, the 17 SDGs and 169 Targets may be blurred through protecting the climate and its surroundings, as illustrated in Fig. 1 [5, 28].

1.2 Interlinkages Between Climate Change, Resilience, and Sustainable Development

The concept has been further expanded to include three dimensions, often referred to as the 3Es: economic, ecological, and social. The economic dimension ensures human well-being through equity and efficiency. The ecological dimension tightly links human activities and associated natural resources. Development should not erode natural capital or exceed the Earth's regenerative capacity. Finally, the social dimension of sustainable development includes poverty reduction and ensuring social justice and security for all people through equitable access to the benefits of both social and economic development [23, 33]. Although these principles are often cited universally, there is frequently a disconnect between theory and practice. Many projects continue to be implemented without considering these core principles. This is especially true of large global problems, such as those attributed to climate change. Such challenges typically require a global solution,however, there is no consensus on the means of achieving it. National interests collide with global concerns, making the realization of truly sustainable development a challenge each time a crisis emerges and often a lost opportunity. Such challenges often require constant attention and an evidence-based approach to achieve a compromise and a globally harmonized response [5, 31]. This chapter aims to initiate a conversation on climate change, utilizing the concept of sustainable development as a lens.

Projected climate change is expected to impact the achievement of sustainable development, particularly in vulnerable developing and least developed countries. However, despite the considerable attention paid to the impact of climate change on mitigation and adaptation, the linkages between climate change, resilience, and the achievement of sustainable development, pertinent to the formulation and implementation of policy and practice, remain underexplored. In the context of the 2030 Agenda for Sustainable Development, this chapter examines the interlinkages between climate change, resilience, and sustainable development [25, 32]. This chapter reviews and discusses key concepts and dimensions of climate change, resilience, and sustainable development, and critically examines the interlinkages between climate change, resilience dimensions, and the 2030 Agenda. Based on an in-depth examination of evidence from research, policy, and practice in both developed and developing countries and a meta-analysis of climate resilience initiatives, we propose a framework that explicitly articulates the pathways and ubiquity of these interlinkages as an analytical lens to understand better and identify the commonalities and differences in the various trajectories toward climate change resilience necessary to achieve sustainable development objectives [14, 40]. Finally, this chapter presents future research needs and possibilities arising from the proposed analytical lens designed to better understand and build climate resilience in the context of the 2030 Agenda.

2 The Science of Climate Change

The scientific foundations of climate change phenomena can be found in the natural sciences. Earth's climate is influenced by a combination of drivers, including solar energy, which shape interactions between different atmospheric systems. Any changes to these drivers, normally over periods longer than 30 years, can disrupt the complex natural balance of climate that has evolved over millions of years. The impacts of climate change on natural ecosystems and human activities have been documented worldwide [3]. Earth's climate has undergone physical changes, with average surface temperatures rising and biogeophysical shifts across various aspects of the atmosphere, hydrosphere, and lithosphere. These changing characteristics of climate are thought to be largely due to human activities. Some researchers have estimated that more than 95% of Earth's natural systems are being affected by localized climate changes, albeit at different levels of severity. In all regions, major contributors to climate change are alterations in how land is used, how energy is sourced and used in human activities, and how materials are consumed and disposed of [17]. Carbon and greenhouse gas emissions from energy production and land use change, in particular, are releasing stored carbon back into the atmosphere, exacerbating the greenhouse effect and the associated warming feedback.

Rates of global and regional orphan or proxy disasters have a close relationship to trends of local and regional populations projected or expected to be forced to flee their homes, lands, and communities. The number of people affected by such disasters has grown rapidly, and evidence shows that economic losses, both absolute and relative to the regional economy, have grown faster than the population in all significant regions of the world [9]. Social and human development, social justice, democratic freedoms, security, and human rights could all be put at risk if this situation continues to deteriorate. Besides the data and evidence, predictions about future climatic events must be acknowledged from a responsible and ethical perspective. These suggest that responsibility becomes even more pronounced beyond a 2 °C rise, where adverse catastrophic impacts would be expected. In such a changed climate system, there is significant consensus from additional research scientists that it will become impossible to adapt the world out of risk [29, 43]. It is for this reason that climate dynamics is a science-oriented discipline, and when assessing alternative approaches to problem-solving, particular attention is given to informed and scientifically derived responses.

2.1 Causes and Impacts

Climate change is fundamentally a problem of excessive carbon in the atmosphere, which acts like a blanket around the Earth, trapping and retaining heat. This, in turn, increases the average temperature on Earth, a phenomenon referred to as the greenhouse effect [30]. The increase in greenhouse gases is primarily due to human activities, mainly the burning of fossil fuels such as coal, oil, and natural gas. However, other factors, including deforestation and land-use changes, also contribute to this increase. The changes are not accounted for as changes in emissions measured, but as emissions from activities beyond the forest and agriculture. The causes of human-induced climate change are crucial in order to formulate strategies to reduce such effects [37, 42]. Although different climate scientists attribute various causes of global warming, it has been noted that recent studies have found that human activities can play a substantial role in climate change. Furthermore, it has also been added that scientists are convinced that the accumulation of greenhouse gases from human economic activities is the main cause of global warming and climate change, as there are complex relationships between these gases [7]. Some of the impacts of climate change are related to national issues, while others require international cooperation, as the full range of impacts on local economies, communities, and individuals is much greater. The local, national, and global factors causing such economic and adverse well-being include not only the economy and social factors, e.g., biophysical features, population, employment, and residence infrastructure, and economic activity, but also a broad range of resources, e.g., natural, financial, and human ones [6, 8].

2.2 *The Role of Sustainable Development in Addressing Climate Change*

Sustainable development is not achievable without the convergence of social, economic, and environmental resilience. Resilience, such as that to climate change, is evident in social relations that can be upheld in challenging circumstances, an economy that generates sustainable livelihoods, and environmental health and renewal. Advancing a sustainable trajectory in each of these dimensions requires a holistic approach that encompasses a systems approach and considers the integration of social, human, and financial capital with human production, natural profit, and one-planet living [42]. When progressives and radical scientists say we need system change, it is because we need to balance all these considerations, which, when out of balance, negatively affect performance unmatched since World War II. Insights included the need for commitments at global, national, and local levels, supported by partnerships, innovation, and data, exploration of local successes that enable a circular and shared economy,and commitment to sustainable behaviors at every level. The framers of the original SDGs, as well as any entity that echoes their vision and wisdom, recognize that human-made systems must align with foundational precepts. These include ecological integrity, the values of self-renewability and circularity, acknowledging that growth cannot be unending and that planetary boundaries exist [4]. Such policy pronouncements then integrate fundamental resilience to climate variability into everything we do through sustainability approaches, as illustrated in Fig. 2.

Fig. 2 Sustainability

2.3 Sustainable Principles and Goals

Principles of sustainable development in the context of climate change. It is stated that the emerging regime changes and improves the viability and efficiency of our production and consumption systems, and over the longer term, will help to ensure greater investment security. The approach adopted is consistent with these objectives. In specific climate change terms, the principles involved in change—resilience and adaptability can be understood as sustainability [24]. This aligns with statements in many countries' national climate change adaptation strategies, such as protecting and conserving ecosystems, the objective is increasingly to promote their sustainability. The climate change solutions we propose aim to generate broad-ranging benefits over time and in various contexts beyond the climate system, thereby fulfilling the key requirement of addressing the trade-offs. The first thing that this means is that how and how much we do any of the things that we propose depends on the local context [34]. However, what follows is that, in practice, what we do, and especially the principles that underpin what we do, particularly at the local level, must be brought into alignment with the principles that are considered by human society to underpin economic, environmental, and ethical sustainability in a broader sense [7]. Trying to find alignment between these different, equally valid principles is the 'real action' that we ultimately want our 'interventions' to be part and parcel of, and are the foundations of effective direct and indirect community and environmental resilience. Portfolio of Adaptation Options—mainstreaming local interventions: case studies, comments, and threats. This manual comprises eight case studies and one study in the form of a comment, demonstrating a variety of approaches and explorations consistent with these principles [5]. They demonstrate localization of objectives, stakeholder involvement, and sustainability.

3 Building Resilience to Climate Change

A common theme throughout all four of these scenarios is the importance of building resilience to climate change. Resilience is the ability to "bounce back" or adapt to new conditions or a shock. In particular, resilience strategies enhance the ability to adapt to changes within the range of historical variability or to resist those changes altogether. Building resilience strategies should lead to a decrease in community vulnerabilities and ultimately empower individuals, communities, and institutions through a localized and proactive approach [39]. Just as with ecosystems and infrastructure, various strategies can enhance the capacity to adapt to a changing climate and sustain developmental gains. Many of those strategies are presented in the Worksheets. Together, the Worksheets outline a complete toolkit for reducing vulnerabilities to current and future climate hazards. Five additional adaptation strategies are presented in this plan and were excluded from the Worksheets because they can be costly and impact food security. The strategies in the Worksheets can be broken down into five key areas of intervention: sustainable agriculture, resource management, local economic development, infrastructure improvement, and collective action (through tools such as community planning, regulating natural resource use, and conflict resolution). This integrated approach addresses the complexity of the causal chain that has led to increasing vulnerabilities and works to build latitude into social-ecological systems. Latitude is the ability to deal with unexpected events and stressors and can be seen as a step beyond building adaptive capacity and resilience. By enhancing the physical, social, and learning dimensions of individuals and communities, we look beyond short-term interventions to ensure a more secure future. Rehabilitation of social and ecological systems must also consider the creation of cooperative and sustainable management systems to guide the distribution and use of resources in support of sustainable development goals [41].

3.1 Climate Change Resilience

Through completing this study, a series of case studies exemplifying successful resilience to climate change is presented. Each case illustrates the practical application of the theory discussed in the preceding chapters. Collectively, the case studies represent three theoretical themes: urban resilience, governance, and rural resilience, and economic development as a resilience strategy. The range of case studies illustrates successful resilience across a mix of social and geographic contexts. The capacity-based framework of resilience is a concept still under development, lacking an operational approach [3]. Resilience within Danang is the most practical and sophisticated example,the remaining cases use similar approaches at present. The stakeholders involved in the framing of theory and practice are financial institutions, policymakers, infrastructure firms or business consultants, academia, and international development agencies. The case studies provide an opportunity to link these stakeholders to deliver resilience approaches in practice [45].

The discussion of challenges and solutions presents a series of issues that must be addressed to ensure that financial, environmental, and infrastructure investments contribute to resilience at multiple scales. At the micro level, the key is to ensure that local communities have sufficient resources to engage in initiatives, and a sense of ownership is promoted to become involved. At the middle level of government, which is dependent upon international finance, the difficulty is to demonstrate accountability and visibility in terms of effective governance in leading their people into safer conditions and maintaining economic buoyancy through political terms of office [17]. At the macro level, for example, flood management authorities in Vietnam will also have to demonstrate that the money has been well spent or be prepared to defend their position in the international financial market. In addition, the case studies provide indications of facilitating factors, for example, at the urban level of Vietnam, the government is using regulations to enforce resilience, and at the rural level of Sri Lanka, farmers are allowing others to buy into resilience by cooperating in the actions, and at the government level, donors are willing to commit finance if resilient policies comply with Kyoto. The evidence provided through these case studies suggests that resilience needs to be constructed from the bottom up. These actions have been successful to date in restoring and improving livelihoods [20]. The case studies, therefore, provide valuable lessons in the practice of resilience through the policies, governance, and management of diverse stakeholders.

3.2 Adaptation Strategies

The role of the agricultural sector in responding to ongoing climate change is challenging due to its dual nature as both a contributor and an adapter. This section of the chapter examines how biophysical and socio-economic factors can be leveraged in the short term to enhance resilience. It explores various adaptation strategies, including the broader concept of climate-smart agriculture. Approaches to building climate resilience are diverse, incorporating climate-resilient farm practices, natural resource management, logical thinking, infrastructure updates, or a combination of these points [9]. Hybrid strategies integrate some green manure use with some external nutrient inputs. Finally, all strategies target a suite of sustainability principles, including economic, social, environmental, and conditions of overall development sustainability, to serve as a basis for their promotion and acceptance. Case studies have reported evidence that the strategies may positively contribute to the adaptation potential of the concerned species in question. However, strategic intervention is often necessary to ensure that the local context is suitable [13]. Because these cases do not represent all vulnerable systems, each approach to a strategy is discussed more extensively in a separate section, along with additional evidence and examples. These case studies each contribute to a better understanding of the challenges, opportunities, and impacts associated with the implementation of each potential adaptation strategy. Major barriers are highlighted, and the way forward, as well as lessons learned, are also discussed. It is crucial to inform stakeholders, particularly local smallholder communities that have limited power to influence [21]. Without a common understanding of the most strategic adaptation options, it is harder to build climate resilience for the future.

3.3 Success Stories and Lessons Learned

Since 2012, the I-RESIST project has supported four village communities in the Pacific to address the unique climate, social, and weather-related challenges they face. For most of these villages, the primary climate threat is from increased storm severity, while long-term concerns focus on the impacts of sea level rise, which is rapidly salinizing the water table. The South Lombok village of Orya was initially designed as a fringing reef village, but now finds itself perched on the coastline atop an eroding sandy platform [43]. The other villages involved in the project are

Nukubalavu, Yaro, and Luata, where rising seas also cause problems with food gardens, and in some cases, rising seawater has destroyed an owner's house or part of it. In each project village, communities have decided to focus on either "moving out" of dangerous areas or "staying put" and strengthening their defenses. Taking a participatory and process-oriented approach to hazard and risk assessments, the project villages worked together to identify possible disaster preparedness and risk reduction strategies [18]. These have been simple, active, and visual, and have been added to in the course of the three years. In the months following disaster workshops, community members have invested money and labor in these activities, increasing their resilience to future threats. The project is expected to be completed in January, and we will then officially document and evaluate this body of work. This project demonstrates that where there is a will to 'move out' or 'move up,' there are behavioral responses active in the absence of more complex forms of governance or policy packages [19].

4 Policy and Governance in Climate Change Resilience

Governance and policy frameworks are the milieu that influence the establishment of laws and sectoral policies, which can promote or detract from resilience, or act in complete isolation from such an outcome. Policymaking is a multi-sectoral process; it involves striking a balance of interests among various areas, including agriculture, health, environment, politics, and trade. The ability to integrate and align the objectives, or at the very least ensure they are in harmony, is critical in any policy that sets a target that is not necessarily agreed upon. The argument made by other disciplines that climate-related governance is not only an indicator of destination (climate resilience itself) was made [16]. No single party can bring about transformative climate action, however, governance structures that ensure climate response cannot be, or will not be, proactive (or worse, exacerbate) directly hinder and prevent local organizations and households from achieving climate adaptation in the very sectors that provide them with a livelihood. Despite the legal authority of top-down law on the ground, what is most important are the day-in, day-out, decade-long informal policies, regulations, and permissions granted by the village head. It is this bottom-up, inter- and intra-personal governance that will ensure that larger, sustainable, mutually beneficial outcomes are achieved [15]. As such, this direction and the structures emerging from such an in-depth analysis will ensure that the policy creators and governance architects develop constructive dialogue with local and other relevant stakeholders within the policy and governance establishment process. This knowledge, which resides in the soil of the unusual cadres, is sometimes more exciting for people in white lab coats, where it matters most—on the ground. Stakeholder involvement is likely to have some effect on this outcome, particularly in the Western context, which is sometimes praised for its democratic governance processes.

4.1 International Agreements and Frameworks

Rather, a significant number of global agreements have been adopted to mitigate climate change since the adoption of the United Nations Framework Convention on Climate Change. All these regulated and legally binding global environmental treaty systems have aimed to respond to the common but differentiated development goals of states. Of those, the Kyoto Protocol has somewhat created a formal treaty structure focusing on a few countries for largely controlling global carbon emissions through a system of multilateral regimes [1]. If the drivers and ultimate goals of resilience involve longer-term policies and major capital projects—that, in turn, are heavily influenced and shaped by national-level legislation, policies, and strategies—how does the Paris Agreement contribute to these implications and actions at the national and local levels? The Paris Agreement has significant potential for practical shifts in its implications for development discourses and practices at the national, regional, and local levels. At the highest international level, the Paris Agreement has been argued as signaling the world's commitment to a sustainable development and poverty eradication agenda. While welcomed by nations, not allowing any 'outside' is both a strength and a potential weakness of such a strategy. The effectiveness of the Paris Agreement will, in part, depend upon the collective realization of its ambition [23]. At the production and policy levels, the national goals have been received with much enthusiasm and optimism. The sense of dynamism and dedication to actually realize those goals is palpable.

4.2 Innovations and Technologies for Climate Change Resilience

Innovations and technologies play a significant role in improving human preparedness and proactive approaches to controlling natural disasters and climate change catastrophes. Practical scientific advancements can and should be utilized to develop tools and techniques that reduce and manage issues, ensuring effective community and environmental risk reduction in the event of catastrophes. Furthermore, emerging technologies have shown potential to contribute to sustainable ecosystem sense and responses. There are a number of innovations and technologies that are already available or in the development phase to meet the needs of reducing global environmental crises. The design thinking and innovative educational approach could help provide solutions from a sustainable development perspective. Some examples of these innovations include developing low-carbon "smart" infrastructure, such as roads, buildings, and utilities, that utilize renewable energy sources, which can also harness solar, wind, and treated drinking water while absorbing stormwater runoff. This infrastructure includes renewable energy train stations, smart road stimulation via solar power innovations, and eco-smart modular flexibility for adaptation and resilience of river infrastructure [15]. Additionally, there is a new vehicle-to-grid technology, which not only allows large power demographics to come from the grid by electric cars but also while parking through bi-directional charging, from the automobile to the grid, and conclude their responsibilities such as reducing the occurrence of non-predictive trouble making as well as exchange of power flexibility and reliability monetization through vehicle-to-grid activities. The fast pace of innovation in growing economies spurring technological advancement that can aid the sustainable industrial transition, an air conditioner company that used greenhouse gases, and a car company that developed electric vehicles are examples of implementing affordable, innovative climate change technology policies in cooperation with governments [16]. Cutting-edge technology innovations that aim for a result-driven technique for greater coherence and pressurized sunk costs in R&D can also help.

4.3 Renewable Energy Solutions

Energy is a crucial component in enhancing resilience. Renewable energy sources, such as solar, wind, and hydroelectric power, can help in transitioning from energy systems that emit to non-emitting ones, thereby minimizing air pollution, protecting human health and the environment, and mitigating climate change impacts by reducing CO_2 emissions. What are considered the developing world's most innovative solutions? For example, solar-powered cars, wind energy turbines, thermoelectric generators that can convert waste heat into electricity, a device that produces continuous electric power from water pressure, solar-powered drilling machines for drilling water wells in clean energy mode, and more [19]. Many of these technological advancements are now being utilized by farmers and entrepreneurs in the developing world to enhance productive and energy-intensive agricultural activities. By 2050, renewable energy sources such as solar, water, and wind could meet the majority of Africa's electricity demands. Relevant, sustainable development goals include renewable energy's share in the total final energy consumption and investing in renewable energy, value-added to the GDP. Renewable energy technologies and practices can significantly contribute to mitigating the severity of climate impacts due to their environmental benefits, reducing carbon emissions, facilitating the transition to clean, efficient, and sustainable energy, ensuring energy access, and promoting economic development [43]. There are several barriers to the adoption of renewable energies, including, but not limited to, high upfront costs, complexity, and a lack of technical support, as well as capacity challenges such as financing, regulation, and awareness. Many of these barriers can be mitigated by addressing system constraints through the engagement and capacity development of relevant stakeholders within ecological-economic systems. Some sampled approaches to achieving SDGs are presented in Fig. 3 [13].

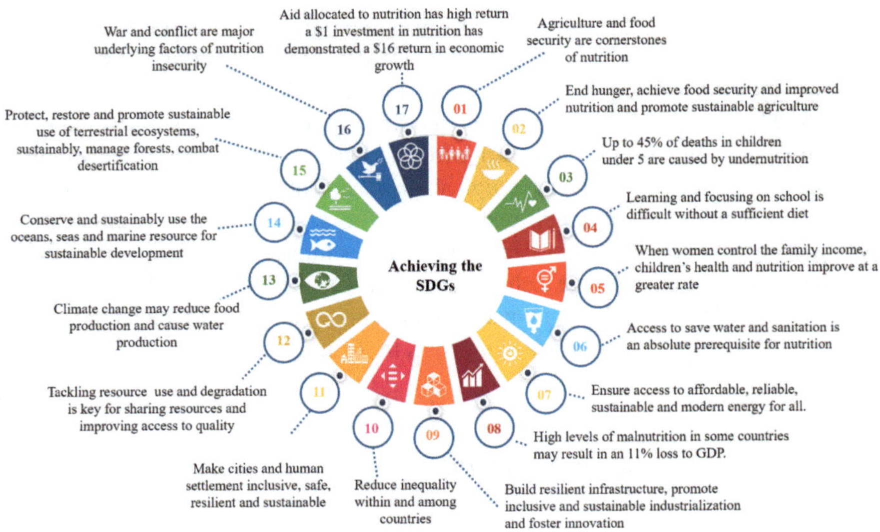

Fig. 3 Achieving the SDGs

5 Community Engagement and Participation

Community engagement is often cited as an important component of resilience-building efforts in relation to social-ecological systems. This is particularly emphasized at the local and national governance levels, both for practical reasons related to the local social-ecological dynamics of resilience concerning climate disruption and development pathways, and in terms of discursive public relations. There is empirical evidence that local community-led initiatives are gradually being supported, especially in developing countries, to be understood as integral parts of national climate change intervention and population-driven development [45]. These initiatives may have been fueled, at least in part, by international grassroots lobbying. There is a need in policy circles to encourage local initiatives that can help ensure the effectiveness of global adaptation and development policies for the water sector. The full scope of opportunities should be explored. Similarly, an increasingly participatory approach to development at the grassroots level is being lobbied for, as the beneficiaries and, indeed, those most affected are also reportedly those most knowledgeable about what will and will not work in terms of adaptability [3].

These bottom-up forces also lend themselves to policy formation by creating public will to move forward. Moreover, community-based activities are in many ways more effective than top-down policy legislation, as they not only create immediate futures of carbon sink conservation but they also create societal change in values, norms, and practices. There is a range of successful projects addressing issues of rapid change that are being generated at the local level in even the most marginalized communities. Being driven by the actual human beings most affected within these communities, these projects are much more grounded and value-adding than many policy projects tend to be [25]. Communities have already had many solutions. It is not possible to solve a problem for a community that did not first perceive the situation as difficult and develop treatment or coping strategies. A solution or remedy is community-driven if it aligns with the conceptual framework of the individuals living, resolving, or dealing with the situation. When those affected are decision-makers on the action, they will be the ones taking responsibility for the consequences and for adapting if change does not occur as expected [6]. Ownership, therefore, will reside at the entry points of the action.

5.1 Local Initiatives and Grassroots Movements

There are many local initiatives and grassroots movements that are working on increasing resilience to climatic variability and change. Many community-driven projects regarding the implementation of the Sustainable Development Goals incorporate activities on climate change by increasing community adaptation strategies in the environment. In improving the local climate change community as an object, it should be noted that many community resources are still untapped. Many local initiatives are born from the awareness of the community itself that is affected by environmental conditions, including climate [39]. The local initiative is typically organized as a grassroots organization led by the community itself. One of the factors for a successful grassroots organization is that it has community leaders who support local initiatives on environmental issues. Several successful local initiatives have been implemented in Indonesia. An initiative in Kotagede, Yogyakarta, has successfully involved local people as heritage activists in developing alternative tourism. Originating from the founder's love of history and heritage, as well as pride in being a resident of an area with a wealth of cultural heritage, working people voluntarily repair, protect, and strengthen the local economy. By utilizing space, fostering local wisdom, promoting cooperation and consensus, and adding a cultural dimension, participation in local economic development is carried out consciously. Local initiatives, carried out by former government administrators who have established affordable and equitable health services, are also found in Malang.

Although it only allows patients to be treated individually, and no free hospitals exist for people, residents continue to visit clinics for easy access. In Mozambique, for example, a local initiative that developed a slum gardening movement built a canal that successfully prevented garbage from entering a river in a household project. In India, local initiatives are undertaken individually in response to the drought problem. Some are building mountains with the resources they control. The success of the activities is driven by the involvement of community leaders in the local initiative. Leaders usually have high social capital, commonly known in the form of networking in social theory [25]. Besides, leaders also have a keen sense and a good heart in the face of local social problems. They have a great concern for development in solving the real problems in the community, starting by creating alternative solutions upon which consensus builds. Some successful domestic initiatives naturally have a legal form or are legalized. Because people believe that the organization is not based on anyone's ego, but as a home to accommodate and realize mutual interests. It shines the dawn of awareness. The community started to see logical differences in reasoning. An elder and a neighborhood figure also visited to discuss. After a long, complex, and tiring process, an agreement was finally reached. An agreement that ultimately contains their desire to have a worthy local legislative body. Building the same and common love that has united them from the beginning [41]. And finally, in May, the people's wishes were fulfilled through local regulations that established the village. Besides the local law, other legal products also emerged if they were desired and contained values as a noble means to solve various social problems in the village. The establishment of this new village is in accordance with the principle of autonomy and local regulatory principles, which have been carried out with citizenship principles in the interpretation of pluralism, enlightened [8]. The communal aspects have been responsible for active society strategies, as presented in Table 1.

5.2 Economic Implications of Climate Change Resilience

While climate adaptation and resilience measures may require substantial initial investment, the costs of inaction can be strikingly higher. Costs of even conservative climate action or defensive practical countermeasures could be offset in part by comprehensive cost–benefit analyses. Over the long term, these studies show that the undeniable benefits often more than recoup the initial costs. Many commentators believe that adopting a more sustainable way of living could trigger an economic boom if implemented on a large scale, thereby alleviating some concerns about cost [13]. Already, considerable sums have been made available to provide immediate support to those regions most adversely affected and to prepare them for the impact of current measures. In the long term, certain indicators suggest that climate adaptation could contribute to job creation and economic growth. Adaptation and resilience include early action strategies to manage climate change risks. The opportunity to reap benefits from large-scale preventive interventions increases with the risks. However, existing barriers make it difficult to make large investments in advance, necessitating the integration of precautionary and disaster risk management measures. Especially in high-risk and low-income areas, short-term planning rightly focuses on immediate, more pressing concerns [21]. Investments in more climate-resilient infrastructure will require incentives or other safeguards, as well as a blend of small and incremental approaches. Their economic rationale must be clarified to encourage their production among policymakers, planners, and stakeholders. To enhance support for focusing and ensuring sustained action within this multifaceted strategy, political actors, governmental and commercial decision-makers, practitioners, civic groups, and other stakeholders require intelligence about the scale of these benefits to ensure that these interventions are worthwhile [20].

Table 1 Climate change impacts and resilience strategies from a sustainable development perspective

Climate change impact	Affected sectors	Resilience strategies	Areas of application	Impact on SDGs	References
Rising sea levels	Coastal infrastructure	Coastal defenses managed retreat	Netherlands' Delta Works	SDG 11 (Sustainable Cities), SDG 13 (Climate Action)	[23]
Vector-Borne disease spread	Public health	Disease surveillance, vaccines	Malaria prevention programs	SDG 3 (Good Health and Well-being)	[11]
Extreme heat	Public health	Urban cooling, heat action plans	Tree planting in urban areas	SDG 3 (Good Health), SDG 11	[1]
Increased storm intensity	Housing and infrastructure	Storm-resistant buildings	Hurricane-resistant Roofs in Florida	SDG 9 (Industry and Infrastructure), SDG 11	[2]
Ocean acidification	Fisheries	Sustainable fishing policies	Coral reef restoration projects	SDG 14 (Life Below Water)	
Water scarcity	Agriculture and water supply	Efficient irrigation, desalination	Drip irrigation in Israel	SDG 6 (Clean Water), SDG 2 (Zero Hunger)	[15]
Agricultural disruptions	Food security	Climate-resilient crops	Drought-resistant maize varieties	SDG 2 (Zero Hunger), SDG 13	[16]
Biodiversity loss	Ecosystems and wildlife	Protected areas, reforestation	Amazon rainforest restoration	SDG 15 (Life on Land), SDG 14 (Life Below Water)	[18]
More frequent wildfires	Forests and communities	Fire-resistant structures	Firebreaks in California Forests	SDG 15 (Life on Land), SDG 11	[43]

5.3 *Cost–Benefit Analysis*

From evaluating disaster risk reduction projects, it is recommended to evaluate the coherence of intervention design from a sustainable development perspective; hence, a cost–benefit analysis (CBA) is performed. The objective is to provide a process of comparing costs and inputs with the benefits and outputs of climate resilience activities. The use of CBA allows for comparing the desirability of different projects as well as the contributions of various stakeholders. A framework for a broader 'extended CBA' can assess the impacts of a climate change adaptation and/or disaster risk reduction project with a sound link to basic development indicators and a wider look at the costs of impacts [3]. The first step of CBA involves estimating project impacts.

The second step of the CBA involves valuing project impacts. Project impacts are valued in monetary terms. The third step of CBA is the estimation of the opportunity costs of project inputs, which are the benefits forgone by allocating resources to a particular project. As a result of these steps, the fourth step of CBA compares costs and benefits by direct comparison and through present-value techniques to account for the time element. An outline of a possible methodology tool for CBA of climate-resilient sanitation projects is developed. Despite a few different statements, the same general steps apply for a general CBA focusing on providing resilience to the promoted development project [9]. Therefore, they can be applied to a range of climate-resilience-building projects.

6 Ethical and Social Justice Considerations

Social and Environmental Inequalities Climate change is already affecting diverse social groups within both developed and developing countries; however, it is considered that the poor, women, marginalized, and vulnerable groups are especially at greater risk. Climate-driven impacts, such as droughts, floods, and typhoons, can be particularly devastating for already impoverished people who are highly vulnerable due to their high dependence on climate-sensitive occupations, including rain-fed cultivation, forest product collection, and fisheries. Women, in particular, contribute significantly to low-income households, and, being largely dependent on natural resources, face the adverse impacts of climate stress [45]. The consequences of these impacts are reflected in terms of food insecurity, income reduction, ill health, socio-cultural changes, and liquefaction. It especially deprives women of accessing resources, with multiple negative impacts on their lives and livelihoods. Moreover, within the same community, households with more social connections can sometimes better manage climate-induced hardships. Social networks are, however, often fragmented by class, caste, ethnicity, and gender, and this further narrows the coping block. The increasing vulnerabilities of poor people, women, marginalized communities, and other vulnerable groups lie in their limited access to resources, weak social networks, insufficient human capital, political redundancy, livelihood insecurity, and the least responsibility for the emission of greenhouse gases [9]. Last but not least, these groups possess the least scientific and technical knowledge that would enable them to understand the existing climate threats.

6.1 Equity and Vulnerable Populations

Climate change adaptation strategies center on the appropriate actions in response to climate effects. Vulnerabilities are defined as the inherent traits particular to vulnerable populations that cause climate risks to act as threats. Vulnerable populations face increased risks and reduced capacities due to preexisting systemic inequities that make social, economic, and environmental resources less accessible and more fragile. Strategies targeting marginalized populations are often required to build adaptive capacities [21]. Targeted strategies are often essential not only because inequities mean different groups will have varying priorities and need different types of help, but also because systemic biases at every level of policy mean the impact of interventions will vary widely. Stratified vulnerability approaches grapple with different sets of social justice considerations that aim to integrate values related to social diversity. Adaptive capacity-building efforts often include improving social justice circumstances. These strategies invariably argue that enhancing capacities in highly vulnerable settings is beneficial, thus addressing preexisting social injustices [18]. Support for policies that protect and empower marginalized people is known most broadly as the human rights approach.

The climate justice adaptation method involved modifying major climate adaptation mechanisms for use in developing nations to ensure cultural equity. These values have led to climate change adaptation initiatives that provide new opportunities for international development mechanisms. Some of the most successful climate change resilience-building initiatives are those designed primarily to address social justice challenges. They were implemented by operating outside of the formal environmental regimes [16]. Support for the approach has grown over recent years, but a significant stumbling block to implementation is the lack of capacity of municipal council officers and politically elected officials to incorporate social justice and climate justice values into their mainstream work. This appears linked to the political priorities set by non-vulnerable decision-makers. Vulnerable populations can sometimes be resistant to these strategies because of their inability to engage with the social injustices and general benefits they offer if they are rejected. Several major barriers stand in the way of marginalized communities receiving the resources they need to address their vulnerabilities.

A focus on individual needs to change people's economic or emotional state without addressing the social and economic system. A reliance on appropriate technologies or neoliberal values to facilitate change, ignoring the extent of existing prejudice against populations. Perhaps most importantly, global and national governments have limited the capacity of local-level actors by sacrificing funding for local communities and community-led solutions that may make a bigger impact. Suitable community-led programming and coaching can be transformative, encouraging marginalized populations to find solutions to their challenges [5]. Many communities rely on social networks and communal activities for support, practical assistance, advice, and access to savings. Solidarity can be increased by facilitating these networks in a way that undermines the current unjust structures of exclusion. In doing so, this investigation suggests general improvement in adaptive capacity and, therefore, resilience. For effective climate resilience, it is possible that social justice is an important starting point to understand and address the underlying unequal vulnerabilities in place [8].

6.2 Education and Capacity Building for Climate Change Resilience

Educational programs and capacity building enhance resilience by raising awareness and understanding of climate change and its risks. At regional and local levels, education and communication are essential in fostering community knowledge, skills, attitudes, and responsibilities that improve actors' abilities to make effective adaptations and engage in resilient actions. This is especially true in sectors sensitive to climate change, such as agriculture, fisheries, forestry, and water resource management. Educating individuals about actual and potentially occurring climate changes, as well as creating understanding and awareness about these changes, fosters understanding of the need for adaptive responses [6]. Key topics are likely to include insights into anticipated change, its implications, options for responding, who will be affected and how, vulnerability assessments, and possible cost–benefit analyses.

Climate change is not a standalone issue. It is an integrating issue and needs to be understood as part of a wider concept, such as sustainable development. Social learning—encompassing communication, learning, and capacity building within communities and among stakeholders—should be a continuous process to mainstream environmental precautions into the operating cultures of authorities, industry, commerce, society, and individuals. Training programs that educate people at the local level and develop their skills are crucial to overcoming social weaknesses and insecurity. Such programs should target a broad audience that can then reach diverse social groups and individuals [39]. Workshops with stakeholders deserve particular mention, as they often emerge as an essential tool for empowerment and supporting the engagement of concerned and interested individuals. The importance of research and action in addressing climate change should not be overstated. More educated countries are more likely to mitigate and adapt to the impacts of climate change. Adaptive capacity is often enhanced by education, particularly through investments in educational infrastructure [25].

6.3 Training Programs and Workshops

Training and workshop programs have been conducted to enhance the capacity and build networks within local communities to respond to climate change effectively. They have been useful, whether they are dealing with integrating climate change adaptation into disaster risk reduction, ecological conservation measures for climate change, incorporating traditional environmental knowledge for climate-smart community regeneration, climate change impact mitigation strategies at the local stakeholder level, or for best practices in adaptation measures leading to more productive and fruitful outcomes [41]. Capacity-building initiatives have evolved and been tailored to the local community context. In certain regions, more quest-oriented approaches to integrating climate change and disaster risk reduction are applied. The community's questions and concerns are integrated into the group training schedule. Because of the heterogeneity of individual and group characteristics, the approaches and methods used for multidimensional capacity-building events are also more diverse [3].

Several case presentations demonstrate the empowerment of various skills and expertise within the local community during the process, including training in transharvesting and seeding techniques for local birds. In certain areas, communities raise the issue of ecological conservation strategies in response to climate change. In certain contexts, knowledge and skills related to the five senses are taught to encourage more community awareness and understanding. Educational approaches for the most vulnerable communities include ecovillage development, food and nutrition resilience, seed exchange skills, preservation of traditional plants, sustainable community-based tourism, waste management, and family economics [17]. The training approach for participants primarily involves practice through simulations, station rotations, and other community empowerment methods. Transformative learning is realized through income-generating training for communities and all stakeholders involved in these initiatives. Training is often initiated with the formation of local community groups or in conjunction with existing groups. Currently, the activities related to addressing climate change are more focused on community education and planting [9].

7 Monitoring and Evaluation of Climate Change Resilience Programs

Monitoring and evaluation (M&E) in climate resilience programs constitutes an important means of examining the effectiveness of resilience initiatives. The development of effective monitoring and evaluation (M&E) methodologies for assessing the effectiveness of climate resilience programs is a growing field of literature. M&E systems are critical components of climate resilience planning, as they facilitate proactive management. Before the development of resilient indicators and metrics, the program's objectives and goals must be clearly defined. The values and principles that guide climate-resilient planning and action should also be clear [21]. Monitoring will inform actions and results in various sections of the program, such as resource allocation, identifying effective strategies, documenting outcomes and impacts, facilitating management, and evaluating monetary values.

Effective assessment will identify sources of successes and constraints, the probability of climate hazard occurrences, and investments worth replicating or avoiding. Robust evaluations of climate resilience and climate-smart investments should be built on the following principles of resilience. M&E systems must cater to various types of evaluations, including continuous monitoring, end-of-program assessments, and one-off evaluations. M&E systems typically have robust feedback mechanisms to implement lessons learned and utilize the information to adjust strategies and management, thereby improving performance. The quality and robustness of the database also inform the robustness of the monitoring and evaluation (M&E) systems. It should be able to balance summative types of evaluation and formative types to cater to learning [43]. M&E systems aim to build accountability, promote learning, and inform policy and practice. It also seeks to do justice in terms of providing clear and good examples of successes and failures.

Indicators and metrics, or some means of measuring resilience, are often incorporated into monitoring, evaluation, and reporting systems. The need for this is an underlying assumption of this text, in that it is only if progress toward the resilience objective can be measured that an adaptive management framework can work properly. But to construct suitable indicators, previous sections suggest the need for clear benchmarks in order to give form to an otherwise non-linear path toward the direction of change; hence, the identification of scenarios [15]. That is, for any given project or set of projects, indicators are designed to reveal not only whether progress is happening but also what form of change it represents within the long-term trajectory of the complex system in question. The earlier parts of the text, which deal with adaptive management, have further explored these aspects. The development of indicators is often belatedly recognized in writing development projects because of an apparent institutional diffusion of the recipe for Logical Framework Analysis [1].

Indicators of resilience, however, are difficult to identify and measure due to the latent effects that are relevant. In this case, indicators can include proxies where there is only a scant possibility of correlation with effect, and residual variability in the data could suggest a direct operationalization of a resilience component within a large system that is impervious to the project. Indicators can be used in a diagnostic phase. This uses both qualitative and quantifiable indicators of the strength of resilience in order to suggest some feasibility scenarios [11]. In any event, the indicators are used to reveal and describe desired paths of change. Indicators need not reflect the only possible roles of an intervention, for example, to improve the legal and policy environment. Stakeholders typically document the use of qualitative indicators because the approach to resilience and its shaping mechanisms is conceptual and nuanced. It can also indicate the proposed boundaries of the project, i.e., the relative empowerment of stakeholders based on the focus of other development agencies on empowerment. Indicators are cheap to effect because they are constructed from a range of client-specific background analyses [5, 6].

There needs to be a context of widely supported scenarios for success, including a silent stakeholder analysis if investing in potential conflict scenarios. Monitoring and evaluation criteria or metrics can be established to help signal when resilience may begin to erode, particularly when project resources are no longer available. This outlines the strategy of treating boundaries, discontinuities, and system-determining states as validity bottom lines. These monitoring processes examine what is actually happening in comparison with projected compliant scenarios to identify genuine risk and consequently improve practice and change the outcomes. These metrics can also allow us to assess trade-offs with other decision factors. In such cases, remember that investment in resilience indicators and metrics is only a proportion of the overall investment that may not be recouped when key stakeholders are sympathetic and thoughtful about the decision criteria [20]. Such surveillance views of resilience anticipate and observe variability interacting thickly and horizontally with a wider range of indices of dependence, some critical resilience technologies, and their application, as presented in Table 2.

Table 2 Key climate resilience technologies and their applications from a sustainable development perspective

Technology	Function	Application areas	Example projects	Impact on SDGs	References
Renewable energy (Solar, Wind)	Reducing carbon emissions	Power generation	Solar farms in India	SDG 7 (Clean Energy), SDG 13	[8]
Green roofs and walls	Reducing urban heat islands	Cities, buildings	Rooftop gardens in Singapore	SDG 11 (Sustainable Cities), SDG 3	[42]
Smart water management	Water conservation and distribution	Agriculture, urban planning	Smart irrigation in Australia	SDG 6 (Clean Water), SDG 2	[4]
AI-Based climate modeling	Predicting climate impacts	Disaster preparedness	AI-driven flood forecasting systems	SDG 9 (Industry and Infrastructure), SDG 11	[24]
Climate-resilient crops	Ensuring food security	Agriculture	Drought-resistant wheat in Africa	SDG 2 (Zero Hunger), SDG 13	[14]
Carbon capture and storage	Reducing atmospheric CO_2 levels	Energy, industry	Carbon storage projects in Norway	SDG 13 (Climate Action), SDG 9	[7]
Electric and hybrid transport	Reducing reliance on fossil fuels	Transportation	Widespread EV adoption in Norway	SDG 11 (Sustainable Cities), SDG 7	[6]
Resilient infrastructure	Enhancing disaster resistance	Transportation, housing	Elevated roads in flood-prone regions	SDG 9 (Industry and Infrastructure), SDG 11	[39]
Early warning systems	Disaster risk reduction	Weather forecasting	Tsunami Warning Systems in Japan	SDG 13 (Climate Action), SDG 11	[45]

8 Future Trends and Emerging Issues

Climate resilience to climate change is expected to be linked with future trends and issues. The future may signal a devastating scenario for technology, instruments, capacity development, and policy direction. To be prepared for these sources of tension, an adaptive approach that incorporates scenario analysis is essential. Proposals were made under the umbrella of future analysis to help stakeholders and community resilience planning. The importance of climate-resilient strategies, including infrastructure and technology, is recognized [9]. If current instruments and technologies are used, different sanctions may arise. The plan has begun in earnest. Energy, vitality, and ecosystems may no longer depend on fossil fuels. Spatial planning and building design might imply adaptation rather than transformation. There are concerns that technology will be too advanced or incapable of stopping climate change without causing more harm. Some of the latest geoscale technologies are included. Biodiversity resilience can benefit from a form of assisted colonization that involves moving species into new areas. Natural resources can be removed, and tactically targeted CO_2 absorption and performance with particulate soiling might be employed, allowing for local information related to urban areas [18]. Prevention and negotiation strategies can help reduce the likelihood of clashes between stakeholders and the public with opposing views. Members gathered issues for one region. For example, some of the regional participatory missions were defined for practical reasons. Resilience planning values and adaptive strategies require accessibility to emerging issues and predictable tools. Quantifying changes in future demand, investment costs, and survival issues is important. To expand upon the importance of climate resilience in addressing these issues, we pose four major 'What if?' questions [2, 23]. They created a critical angle, prompting a strategic response.

9 Computational Techniques in Climate Resilience

Computational techniques have revolutionized traditional practices, enabling people to obtain new analytics and decision-making aids. Several computational techniques are being employed to enhance climate resilience. Remote sensing techniques can be used to map climate risk for identifying hotspots in various sectors. Climate modeling and statistical techniques are now capable of generating various scenarios for climate variables and their associated risks. Similarly, disaggregation techniques can downscale global climate predictions into local micro-scale predictions [13, 23]. Other computational techniques, like machine learning applications, big data analytics, and blockchain technologies, also have great potential for contributing solutions to identified roadblocks in the perspective of climate change [22]. In fact, these techniques have been successfully deployed to transform various nature-based responses to climate change and have proven to be advantageous for enhancing climate resilience at different levels. State-of-the-art technology developments have enabled us to enhance climate resilience for various systems with improved decision-support requirements, utilizing techniques from diverse computational methods, including remote sensing, climate modeling, data analytics, and data science inspiration. Remote sensing and geographical information systems are used for predicting climate risk in agriculture, urban climate hazards, disaster vulnerability, urban growth assessment, and even in biodiversity assessments [18]. Climate models and statistical techniques are used to simulate past behavior and generate future predictions. Downscaling is also a statistical technique used for this purpose, while data science methods are employed to generate strategies and frame policies based on the quantification and documentation of aggregated multidimensional and multisource data. Since spatial climate information is complex in terms of coverage area, resources, and interdisciplinary knowledge, using a combination of the above methods can provide us with considerable advances in preparation, reduction of budget and time, and assessment of the locally focused systems to climate impacts, enhancing the decision-making capability for climate resilience [21].

9.1 Remote Sensing and Earth Observation

Early Warning Early Action Systems are gaining traction as climate extreme events are becoming more frequent and severe. This requires gathering a myriad of data streams in a timely manner and presents satellite Earth observation and remote sensing as prime assets for the resilience data toolbox. These tools collect data from space via satellite measurement of light, with the data then processed and transmitted, and regions of certain habitats that are described as optical or rich in light from the electromagnetic spectrum [43]. The most commonly used satellites in this particular research are from the Landsat series, which has provided consistent and high-quality data since the early 1970s. The images used in this research, taken by Landsat 8 satellites, are part of the USGS space mission. One of the advantages of using this data as a key factor is that it has not reduced the accuracy of the model in any way. The use of more recent satellite images would project better image quality because of improvements in sensor technology [44]. These images were smaller in file size and lighter to carry when accessed off-cloud, and as a result, they were preferable for the research due to the brevity of processing. Remote sensing and Earth observation technologies provide decision-makers with real-time, accurate, and regularly updated assessments of trends in geographical areas, ecosystems, vegetation, and built environments. The technologies of remote sensing and Earth observation have improved over time as sensor technology continues to evolve. Furthermore, remote sensing is a vast field with applications spanning nearly all geographies, disciplines, and sectors. It encompasses the assessment of the extent of crop damage due to a cyclone event, debris flow measurement for environmental models, satellite-radar monitoring of floods, and observations to map and monitor any changes in ecosystems [3]. Despite their depth in usage, one of these, which mostly requires timely data, is environmental monitoring of disaster impact assessment, used in predicting and preparing for events as they happen because it provides help in forecasting. Remote sensing is employed to monitor disasters and their consequences.

9.2 Climate Modeling and Simulation

The climate system cannot be fully understood without advanced computational techniques, which are now available in the era of high-performance computing. Climate models, particularly due to their computational capabilities for simulation, are recognized for their significant contributions to the scientific understanding of climate dynamics. Predictive models are essential and useful for investigating future changes in various climate components, such as temperature, rainfall, and sea-level rise, over periods of ten, twenty, fifty, or even a hundred years [35]. These models can be applied in several development-related fields, such as water resources, agriculture, transportation, energy, and health, because they are essential tools for predicting the impacts of climate change. For scientific and computational reasons, various modeling approaches, categorized into global and downscaled models, have been employed to describe future climate scenarios and assess associated risks and vulnerabilities. Over time, climate models have evolved and expanded in their ability to predict climate variability and changes, from regional to global scales, primarily due to an improved understanding of physical processes and the rapid increase in computational power [42]. One of the values of climate models is that they can consistently handle climate components under simulations both as a part of the Earth's climate system and as a coupled part of wider Earth systems, such as the environment, society, and development institutions.

A substantial body of research has utilized climate modeling as a primary input in climate policy formulation, strategic planning, and vulnerability and adaptation studies. A number of significant advances in the methodological underpinning of climate models have allowed climate scientists to develop more realistic models of the Earth system, with both higher resolution and better ability to represent sub-grid-scale physical processes, so that the number of representative processes has increased and the linkages between processes have been better represented [23]. These improvements have led to models that are computationally much more expensive. The availability of increasingly powerful supercomputers has helped to compensate for the higher computational cost, enabling larger ensembles of stable simulations and more detailed, high spatial resolution atmosphere and ocean simulations, now with the possibility of coupling to higher resolution model components for sea ice, land surface, etc., on increasingly powerful computer resources. This breakthrough in computing has enabled the latest suite of climate models to conduct experiments that were previously impossible. More precise and high-resolution climate scenarios allow more regional development institutions to have better representation of key climate processes relevant to policy-relevant adaptation scenarios [11]. Currently, climate modeling techniques can also represent complex interrelationships among social, economic, and environmental sectors to define future sustainable development scenarios, thereby addressing the cross-disciplinary and cross-stakeholder resilience agenda at regional and local levels.

9.3 Big Data Analytics and Machine Learning

Big data analytics has become indispensable for managing and processing large datasets, as observed in social media, search engines, and genomic products. These technologies are capable of processing huge volumes of environmental data that are difficult for traditional systems and techniques. Big data analytics is also an extension of the technological data management zone. To gain a deeper insight into data and generate effective data-driven solutions, machine learning algorithms are now commonly used in tandem with big data analytics technologies [46]. There are various domains in which the blending of machine learning and big data analytics is beneficial. Several applications are also underway that utilize these technologies in the domain of climate resilience. The machine learning algorithms enable the discovery of hidden patterns and correlations in the data with greater precision. This will help in forecasting the extent of loss or damage that is likely to occur due to extreme weather events and in taking necessary actions [13]. When big data analytics and climate science intersect, and advanced machine learning and analytics tools and techniques are available to build models for impact assessment and resilience building, it will result in more accurate climate predictions and precise actions for improving climate resilience. All the above-discussed techniques and processes are a small part of the picture of how big data analytics and machine learning could help in climate resilience. Various other applications are also underway based on these cutting-edge techniques for cost-effective climate resilience. It is sufficiently clear from these studies that machine learning and big data analytics could offer further insights into local adaptation strategies for climate risk management [11]. Little research is available that directly demonstrates these innovative approaches, but it is evident that these advanced technologies are the need of the hour for issues like climate resilience. Otherwise, climate challenges will not have sustainable solutions, which in turn could have significant impacts on the SDGs.

10 Conclusion

Technological advancements are widely recognized for their potential to impact development and build resilience to climate change in both current and future climate scenarios, i.e., the process of adaptation. Several case studies report on the innovative developments and harnessing of technology for climate change resilience in various sectors. The deployment of electric vehicles is enhancing transportation systems' ability to cope with extreme weather events. A case study in China is looking to develop drought forecasting systems. The deployment of solar photovoltaic and wind technologies has significantly enhanced the resilience of building infrastructure in the face of extreme weather events. A forward-looking perspective on how technological innovation can lead to geological, social, cultural, and economic advancements that can result in the formation and invigoration of strong and sustainable communities for the long term. A significant portion of the dialogue attributes this potential to the use of new and renewable energy resources in mitigating the effects of global climate change. Other advances and changes in technology, however, have not been widely discussed in the context of global social change. With a notable exception in tech-enthusiast circles, we have hardly begun to realize their implications and transformational potential. This narrative is of particular interest due to the recent surge in political, scientific, and popular discussions about the potential role these energy technologies can play in mitigating global climate change. To integrate technological, social, and institutional capacities across communities and places, we must also integrate the seemingly diverse body of literature concerned with its study. In this chapter, we present a framework for integrating these multilayered studies of transformative capacity. This framework is intended to outline a set of multidisciplinary, interdisciplinary, and place-based studies, bringing a heightened sense of urgency to the question of how we can harness the tools at our disposal to mitigate the social and ecological consequences of poor development. While the focus of our research, therefore, is at the state and federal levels of government, increasingly, it is clear to us that the work of civil society and marginalized communities is the area demanding the greatest attention at present and that the results of our research may have the potential to be most handily deployed in these areas.

References

1. Abdillah, A., Buchari, R.A., Widianingsih, I., Nurasa, H.: Climate change governance for urban resilience for Indonesia: a systematic literature review. Cogent Soc. Sci. **9**(1) (2023). https://doi.org/10.1080/23311886.2023.2235170
2. Awazi, N.P.: Agroforestry for climate change adaptation, resilience enhancement and vulnerability attenuation in smallholder farming systems in cameroon. J. Atmos. Sci. Res. **5**(1) (2022). https://doi.org/10.30564/jasr.v5i1.4303
3. Bahadur, A.V., Tanner, T.: Policy climates and climate policies: analysing the politics of building urban climate change resilience. Urban Clim. **7** (2014). https://doi.org/10.1016/j.uclim.2013.08.004
4. Banwell, N., Gesche, A.S., Vilches, O.R., Hostettler, S.: Barriers to the implementation of international agreements on the ground: climate change and resilience building in the Araucanía Region of Chile. Int. J. Disaster Risk Reduction **50** (2020). https://doi.org/10.1016/j.ijdrr.2020.101703
5. Batung, E.S., Mohammed, K., Kansanga, M.M., Nyantakyi-Frimpong, H., Luginaah, I.: Credit access and perceived climate change resilience of smallholder farmers in semi-arid northern Ghana. Environ. Dev. Sustain. **25**(1) (2023). https://doi.org/10.1007/s10668-021-02056-x
6. Brown, A., Dayal, A., Rumbaitis Del Rio, C.: From practice to theory: Emerging lessons from Asia for building urban climate change resilience. Environ. Urbanization **24**(2) (2012). https://doi.org/10.1177/0956247812456490
7. Cardoso, M.A., Brito, R.S., Almeida, M.C.: Approach to develop a climate change resilience assessment framework. H2Open J. **3**(1) (2020). https://doi.org/10.2166/h2oj.2020.003

8. Davies, H.N., Beckley, L.E., Kobryn, H.T., Lombard, A.T., Radford, B., Heyward, A.: Integrating climate change resilience features into the incremental refinement of an existing marine park. PLoS ONE **11**(8) (2016). https://doi.org/10.1371/journal.pone.0161094
9. Dogru, T., Marchio, E.A., Bulut, U., Suess, C.: Climate change: Vulnerability and resilience of tourism and the entire economy. Tourism Manag. **72** (2019). https://doi.org/10.1016/j.tourman.2018.12.010
10. Feldmeyer, D., Wilden, D., Kind, C., Kaiser, T., Goldschmidt, R., Diller, C., Birkmann, J.: Indicators for monitoring urban climate change resilience and adaptation. Sustainability (Switzerland) **11**(10) (2019). https://doi.org/10.3390/su11102931
11. Haque, A., Jahid, A.: Climate-change beliefs and resilience to climate change in Bangladesh: is leadership making any difference? J. Environ. Stud. Sci. **11**(4) (2021). https://doi.org/10.1007/s13412-021-00706-0
12. Izuogu, C.U., Onyeneke, R.U., Njoku, L.C., Azuamairo, G.C., Atasie, M.C.: Repositioning Nigeria's agricultural extension system towards building climate change resilience. Sarhad J. Agric. **37**(1) (2021). https://doi.org/10.17582/JOURNAL.SJA/2021/37.1.180.189
13. Karamidehkordi, E., Hashemi Sadati, S.A., Tajvar, Y., Mirmousavi, S.H.: Climate change vulnerability and resilience strategies for citrus farmers. Environ. Sustain. Indic. **20** (2023). https://doi.org/10.1016/j.indic.2023.100317
14. Karunarathne, P., Zhou, Q., Lascoux, M., Milesi, P.: Hybridization mediated range expansion and climate change resilience in two conifers. BioRxiv (2023)
15. Kernaghan, S., da Silva, J.: Initiating and sustaining action: experiences building resilience to climate change in Asian cities. Urban Clim. **7** (2014). https://doi.org/10.1016/j.uclim.2013.10.008
16. Lassa, J.A.: Negotiating institutional pathways for sustaining climate change resilience and risk governance in Indonesia. Climate **7**(8) (2019). https://doi.org/10.3390/cli7080095
17. Liu, L., Lei, Y., Zhuang, M., Ding, S.: The impact of climate change on urban resilience in the Beijing-Tianjin-Hebei region. Sci. Total Environ. **827** (2022). https://doi.org/10.1016/j.scitotenv.2022.154157
18. Mahmood, R., Zhang, L., Li, G., Ranjon Roy, N., Rawnaq, N., Yan, M., Dong, Y., Chen, B.: Geospatial assessment of intrinsic resilience to the climate change for the central coast of Bangladesh. Clim. Risk Manag. **40** (2023). https://doi.org/10.1016/j.crm.2023.10052
19. Mngumi, L.E.: Exploring the contribution of social capital in building resilience for climate change effects in peri-urban areas, Dar es Salaam, Tanzania. GeoJournal **86**(6) (2021). https://doi.org/10.1007/s10708-020-10214-3
20. O'Connor, D., Zheng, X., Hou, D., Shen, Z., Li, G., Miao, G., O'Connell, S., Guo, M.: Phytoremediation: Climate change resilience and sustainability assessment at a coastal brownfield redevelopment. Environ. Int. **130** (2019). https://doi.org/10.1016/j.envint.2019.104945
21. Rajkovich, N.B., Okour, Y.: Climate change resilience strategies for the building sector: Examining existing domains of resilience utilized by design professionals. Sustainability (Switzerland) **11**(10) (2019). https://doi.org/10.3390/su11102888
22. Samuel, J., Rama Rao, C.A., Pushpanjali, Anshida Beevi, C.N., Raju, B.M.K., Amarender Reddy, A., Nagarjuna Kumar, R., Reddy, A.G.K., Singh, V.K., Prabhakar, M., Siva, G.S., Teggelli, R.G.: Enhancing farm income resilience through climate smart agriculture in drought-prone regions of India. Front. Water **6** (2024). https://doi.org/10.3389/frwa.2024.1327651
23. Sanson, A.V., Masten, A.S.: Climate change and resilience: developmental science perspectives. Int. J. Behav. Dev. **48**(2) (2024). https://doi.org/10.1177/01650254231186332
24. Sen, L.T.H., Bond, J., Winkels, A., Linh, N.H.K., Dung, N.T.: Climate change resilience and adaption of ethnic minority communities in the upland area in Thừa Thiên-Huế province, Vietnam. NJAS—Wageningen J. Life Sci. **92** (2020). https://doi.org/10.1016/j.njas.2020.100324
25. Shade, C., Kremer, P., Rockwell, J.S., Henderson, K.G.: The effects of urban development and current green infrastructure policy on future climate change resilience. Ecol. Soc. **25**(4) (2020). https://doi.org/10.5751/ES-12076-250437

26. Shafik, W.: Environmental conservation and sustainable development. In: Developing Digital Inclusion Through Globalization and Digitalization, pp. 142–176. IGI Global (2024a). https://doi.org/10.4018/979-8-3693-4111-7.ch008
27. Shafik, W.: Factors affecting green purchase behaviors for small and medium enterprises (SMEs) in Uganda. In: Entrepreneurial Strategies for the Internationalization and Digitalization of SMEs, pp. 198–224. IGI Global (2024b). https://doi.org/10.4018/979-8-3693-3518-5.ch009
28. Shafik, W.: Generative AI for social good and sustainable development. In: Generative AI: Current Trends and Applications, pp. 185–217. Springer (2024c)
29. Shafik, W.: Incorporating artificial intelligence for urban and smart cities' sustainability. In: Maintaining a Sustainable World in the Nexus of Environmental Science and AI, pp. 23–58. IGI Global (2024d). https://doi.org/10.4018/979-8-3693-6336-2.ch002
30. Shafik, W.: Industry 4.0 technologies' opportunities and challenges for realising net-zero economy. In: Net Zero Economy, Corporate Social Responsibility and Sustainable Value Creation: Exploring Strategies, Drivers, and Challenges, pp. 19–41. Springer (2024e). https://doi.org/10.1007/978-3-031-55779-8_2
31. Shafik, W.: Industry revolution 4.0 and beyond: abilities and perils for sustainable zero net economy development. In: Powering Industry 5.0 and Sustainable Development Through Innovation, pp. 292–316. IGI Global (2024f)
32. Shafik, W.: Industry revolution 4.0 and beyond: abilities and perils for sustainable zero net economy development. In: Powering Industry 5.0 and Sustainable Development Through Innovation, pp. 292–316. IGI Global (2024g)
33. Shafik, W.: Sustainable development. In: Viana Hassan, A.S.S.B. (ed.) Building Community Resiliency and Sustainability With Tourism Development, 1st edn., vol. 1, pp. 1–30. IGI Global (2024h). https://doi.org/10.4018/979-8-3693-5405-6.ch001
34. Shafik, W.: Sustainable development goal 14: Explainable AI (XAI) for ocean health. In: Artificial Intelligence and Edge Computing for Sustainable Ocean Health, pp. 167–198. Springer (2024i). https://doi.org/10.1007/978-3-031-64642-3_8
35. Shafik, W.: Ecological Health Impact in the Marine World, pp. 307–335 (2025a). https://doi.org/10.1007/978-3-031-82403-6_10
36. Shafik, W.: Pathways to climate neutrality: understanding ice cap reduction, glacial melt, and sea level change and adoption. In: Climate Neutrality Through Smart Eco-Innovation and Environmental Sustainability, pp. 53–67. Springer (2025b)
37. Shafik, W.: Zero carbon industry challenges and opportunities in enhancing urban water sector for environmental sustainability and innovation. In: Zero Carbon Industry, Eco-Innovation and Environmental Sustainability, pp. 269–289 (2025c)
38. Shafik, W., Azrour, M.: Building a greener world: harnessing the power of IoT and smart devices for sustainable environment. In: Technical and Technological Solutions Towards a Sustainable Society and Circular Economy, pp. 35–58. Springer (2024). https://doi.org/10.1007/978-3-031-56292-1_3
39. Singh, P., Tabe, T., Martin, T.: The role of women in community resilience to climate change: A case study of an Indigenous Fijian community. Women's Stud. Int. Forum **90** (2022). https://doi.org/10.1016/j.wsif.2021.102550
40. Singh, R., Shafik, W., Crowther, D., Kumar, V. (eds.).: Transforming Healthcare Sector Through Artificial Intelligence and Environmental Sustainability, 1st edn., vol. 1. Springer Nature, Singapore (2024). https://doi.org/10.1007/978-981-97-9555-0
41. Uduji, J.I., Okolo-Obasi, E.N.: The impact of CSR on rural women custodians of seed, food and climate change resilience in Nigeria's Niger Delta region. J. Enterprising Communities **18**(3) (2024). https://doi.org/10.1108/JEC-09-2022-0136
42. Wang, L.: Exploring a knowledge map for urban resilience to climate change. Cities **131** (2022). https://doi.org/10.1016/j.cities.2022.104048
43. Woodruff, S.C., Meerow, S., Stults, M., Wilkins, C.: Adaptation to resilience planning: alternative pathways to prepare for climate change. J. Plann. Educ. Res. **42**(1) (2022). https://doi.org/10.1177/0739456X18801057

44. World Bank.: Climate-smart agriculture: increased productivity and food security, enhanced resilience and reduced carbon emissions for sustainable development—opportunities and challenges for a converging agenda: country examples. In: Working Paper (2011)
45. Yanda, P.Z., Mabhuye, E.B., Mwajombe, A.: Linking coastal and marine resources endowments and climate change resilience of tanzania coastal communities. Environ. Manag. **71**(1) (2023). https://doi.org/10.1007/s00267-021-01553-z
46. Yang, Z., Jianjun, L., Faqiri, H., Shafik, W., Talal Abdulrahman, A., Yusuf, M., Sharawy, A.M.: Green internet of things and big data application in smart cities development. Complexity (2021). https://doi.org/10.1155/2021/4922697

Climate Change and Resilience: Need, Challenges and Future

Rajnish Srivastava, Geetanjali Mehara, Akash Sharma, and Mandeep Kumar Gupta

Abstract Climate change poses a unique global threat, affecting ecosystems, economies, and societies with increasing speed. The increasing frequency of rising sea levels, extreme weather events, and shifting climatic patterns emphasize the need for resilience-building. This chapter describes the concept of resilience in climate change, with a focus on its role in minimizing adverse impacts and ensuring sustainable development. It explores resilience planning challenges, such as financial limitations, policy lacunas, socio-economic inequalities, and technological limitations. It also explores governance, community participation, and scientific invention role in increasing adaptive capacity. The discourse also touches on sector-specific resilience in agriculture, water supply, urban planning, and disaster management, highlighting integrated strategies towards climate adaptation and mitigation. Additionally, the chapter contains best-practice climate resilience programs from around the globe and makes conclusions on how to replicate them. Policymaking for climate resilience is dependent on the future technical developments and the input of businesses, governments, and societies. Climate change needs to be addressed through a shift in paradigms toward sustainable development, equitable resource sharing, and robust institutional arrangements. The chapter ends by highlighting the imperative of a whole-of-society, cross-disciplinary approach to climate resilience aligned with universal sustainability objectives. With increasing intensity of climate problems, building resilience is no longer a choice but an imperative for ensuring a habitable future for everyone.

Keywords Climate change · Resilience · Urban planning · Disaster management · Climate adaptation

---

R. Srivastava
Chitkara University School of Pharmacy, Chitkara University, Baddi, Himachal Pradesh, India

G. Mehara · A. Sharma · M. K. Gupta (✉)
Moradabad Educational Trust Group of Institutions Faculty of Pharmacy, Moradabad, Uttar Pradesh, India
e-mail: mandeepkrgupta@gmail.com

A. L. Srivastav et al. (eds.), *Climate Resilience: Impact of Quantum Computing and Artificial Intelligence on Urban Planning*, Studies in Systems, Decision and Control 637, https://doi.org/10.1007/978-3-032-06791-3_3

1 Introduction

"Climate change" is the most often used environmental word nowadays, and it refers to the modern climate shift that is mostly caused by human activity. The world's population is now dealing with one of the most severe environmental problems [68]. Even so, the problem is not new. The problem was only debated inside the scientific community from the time it first arose in the early nineteenth century until the late twentieth century. It initially appeared on the public agenda in the mid to late 1980s [65, 83]. Since then, believers have expressed the belief that the effects of human activity on the global climate have become worse and are seriously endangering physical and socioeconomic institutions. However, the sceptics have provided sufficient data to rule out the human aspect of climate change [42]. Once again, the proponents of climate change have argued about the best ways to deal with this pressing problem. As a result of growing public interest in the topic of climate change and the resulting understanding of the dangers and uncertainties associated with it, the topic has been discussed and analysed from a variety of angles. Although there is ongoing discussion on how to manage the new hazards and the extent and cause of the shift, there is also no denying that the climate has been changing since the dawn of time [41]. Even if the effects of climate change are still up for debate, they are probably going to include severe weather events, including heat waves, droughts, starvation, ocean acidification, and the extinction of species [86]. However, temperature increases will have an impact on the agricultural sector, which is primarily responsible for food security, and impose additional burdens on people, potentially raising the risk of food, water, and vector-borne illnesses such as malaria, giardiasis, and typhoid [88]. Consequently, there is a complex link between climate change and human health [65]. Addressing and managing the cause of climate change and taking action to lessen its potential effects should be a top concern for society and politics [59]. Resilience is thus one of the primary approaches to climate change management. When a community or society is exposed to risks, resilience is defined as its capacity to confront, adapt to, and recover from such impacts in a reasonable and well-organised manner by preserving and reestablishing its basic composition and functions. Resilience is thus one of the primary approaches to climate change management [20]. When a community or society is exposed to risks, resilience is defined as its capacity to confront, adapt to, and recover from such impacts in a reasonable and well-organised manner by preserving and reestablishing its basic composition and functions [89]. The impact of global climate change has significant negative effects on environmental safety in addition to social welfare, inconsistency, and cross-border relocations, especially in dry regions. Yet, these changes are severe enough to allow for changes in land use, which make the area vulnerable to droughts and provide challenges for disaster management [73]. The region's drought has resulted in tragedies, raising concerns about the quality of the local water supply [73] and food security.

1.1 Historical and Characteristic Features of Climate Change

'Climate Change' literally refers to a long-term shift in the statistical distribution of weather patterns (temperature, precipitation, etc.) over periods ranging from decades to millions of years. The earth's climate has changed throughout all time spans, even before human activity could have contributed to this shift [61]. However, according to UNFCCC (United Nations Framework Convention on Climate Change), climate change is "a change of climate which is attributed directly or indirectly to human activity that alters the composition of the global atmosphere, and which is in addition to natural climate variability observed over comparable time periods". But according to the IPCC, climate change encompasses both human activity and changes brought about by natural variability [46]. Climate Change was explained by the Australian Government's DCCEE on its website- "our climate is changing, largely due to the observed increases in human produced greenhouse gases. Greenhouse gases absorb heat from the sun in the atmosphere and reduce the amount of heat escaping into space. This extra heat has been found to be the primary cause of observed changes in the climate system over the twentieth century". Therefore, several stakeholders have defined climate change in the context of the environment as primarily the alteration of the modern climate brought about by human activity. Additionally, destructive human actions like burning fossil fuels and deforestation are thought to have the potential to alter various climatic features, which are briefly listed in Table 1. These are the characteristics that the phrase mostly refers to when discussing climate change.

However, according to definitions and policy dissemination, the word "climate change" qualifies as a negative manmade climate change in its current sense, although its detrimental impact was first not fully recognized [97]. The discussion is said to have begun in 1824 when French physician and mathematician Jean Fourier wrote an essay for the "Annales de la Chimie et de Physique" that defined the greenhouse effect, which is really at the centre of the climate dispute [22]. Although Keeling was the first to precisely quantify carbon dioxide in the Earth's atmosphere in 1960, Arrhenius [9] presented the first computation of global warming resulting from human emissions of the gas around fifty years later [4]. Vlassopoulos [97] notes that the discussion was mostly limited to the professional community of climatologists and pertinent study was dispersed across several academic endeavours, as climatic fluctuations were not exactly regarded as a scientific concern until 1970. According to AIP, the emergence of environmentalism in the early 1970s caused people to question the advantages of human activity for the earth, which in turn transformed their interest in climate into an anxious concern. Since then, awareness of anthropogenic global deterioration has grown, sparking many initiatives, meetings, and collaborations amongst interested parties, including members from the community outside scientists. It seems that in the 1970s, programs and gatherings were held to examine and recognize the magnitude of human-caused climate change. An example in this respect is the 1974 Global Atmospheric Research Program (GARP), which was arranged by the International Council of Scientific Unions (ICSU) and the World Meteorological Organization (WMO). Examining the very complicated issue of the physical foundation of climate

Table 1 Characteristics of climate change and its alleged effects

Climatic features		Implications of change
Global warming	GHG concentration	The concentration of greenhouse gases in the atmosphere is rising due to emissions from industrialization, travel, and other activities. The concentration of CO_2 is now at 393 ppm, the highest level in 650,000 years [24]
	Change in world temperature	GHG concentration causes the earth to wane along with a few other problems. Since 1880, Earth has warmed. Most of this warming has happened during the 1970s; the last 20 hottest years have happened since 1981, and the last 12 years have seen all 10 of the warmest years. It's being primarily at the heart of the problem. Changes in several aspects of the human environment are brought about by global desire [21]
Ozone layer depletion		It is predicted that since the late 1970s, the total amount of ozone in Earth's stratosphere (the ozone layer) has been steadily declining at a rate of around 4% per decade [52]
Shrinking ice sheets		The Gravity Recovery and Climate Experiment (GRACE) satellites (2002–2017) and GRACE Follow-On (since 2018) show that from 2002 to 2020, Antarctica lost around 150 gigatons of ice year, contributing to a global sea level increase of 0.4 mm per year [5]
Increasing sea level		The sea level risen by almost 17 cm (6, 7 inches) in the last century. If it keeps growing, it will probably flood numerous island nations and low-lying delta areas, depriving the people of any land on which to live [30]
Ocean acidification		Ocean acidity has risen by over 30% since 1750 because of the Earth's oceans' growing CO_2 concentration, which is now rising by approximately 2 billion tons per year [72]
Warming oceans		With the top 700 m (about 2300 ft) of ocean showing warming of 0,16 °C since 1969 due to increased heat absorbed by the Earth. These two changes are likely to bring massive change/destruction in ocean habitats [64]

was its goal [23]. The inaugural World Climate Conference (WCC), which was hosted by the WMO in 1979, may serve as another example. However, it may be argued that several noteworthy events in the 1980 and 1990s prompted the development of solutions. For instance, worldwide limits on the release of gases that deplete ozone are enforced by the 1987 Vienna Convention's Montreal Protocol. Two significant events in the 1990s were the Rio de Janeiro meeting in 1992, which produced the UN Framework Convention on Climate Change, and the International meeting in 1997, which produced the Kyoto Protocol, which went into force in 2005 and established goals for developed countries to cut their greenhouse gas emissions.

1.2 Climate Change Discourse: Key Perspectives and Assumptions

The Climate Change discourse has been disseminated over three centuries, but it has encountered some crises because of the perspectives of the concerned community (e.g., scientific) of the time, which were based on pertinent assumptions. This initiated investigations, which were conducted after research knowledge had been accumulated and established in society, resulting in the development of the standard science. Subsequently, a crisis arose, resulting in a shift in the conventional science that persisted for a period. Consequently, the discourse continues with a revolution of the established conventional science. Based on the historicization in the preceding section of the essay and a review of other relevant literature, we would like to propose that the discourse surrounding climate change has been classified into three distinct paradigms, each of which has evolved by the development of specific perspectives based on pertinent assumptions (Fig. 1). The 'pre-science' period, as Kuhn defines it, is the period preceding the initial 'crisis' [54]. The initial "crisis" that initiates the climate study can be characterised as a "lack of knowledge" regarding the changing climate, or in other words, a "curiosity" regarding the climate, as noted by AIP [72]. Thus, the suggestion made by Foe that Benjamin Franklin was the pioneer of the discourse (as he initiated climate studies in 1763) appears more plausible to me in the years following of that crisis. The discourse originated in the nineteenth century with Fourier. Despite this debate, the scientific community, particularly climatologists, were exclusively drawn to the investigation of climate change science from the very beginning, as it was motivated by "curiosity" or the pursuit of climate change knowledge. This trend continued until the next crisis in the 1970s. The negative impact of increased CO_2 emissions and the effects of burning coal on the Earth's temperature was not appreciated by Fourier, after Arrhenius's research. It illustrated that Climate Change was a state that was governed by a 'objective view' during this period, which was neutral to any Eco-centric or Anthropocentric perspective. This view was influenced by the notion that it was entirely a scientific issue [97].

1.2.1 Intergovernmental Panel on Climate Change (IPCC)

Since its inception in 1988, the Intergovernmental Panel on Climate Change (IPCC) has been tasked with presenting decision-makers with the most recent state of knowledge. Specifically, the IPCC is responsible for "providing a comprehensive summary of what is known about the drivers of climate change, its impacts and future risks, and how adaptation and mitigation can reduce those risks [53]." This task is characterized by a significant amount of uncertainty in terms of knowledge. This is due in part to the fact that it pertains to the highly intricate climate system, which complicates the determination of the consequences of climate change, and in part because it pertains to the (largely unknown) societal responses to climate change. The climate issue is future-oriented, and the knowledge uncertainty is also associated with the future

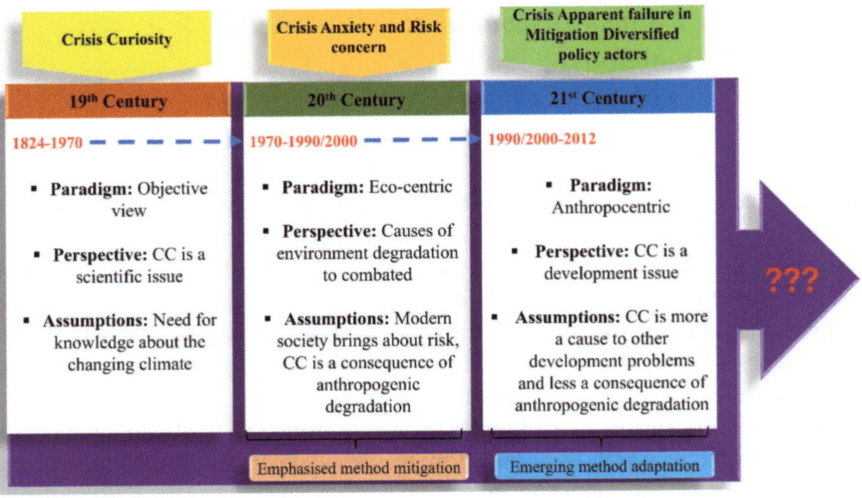

Fig. 1 Structure of the revolution of the climate change discourse

social and environmental consequences of today's emissions [50]. This is a typical risk issue, as it pertains to the potential for severe negative repercussions. The climate issue is further complicated by the fact that it is not solely a matter of knowledge uncertainty, but also of normative uncertainty. Climate change, like all environmental challenges, is a hazard to a societal value that necessitates compromises with other societal objectives and values [51, 91]. The IPCC has developed various strategies to manage uncertainty and risk, including guidelines, a language of uncertainty and confidence, a risk framework, and organization-wide guidelines. These strategies aim to communicate the unknowns about adverse impacts, understand and manage climate extreme risks, and ensure a uniform use of risk terminology. However, implementing these guidelines and regulations is crucial, as they may not align with the way scientists are trained to manage these factors in their respective fields of research. Therefore, it is important to investigate the IPCC's approach and understanding of risk and uncertainty to effectively manage these issues [49].

2 Key Concepts in Climate Science

Many of the important concepts in climate science are briefly described in this section; however, they were previously summarised in more comprehensive detail in previous IPCC assessments [96]. Initially, it is crucial to differentiate between the terms "weather" and "climate". The term "weather" refers to the conditions of the atmosphere at a specific location and time, including temperature, pressure, humidity, wind, and other key parameters (meteorological elements). It also encompasses the presence of clouds, precipitation, and the occurrence of special phenomena, such as thunderstorms, dust storms, and tornadoes. The average weather is typically the narrowest definition of climate. The World Meteorological Organization defines climate as a long-term shift in weather patterns, characterized by variations in the mean and variability of variables. Climate change, a term used to describe such changes, is a shift in weather patterns that can be detected through statistical tests, often characterized by factors such as frequency, amplitude, persistence, and trends. Solar radiation controls the Earth's climate system (Fig. 2). The visible portion of the electromagnetic spectrum contains about half of the Sun's energy. The entering solar energy must be almost equal to the exiting radiation since the Earth's temperature has remained mostly stable over countless generations. The Earth's surface absorbs almost half of the incoming solar shortwave radiation (SWR). About 20% of SWR is absorbed in the atmosphere, while the remaining 30% is reflected to space by gases, aerosols, clouds, and the Earth's surface (albedo). Most of the energy flow that leaves the Earth is in the infrared portion of the spectrum, which is determined by the temperature of the Earth's surface. Some atmospheric components, such as water vapour, carbon dioxide (CO_2), methane (CH_4), nitrous oxide (N_2O), and other greenhouse gases, absorb a large portion of the longwave radiation (LWR, also known as infrared radiation) emitted from the Earth's surface. Clouds also emit LWR in all directions. The Earth's surface and lower atmosphere get additional heat from this LWR's downward-directed component (greenhouse effect). Higher troposphere levels account for the majority of the Earth's infrared radiation energy loss. The tropics and subtropics get the majority of the Sun's energy, which is then partly transferred to the middle and high latitudes via marine and atmospheric transport mechanisms. Changes in either the net incoming solar radiation or the outgoing longwave radiation (OLR) cause variations in the global energy balance [44].

3 Impacts and Challenges of Climate Change

Climate change increases acute catastrophes like floods and wildfires, while causing longer-term harm like droughts and rising sea levels, impacting regional, national, community, family, and individual systems [19]. There is now enough evidence of the direct effects of climatic catastrophes on physical health, including illnesses, respiratory disorders, malnourishment, fatalities and injuries [98]. Climate change catastrophes disrupt human systems such as waste disposal, transportation, health, education, and food production, causing long-term effects like prolonged hunger. Famine can lead to physical disability, cognitive decline, and neurological damage, highlighting the urgent need for action to mitigate these impacts [3]. The UN Security Council identifies climate change as a "threat multiplier" due to its potential to cause political and economic instability by destroying resources like habitable and arable land [78]. By 2050, the World Bank estimates climate change would force hundreds of millions of people to relocate [77]. Climate refugees are fleeing their homes due to extreme temperatures, poor food production, lack of clean water, loss of infrastructure, and wars. Rising temperatures and altered rainfall patterns are

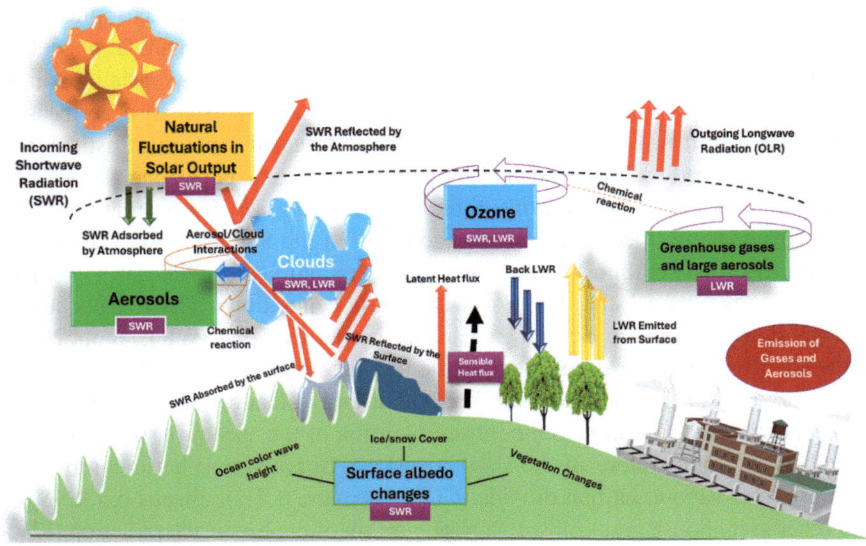

Fig. 2 Solar radiation controls the Earth's climate system

linked to increased conflict and violence among people, communities, and nation-states, according to numerous studies [35]. Following a tragedy, there may be an upsurge in interpersonal and communal violence, and some parents may resort to harsher parenting techniques [63]. Damage and loss from climate change are more likely to affect those who are already vulnerable. For instance, people with disabilities, who make up 15% of the world's population, are particularly vulnerable to harm or death in the event of a catastrophe [11]. Despite years of neglect, efforts to reduce disaster risk have recently emphasised addressing special needs associated with disabilities, as seen in the Sendai Framework for Disaster Risk Reduction 2015–2030, adopted at a UN world conference [87]. Many still lack the climate funding required to cover mitigation and adaptation, given recognised drawbacks confronting poor nations, and many still cannot reach the existing climate finance reservoirs. For these nations, many of whom are Commonwealth poor countries, this is a major handicap as they need this money to improve their infrastructure, decarbonise and move to green technology and renewables, and eliminate poverty [28]. Advanced age increases vulnerability to infectious illnesses, deteriorating air quality, and severe weather events. Due to increased dependence, lower socioeconomic status, and multimorbidity, older adults are less likely to survive injuries and escape disasters [40]. Perhaps the age group most at risk from climate change is young people, for several reasons [93]. Younger children and adolescents often rely on adults for safety and care, often under stress or unavailable due to separation, injury, or death. Poverty, especially in poor countries, increases susceptibility to catastrophic impacts, insufficient nutrition, and disasters [25]. Additionally, children's immunological and neurological systems are less developed, and they are less able to recognise and react to threats (American Public Health Association [APHA], 2019) [14]. Children's shorter stature, increased ground-based activities, and increased hand-to-mouth movements contribute to increased exposure to air, water, soil pollutants and disease vectors [2]. The World Health Organisation predicts that children will experience over 80% of the physical health effects of climate change due to their increased exposure to these pollutants and disease vectors [13]. Additionally, children of today are expected to encounter a variety of climate-related stressors that will have an accumulation effect throughout the course of their lifetimes. A child born in 2020 is on average likely to experience up to 24 times as many climate-induced extreme weather events (heat waves, wildfires, crop failures, droughts, river floods, and tropical cyclones) in their lifetime as someone born in the 1960s, according to [92], even if warming is kept to 1.5 °C [92]. The Children's Climate Risk Index (CCRI) by UNICEF assesses countries based on 57 factors, including direct exposure to environmental and climate disasters, poverty, education, sanitation, health, nutrition systems, and social protection (Table 2). Nearly half of all children live in countries at extreme risk from climate change, mainly in the Global South. The CCRI highlights the devastating impact of climate change on children's well-being, with nearly every child exposed to at least one serious climate or environmental hazard, and 850 million children exposed to four or more hazards [82]. Desperation, hopelessness, and sadness are common among young individuals, with high rates of post-traumatic stress disorder (PTSD) symptoms among children and teens after climate-related disasters. For example, 73% of children in Pakistan aged 10–19 showed significant PTSD symptoms four months after floods, with displaced females being disproportionately affected [66]. Yelland and colleagues (2010), on the other hand, found that among students aged 8 to 18 who had been exposed to bushfires in Australia, just 10% had severe or very severe PTSD symptoms a year after the flames, 17% had moderate symptoms, and the majority had mild symptoms. Researchers found that younger children reported higher symptoms of disasters, indicating that the design of the study may influence the main conclusions, as the consequences of disasters vary depending on developmental timing and exposure [57]. Additionally, children who endure a series of climate-related catastrophes may face cumulative negative health impacts, particularly if they do not receive enough assistance [47]. Anticipating escalating climate consequences gives birth to a different class of climate change psychological effects [15, 16]. Pew Research Centre study found that 75% of respondents in 19 North American, European, and Asia–Pacific countries believed climate change was a danger to their country, while only 5% did not. A recent poll of 10,000 young people aged 16–25 from ten Global North and Global South countries also showed similar results [33].

Table 2 Major impacts and key challenges of climate change [28, 90]

Category	Impact	Challenges
Environment	Rising temperatures, melting glaciers, sea-level rise	Preserving biodiversity, protecting natural habitats
Weather patterns	Increased frequency of extreme weather events (storms, droughts)	Improving early warning systems, disaster preparedness
Agriculture	Reduced crop yields, soil degradation	Developing climate-resilient crops, ensuring food security
Water resources	Water scarcity, contamination of freshwater supplies	Managing water resources sustainably
Human health	Spread of diseases, heat-related illnesses	Strengthening healthcare systems, public awareness
Economy	Damage to infrastructure, increased costs for recovery and adaptation	Securing funding for mitigation and adaptation efforts
Social systems	Migration and displacement, increased inequality	Creating social safety nets, managing climate refugees
Energy sector	Disruption to energy production and supply	Transitioning to renewable energy sources
Political stability	Conflicts over resources, national security risks	Strengthening international cooperation and policy responses

4 Building Resilience of Current and Future Generations to Climate Threats

The multisystem processes that safeguard people in the short term and foster future resilience must be considered to address the numerous challenges that climate change poses to human development. Resilience and positive development models provide valuable theoretical insights, empirical support, and actionable advice on how to avoid, lessen, or minimise the hazards that climate change poses to human development, especially for children and young people [55]. Human development and human adaptation are influenced by protective and promotional variables, particularly in the face of adversity. In climate crises, human adaptation is influenced by threat exposures and vulnerabilities within the community, family, and individual. Resources, protections, and capabilities at multiple system levels are also important for human survival and development. Research on responses to climate-related disasters and multi-system strategies for risk reduction or resilience enhancement aligns with decades of prior research on resilience among people who overcome threats associated with acute and cumulative adversity [56, 95]. Chronic risks like child abuse and climate-related catastrophes can improve outcomes, and individuals who recover from adversity have critical assets or resources that are essential for their well-being and overall development, according to research [31, 48, 70]. Interpersonal resources and support from relationships with family, friends, mentors, and teachers; engagement with broader social resources and supports offered by successful schools, religious and cultural traditions, and safe communities invested in the well-being of children, youth, and families; and individual skills, behaviours, and mindsets like problem-solving, self-regulation skills, optimism, hope, self-efficacy, a sense of belonging, and a sense of meaning or purpose are all examples of the capacities and resources associated with resilience in development. Numerous of these often-noted indicators of successful adaptation in adversity conditions are supported by systematic studies [26, 60]. Communities, schools, and families prepare for future calamities and foster resilience by providing resources and protection against current adversity. Policies at local and societal levels support ecological systems that benefit children, youth, and families, such as clean water, childcare, tax credits, and healthcare. Funding for community initiatives enhances schools, safety, and youth leadership opportunities, increasing resilience capacity. Benefits of these elements are generally promotive, regardless of risk or adversity exposure, and are particularly significant when danger levels are high [55]. According to new research, several well-known resources, including optimism, agency, and belonging, will be helpful for tackling climatic adversity [81]. It has been shown that meaning-focused coping mechanisms safeguard psychological health in young people in Sweden dealing with climate change issues [69]. Spending time in nature and participating in environmental advocacy are two activities that may help prepare young people to handle the growing issues associated with climate change [69, 93]. Engaging in collective action, like attending rallies, is associated with a feeling of agency or self-efficacy that seems to offset the helplessness and anxiety brought on by worries about climate change [80]. A feeling of purpose, the capacity for constructive reconsideration, and faith in scientific institutions are further protective and motivating qualities. Scholars have observed remarkable similarities in resilience factors found at several system levels crucial to human development, such as communities, schools, and families [55]. The World Bank emphasizes the significance of resilience in socio-ecological systems, which are adaptive qualities resulting from beneficial interactions between systems and inhabitants. Governments, humanitarian organizations, and the World Bank have implemented programs integrating various sectors to promote children's healthy development worldwide [7]. However, assessing the effectiveness of multi-system and multisector interventions that are adapted to situational, individual, and cultural variables is difficult.

5 Strategies for Enhancing Climate Resilience

A comprehensive, multi-level strategy that incorporates environmental, technical, social, and policy-based tactics is needed to build climate resilience [36]. Societies everywhere must take proactive measures to safeguard people and ecosystems as the effects of climate change worsen and occur more often [1]. These effects include rising sea levels, harsh weather, and food insecurity. The use of nature-based solutions is one of the most popular tactics [6]. Ecosystems can serve protective functions by restoring mangrove forests to reduce coastal erosion and floods, reforesting degraded areas to stabilise soils and absorb carbon, and maintaining wetlands as organic water storage systems [38]. These interventions are a sustainable and affordable method of promoting resilience since they not only lessen vulnerability but also boost biodiversity, increase air and water quality, and assist local livelihoods [34]. Innovations in infrastructure and technology are crucial in addition to natural methods. These include drought-tolerant agricultural types, climate-resilient infrastructure like flood-resistant buildings and roads, and renewable energy sources that provide energy security while lowering reliance on fossil fuels [27]. Communities may better plan for and react to catastrophes with the use of smart irrigation systems, early-warning technology, and geographic information systems (GIS) for climate monitoring [45]. Cities are implementing climate-resilient urban design concepts like permeable pavements, green roofs, and increased vegetation to mitigate heat island impacts and manage stormwater, emphasizing the importance of governance and policy [17]. Resilience is being progressively incorporated into legal and policy frameworks by national governments, local governments, and international organisations. Adaptation initiatives in sensitive areas are supported by climate financing instruments, including the Green Climate Fund, the Paris Agreement, and National Adaptation Plans (NAPS) [58]. Effective governance promotes cross-sectoral cooperation and guarantees that local initiatives are in line with larger national and international objective [12]. Decentralisation of planning, which gives communities and local governments the tools and freedom to put solutions in place that are specific to their climate threats, is equally crucial [32] Public awareness and education are the cornerstones of resilience. Increasing awareness and comprehension of climate hazards enables people and communities to respond with knowledge. Social cohesion is improved, and culturally and regionally relevant solutions are guaranteed via community-based adaptation projects that include residents in planning and decision-making [43]. Long-term sustainability is promoted by capacity-building initiatives that teach farmers, young people, and local leaders' climate-resilient techniques. Additionally, inclusive participation guarantees that disadvantaged groups', women's, and indigenous peoples' views are heard, which leads to more equitable results. In locally driven, context-specific solutions backed by both domestic and international collaboration [94]. A strong foundation for lowering vulnerability and boosting adaptive capability is provided by the combination of nature, technology, policy, and education. Investing in resilience now is crucial to ensuring a sustainable and equitable future for future generations as climate risks continue to change [29].

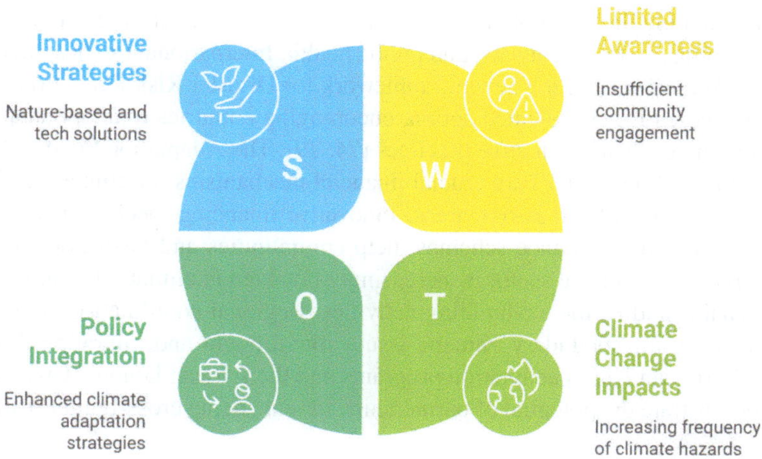

Fig. 3 SWOT (Strengths, Weakness, Opportunities and Threats) analysis on climate resilience

6 Future Perspectives on Climate Resilience

6.1 Emerging Trends in Climate Adaptation

As climate change intensifies, new climate adaptation trends are important to increase resilience in communities all over the world. One of the most exciting trends is nature-based solutions, which include using ecosystems to buffer against climate effects (Fig. 3). Wetland restoration, afforestation of degraded areas, and the establishment of urban green spaces enhance flood management, mitigate heat stress, and promote biodiversity. Climate-smart agriculture (CSA) is gaining increased popularity. This approach employs innovative techniques such as regenerative agriculture, drought-tolerant crops, and precision farming to enhance food security while minimizing environmental impact [67].

Technological innovation, including Artificial intelligence and GIS, is transforming systems to prepare for extreme weather. These technologies enable efficient resource allocation, real-time environmental tracking, and informed decision-making. Remote sensing and GIS enable real-time tracking of environmental change, while decentralized renewable energy systems like microgrids and community solar programs provide energy security and minimize fossil fuel dependence [18, 37].

6.2 Policy Frameworks and International Cooperation

Effective climate adaptation necessitates strong policy frameworks and international collaboration (Fig. 3). Governments are integrating climate resilience into national planning through policies aimed at addressing vulnerabilities in infrastructure, agriculture, water resources, and public health. International treaties, including the Paris Agreement and the Sendai Framework for Disaster Risk Reduction, establish frameworks for coordinated action, encouraging countries to adopt adaptation strategies and exchange effective practices [74, 79]. The Adaptation Fund (AF) and Green Climate Fund (GCF) are crucial financial mechanisms for climate resilience initiatives, particularly in at-risk areas. Innovative financing mechanisms, such as climate bonds and insurance schemes, help communities and businesses mitigate climate risks. The private sector is increasingly involved in climate risk appraisal in supply chains and resilience-building activities. Regional coordination is essential for addressing transboundary climatic issues like drought and migration. Policies like the European Union adaptation programs and the African Union's Great Green Wall demonstrate the potential of partnerships for enhancing cross-border resilience [16].

6.3 Integration of Indigenous Knowledge and Modern Science

Traditional knowledge systems honed over several centuries offer penetrating insights in response to climate change. Traditional Australians possess deep environmental knowledge, deploying conventional methods such as agroforestry, water harvesting, and rotational grazing to maintain balanced ecosystems. Combining this with modern scientific methodology can produce ecologically sustainable and culturally suitable adaptation strategies [75]. Participatory strategies for climate resilience planning are on the rise, focusing on integrating Indigenous perspectives into policy development and decision-making. Joint research efforts that merge Indigenous methods with modern science have been effective in building resilience at the community level. For example, in the Arctic, Indigenous observations of ice patterns complement satellite data to improve the accuracy of climate models [84]. Cross-sectoral learning between scientific institutions and Indigenous peoples is important to build resilience through mutual respect and knowledge sharing. Empowering Indigenous land rights and self-determined adaptation management also supports resilience by ensuring that local appropriateness and social equity of adaptation solutions are both achieved [71].

6.4 Towards a Sustainable and Resilient Future

Creation of a resilient and sustainable future will require an integrative multi-dimensional approach that merges adaptation, mitigation, and socio-economic development. Increased urban resilience in the shape of climate-resilient infrastructure such as flood-proof houses and heat-resistant roads will become necessary to protect burgeoning populations from the effects of climate change. Enhancing water security by adopting sustainable management practices, such as desalination strategies and rainwater harvesting innovations, will be imperative in the face of the expected increase in global water demand [10, 76].

Climate literacy is crucial for fostering resilience against climate change. It promotes public awareness, empowers youth, and encourages anticipatory measures for climate adaptation. Gender-sensitive approaches, including those for women and minority populations, ensure that these groups are not left behind in the process [39]. A resilient society requires combined action at the global, national, and local levels. This includes governments, businesses, scientists, and communities working together to find solutions to climate change. By integrating emerging technologies, strong policy actions, Indigenous knowledge, and participatory governance, a more equitable and resilient society can be built [8, 62, 85].

7 Conclusion

Climate change presents humanity with an existential threat, for which urgent and consistent efforts to build resilience at individual, community, national, and international levels are needed. With rising climate-related disasters becoming more frequent and intense, resilience has never been more essential. Long-term adaptation and sustainability in the face of changing climate hazards are important aspects of resilience, in addition to surviving immediate climatic shocks. The need for resilience, its attainment, and the routes to a sustainable and climate-resilient future have all been covered in this chapter. The need for resilience is prompted by rising risks attributable to global warming, weather hazards, sea-level rise, and ecosystem changes. The disproportionate consequences will have largest impacts on most vulnerable groups, particularly in poor countries, rendering it essential to establish targeted measures of resilience. Resilience is needed to ensure that infrastructure, agriculture, water resources, and public health infrastructure can absorb climate shocks. Secondly, economic resilience is also important, as climate change endangers livelihoods, deepens poverty, and interferes with global supply chains. With no preventive actions, the negative impacts of climate change might leave millions more in deeper vulnerability.

Policy structures face key challenges in climate adaptation, including financial constraints and inefficiencies. Developing countries struggle to mobilize sufficient budgets, and external finance remains insufficient. Policy structures are hindered by inefficiencies, lack of enforcement, and political obstacles. Scientific research and technological options can enhance climate resilience, but uneven access and incorrect information hinder collective action against climate threats. Interconnected policies are crucial for a future that balances economic development with climate resilience. These policies involve renewable energy, nature-based solutions, and green urban planning. International cooperation, knowledge exchange, and participatory governance are essential for managing climate risks. A shift towards a climate-resilient future requires technological progress, behavioral transformations, learning, and community action. The private sector also plays a significant role, with companies investing in sustainable development. Looking ahead, the future of climate resilience hangs in balance with a multidimensional and collaborative approach. Governments, scientific institutions, industries, and civil society organizations must collaborate to develop adaptive capacities. Climate resilience must be mainstreamed into policy making, development planning, and investment. Also, the integration of traditional knowledge with scientific approaches can yield holistic solutions. Technological advancements like artificial intelligence, climate models, and climate-resilient infrastructure will be at the forefront to predict and absorb climate risks. Behind each of these efforts lies a realization that disaster response is only part of resilience—creating sustainable development patterns that reduce vulnerabilities.

Consent for Publicaton We hereby give our consent to publish the content of this book chapter.

Acknowledgements None.

Conflict of Interest The authors have declared that this book chapter is not associated with any conflicts of interest.

References

1. Abbass, K., Qasim, M.Z., Song, H., Murshed, M., et al.: A review of the global climate change impacts, adaptation, and sustainable mitigation measures. Environ. Sci. Pollut. Res. Int. **29**(28), 42539 (2022). Available at: https://doi.org/10.1007/S11356-022-19718-6
2. Alves, J.G.B., Alves, G.V.: Effects of physical activity on children's growth. Jornal de Pediatria 9572–9578 (2019). Available at: https://doi.org/10.1016/J.JPED.2018.11.003
3. Anderko, L., Chalupka, S., Du, M., Hauptman, M.: Climate changes reproductive and children's health: a review of risks, exposures, and impacts. Pediatric Res. **87**(2), 414–419 (2020). Available at: https://doi.org/10.1038/S41390-019-0654-7
4. Anderson, T.R., Hawkins, E., Jones, P.D.: CO_2, the greenhouse effect and global warming: from the pioneering work of Arrhenius and Callendar to today's earth system models. Endeavour **40**(3), 178–187 (2016). Available at: https://doi.org/10.1016/J.ENDEAVOUR.2016.07.002
5. Anon.: Antarctic Ice Loss 2002–2020: GRACE Tellus. Available at: https://grace.jpl.nasa.gov/resources/31/antarctic-ice-loss-2002-2020
6. Anon.: Climate Change and Agriculture, vol. 9 (2023). Available at: https://doi.org/10.22271/ED.BOOK.2395
7. Anon.: Investing in Early Childhood Development: Review of the World Bank's Recent Experience (2015). Available at: https://hdl.handle.net/10986/20715
8. Argyroudis, S.A., Mitoulis, S.A., Chatzi, E., Baker, J.W., et al.: Digital technologies can enhance climate resilience of critical infrastructure. Clim. Risk Manag. **35** (2022). Available at: https://doi.org/10.1016/j.crm.2021.100387
9. Arrhenius, S.: On the influence of carbonic acid in the air upon the temperature of the ground. Lond. Edinburgh, Dublin Philos. Mag. J. Sci. [Online] **41**(251), 237–276 (1896). Available at: https://doi.org/10.1080/14786449608620846

10. Baraj, B., Mishra, M., Sudarsan, D., Silva, R.M. da, et al.: Climate change and resilience, adaptation, and sustainability of agriculture in India: a bibliometric review. Heliyon **10**(8), e29586 (2024). Available at: https://doi.org/10.1016/j.heliyon.2024.e29586
11. Bennett, D.D.: Five years later: assessing the implementation of the four priorities of the sendai framework for inclusion of people with disabilities. Int. J. Disaster Risk Sci. **11**(2), 155–166 (2020). Available at: https://doi.org/10.1007/S13753-020-00267-W/TABLES/1
12. Boas, I., Biermann, F., Kanie, N.: Cross-sectoral strategies in global sustainability governance: towards a nexus approach. Int. Environ. Agreements: Polit. Law Econ. **16**(3), 449–464 (2016). Available at: https://doi.org/10.1007/S10784-016-9321-1/METRICS
13. Bull, F.C., Al-Ansari, S.S., Biddle, S., Borodulin, K., et al.: World Health Organization 2020 guidelines on physical activity and sedentary behaviour. Br. J. Sports Med. **54**(24), 1451–1462 (2020). Available at: https://doi.org/10.1136/BJSPORTS-2020-102955
14. Charles, Hodges, H., Masten, A.S.: Risk and resilience of somali children in the context of climate change, famine, and conflict. J. Appl. Res. Child. Informing Policy Child. Risk **12** (2021). Available at: https://doi.org/10.58464/2155-5834.1453
15. Clayton, S.: Climate anxiety: psychological responses to climate change. J. Anxiety Disord. 74102263 (2020). Available at: https://doi.org/10.1016/J.JANXDIS.2020.102263
16. Commonwealth Practical Guide, A.: Toolkit to Enhance Access to Climate Finance A Commonwealth Practical Guide (2022)
17. Cuce, P.M., Cuce, E., Santamouris, M.: Towards sustainable and climate-resilient cities: mitigating urban heat islands through green infrastructure. Sustainability **17**(3), 1303 (2025). Available at: https://doi.org/10.3390/SU17031303
18. Diehr, J., Ogunyiola, A., Dada, O.: Artificial intelligence and machine learning-powered GIS for proactive disaster resilience in a changing climate. Ann. GIS 1–14 (2025). Available at: https://doi.org/10.1080/19475683.2025.2473596
19. Evans, G.W.: Projected behavioral impacts of global climate change. Ann. Rev. Psychol. 70449–70474 (2019). Available at: https://doi.org/10.1146/ANNUREV-PSYCH-010418-103023
20. Ferguson, P., Wollersheim, L., Lowe, M.: Approaches to climate resilience. In: The Palgrave Handbook of Climate Resilient Societies, pp. 1–25 (2021). Available at: https://doi.org/10.1007/978-3-030-32811-5_97-1
21. Filonchyk, M., Peterson, M.P., Zhang, L., Hurynovich, V., et al.: Greenhouse gases emissions and global climate change: examining the influence of CO_2, CH_4, and N_2O. Sci. Total Environ. 935173359 (2024). Available at: https://doi.org/10.1016/J.SCITOTENV.2024.173359
22. Fleming, J.R.: Joseph Fourier, the 'greenhouse effect', and the quest for a universal theory of terrestrial temperatures. Endeavour **23**(2), 72–75 (1999). Available at: https://doi.org/10.1016/S0160-9327(99)01210-7
23. Flohn, H.: Climate and energy: a scenario to a 21st century problem. Climatic Change **1** (1), 5–20 (1977). Available at: https://doi.org/10.1007/BF00162774/METRICS
24. Florini, A., Pauli, M.: Collaborative governance for the sustainable development goals. Asia Pac. Policy Stud. **5**(3), 583–598 (2018). Available at: https://doi.org/10.1002/APP5.252;REQUESTEDJOURNAL:JOURNAL:20502680;WGROUP:STRING:PUBLICATION
25. Fothergill, A.: Children, Youth, and Disaster. Oxford Research Encyclopedia of Natural Hazard Science (2017). Available at: https://doi.org/10.1093/ACREFORE/9780199389407.013.23
26. Fritz, J., de Graaff, A.M., Caisley, H., van Harmelen, A.L., et al.: A systematic review of amenable resilience factors that moderate and/or mediate the relationship between childhood adversity and mental health in young people. Front. Psychiatry **9** (2018). Available at: https://doi.org/10.3389/FPSYT.2018.00230
27. Gamage, A., Gangahagedara, R., Subasinghe, S., Gamage, J., et al.: Advancing sustainability: The impact of emerging technologies in agriculture. Curr. Plant Biol. 40100420 (2024). Available at: https://doi.org/10.1016/J.CPB.2024.100420
28. Gomez, D.: Climate change–challenges, issues and Commonwealth responses. Round Table **110**(5), 539–545 (2021). Available at: https://doi.org/10.1080/00358533.2021.1985261;JOURNAL:JOURNAL:CTRT20;WGROUP:STRING:PUBLICATION

29. Haider, L.J., Cleaver, F.: Capacities for resilience: persisting, adapting and transforming through bricolage. Ecosyst. People **19**(1) (2023). Available at: https://doi.org/10.1080/26395916.2023.2240434
30. Hamlington, B.D., Bellas-Manley, A., Willis, J.K., Fournier, S., et al.: The rate of global sea level rise doubled during the past three decades. Commun. Earth Environ **5**(1), 1–4 (2024). Available at: https://doi.org/10.1038/S43247-024-01761-5
31. Hawkins, M.T., Letcher, P., Sanson, A., Smart, D., et al.: Positive development in emerging adulthood. Australian J. Psychol. **61**(2), 89–99 (2009). Available at: https://doi.org/10.1080/00049530802001346
32. Helmsing, A.H.J.: Decentralisation, enablement, and local governance in low-income countries. Environ. Plann. C: Govern. Policy **20**(3), 317–340 (2002). Available at: https://doi.org/10.1068/C0040
33. Hickman, C., Marks, E., Pihkala, P., Clayton, S., et al.: Climate anxiety in children and young people and their beliefs about government responses to climate change: a global survey. Lancet Planet. Health **5**(12), e863–e873 (2021). Available at: https://doi.org/10.1016/S2542-5196(21)00278-3
34. Hollaar, M.H.L., Kemmere, B., Kocken, P.L., Denktaş, S.: Resilience-based interventions in the public sector workplace: a systematic review. BMC Public Health **25**(1), 350 (2025). Available at: https://doi.org/10.1186/S12889-024-21177-2/TABLES/6
35. Hsiang, S.M., Burke, M., Miguel, E.: Quantifying the influence of climate on human conflict. Science **341**(6151) (2013). Available at: https://doi.org/10.1126/SCIENCE.1235367
36. Irepoglu Carreras, Y.: Problem-solving across literatures: comparative federalism and multilevel governance in climate change action. Eur. Policy Anal. **5**(1), 117–134 (2019). Available at: https://doi.org/10.1002/EPA2.1066
37. Jain, H., Dhupper, R., Shrivastava, A., Kumar, D., et al.: AI-enabled strategies for climate change adaptation: protecting communities, infrastructure, and businesses from the impacts of climate change. Comput. Urban Sci. **3**(1) (2023). Available at: https://doi.org/10.1007/s43762-023-00100-2
38. Jiang, Y., Zhang, Z., Friess, D.A., Li, Y., et al.: Restoring mangroves lost by aquaculture offers large blue carbon benefits. One Earth **8**(1), 101149 (2025). Available at: https://doi.org/10.1016/J.ONEEAR.2024.11.003
39. Johnston, J.D.: Climate Change Literacy to Combat Climate Change and Its Impacts, pp. 1–14 (2019). Available at: https://doi.org/10.1007/978-3-319-71063-1_31-1
40. Jonkman, S.N., Maaskant, B., Boyd, E., Levitan, M.L.: Loss of life caused by the flooding of New Orleans after Hurricane Katrina: analysis of the relationship between flood characteristics and mortality. Risk Anal. Official Publ. Soc. Risk Anal. **29**(5), 676–698 (2009). Available at: https://doi.org/10.1111/J.1539-6924.2008.01190.X
41. Kane, J. V., Perry, S.L.: Belief in divine (versus human) control of earth affects perceived threat of climate change. NPJ Clim. Action **3**(1), 1–9 (2024). Available at: https://doi.org/10.1038/s44168-024-00163-9
42. Karakas, L.D., Mitra, D.: Believers vs. deniers: Climate change and environmental policy polarization. Eur. J. Polit. Econ. 65101948 (2020). Available at: https://doi.org/10.1016/J.EJPOLECO.2020.101948
43. Khatibi, F.S., Dedekorkut-Howes, A., Howes, M., Torabi, E.: Can public awareness, knowledge and engagement improve climate change adaptation policies? Discover Sustain. **2**(1), 1–24 (2021). Available at: https://doi.org/10.1007/S43621-021-00024-Z/FIGURES/5
44. Kopp, G., Lean, J.L.: A new, lower value of total solar irradiance: evidence and climate significance. Geophys. Res. Lett. **38**(1) (2011). Available at: https://doi.org/10.1029/2010GL045777
45. Krichen, M., Abdalzaher, M.S., Elwekeil, M., Fouda, M.M.: Managing natural disasters: An analysis of technological advancements, opportunities, and challenges. Internet Things Cyber-Phys. Syst. 499–109 (2024). Available at: https://doi.org/10.1016/J.IOTCPS.2023.09
46. Kumar, A., Mahajan, N.P., Sorokhaibam, R., Sunthlia, A., et al.: National programme on climate change and human health-India. J. Commun. Dis. **52**(3), 43–48 (2019). Available at: https://doi.org/10.24321/0019.5138.202029

47. Leppold, C., Gibbs, L., Block, K., Reifels, L., et al.: Public health implications of multiple disaster exposures. Lancet Public Health **7**(3), e274–e286 (2022). Available at: https://doi.org/10.1016/S2468-2667(21)00255-3
48. Lerner, R.M., Almerigi, J.B., Theokas, C., Lerner, J.V.: Positive youth development: A view of the issues. J. Early Adolesc. **25**(1), 10–16 (2005). Available at: https://doi.org/10.1177/0272431604273211
49. Lidskog, R.: From climate facts to climate risks. How the IPCC treats risk and uncertainty. J. Risk Res. (2025). Available at: https://doi.org/10.1080/13669877.2025.2488392
50. Lidskog, R., Elander, I., Standring, A.: COVID-19, the climate, and transformative change: comparing the social anatomies of crises and their regulatory responses. Sustainability (Switzerland) **12**(16) (2020). Available at: https://doi.org/10.3390/SU12166337
51. Lidskog, R., Sundqvist, G.: Lost in transformation: the Paris Agreement, the IPCC and the quest for national transformative change. Front. Clim. **4** (2022). Available at: https://doi.org/10.3389/FCLIM.2022.906054
52. Lu, Q.B.: Observation of large and all-season ozone losses over the tropics. AIP Adv. **12**(7), 075006 (2022). Available at: https://doi.org/10.1063/5.0094629
53. Lucas, A.: Risking the earth Part 1: Reassessing dangerous anthropogenic interference and climate risk in IPCC processes. Clim. Risk Manag. **31** (2021). Available at: https://doi.org/10.1016/j.crm.2020.100257
54. Madrueño, R., Tezanos, S.: The contemporary development discourse: Analysing the influence of development studies' journals. World Dev. 109334–109345 (2018). Available at: https://doi.org/10.1016/J.WORLDDEV.2018.05.005
55. Masten, A.S.: Resilience of children in disasters: a multisystem perspective. Int. J. Psychol. **56**(1), 1–11 (2021). Available at: https://doi.org/10.1002/IJOP.12737.
56. Masten, A.S., Lucke, C.M., Nelson, K.M., Stallworthy, I.C.: Resilience in Development and Psychopathology: Multisystem Perspectives. Ann. Rev. Clin. Psychol. 17521–549 (2021). Available at: https://doi.org/10.1146/ANNUREV-CLINPSY-081219-120307
57. Masten, A.S., Narayan, A.J.: Child development in the context of disaster, war, and terrorism: pathways of risk and resilience. Ann. Rev. Psychol. 63227–63257 (2012). Available at: https://doi.org/10.1146/ANNUREV-PSYCH-120710-100356
58. Mazhar, S.S., Khan, F.S., Srivastava, P., Khan, A.S.: Resilience and Vulnerability: Concepts and Policy Contexts, pp. 327–341 (2024). Available at: https://doi.org/10.1007/978-3-031-52561-2_18
59. Mcneal, K.S., Hammerman, J.K.L., Christiansen, J.A., Carroll, F.J.: Climate Change Education in the Southeastern U.S. Through Public Dialogue: Not Just Preaching to the Choir (2014). Available at: https://doi.org/10.5408/13-061.1
60. Meng, X., Fleury, M.J., Xiang, Y.T., Li, M., et al.: Resilience and protective factors among people with a history of child maltreatment: a systematic review. Soc. Psychiatry Psychiatric Epidemiol. **53**(5), 453–475 (2018). Available at: https://doi.org/10.1007/S00127-018-1485-2
61. Misiou, O., Koutsoumanis, K.: Climate change and its implications for food safety and spoilage. Trends Food Sci. Technol. 126142–126152 (2022). Available at: https://doi.org/10.1016/j.tifs.2021.03.031
62. Moallemi, E.A., Malekpour, S., Hadjikakou, M., Raven, R., et al.: Achieving the sustainable development goals requires transdisciplinary innovation at the local scale. One Earth **3**(3) (2020). Available at: https://doi.org/10.1016/j.oneear.2020.08.006
63. Molyneaux, R., Gibbs, L., Bryant, R.A., Humphreys, C., et al.: Interpersonal violence and mental health outcomes following disaster. BJPsych Open **6**(1) (2019). Available at: https://doi.org/10.1192/BJO.2019.82
64. Moore, P.: Ocean "Acidification" Alarmism in Perspective. In: Evidence-Based Climate Science: Data Opposing CO_2 Emissions as the Primary Source of Global Warming: Second Edition, pp. 231–242 (2016). Available at: https://doi.org/10.1016/B978-0-12-804588-6.00013-6c
65. Moser, S.C.: Communicating climate change: History, challenges, process and future directions. Wiley Interdisc. Rev. Clim. Change **1**(1), 31–53 (2010). Available at: https://doi.org/10.1002/WCC.11

66. Murtaza, F., Mustafa, T., Awan, R.: Child health inequalities and its dimensions in Pakistan. J. Family Community Med. **22**(3), 169 (2015). Available at: https://doi.org/10.4103/2230-8229.163036
67. Nalau, J., Verrall, B.: Mapping the evolution and current trends in climate change adaptation science. Clim. Risk Manag **32** (2021). Available at: https://doi.org/10.1016/j.crm.2021.100290
68. Nema, P., Nema, S., Roy, P.: An overview of global climate changing in current scenario and mitigation action. Renew. Sustain. Energy Rev. **16**(4), 2329–2336 (2012). Available at: https://doi.org/10.1016/J.RSER.2012.01.04
69. Ojala, M.: How do children, adolescents, and young adults relate to climate change? Implications for developmental psychology. Eur. J. Dev. Psychol. **20**(6), 929–943 (2023). Available at: https://doi.org/10.1080/17405629.2022.2108396
70. Petersen, A.C., Koller, S.H., Motti-Stefanidi, F., Verma, S.: Positive youth development in global contexts of social and economic change. In: Positive Youth Development in Global Contexts of Social and Economic Change, pp. 1–322 (2016). Available at: https://doi.org/10.4324/9781315307275/POSITIVE-YOUTH-DEVELOPMENT-GLOBAL-CONTEXTS-SOCIAL-ECONOMIC-CHANGE-ANNE-PETERSEN-FROSSO-MOTTI-STEFANIDI-SILVIA-KOLLER-SUMAN-VERMA/RIGHTS-AND-PERMISSIONS
71. Quinn, P., Williamson, B., Gibbs, L.: Indigenous-informed disaster recovery: Addressing collective trauma using a healing framework. Progress Disaster Sci. **16** (2022). Available at: https://doi.org/10.1016/j.pdisas.2022.100257
72. Rahman, M.I.: Climate change: a theoretical review. Interdisc. Description Complex Syst. INDECS **11**(1), 1–13 (2013). Available at: https://doi.org/10.7906/INDECS.11.1.1
73. Rajsekhar, D., Gorelick, S.M.: Increasing drought in Jordan: climate change and cascading Syrian land-use impacts on reducing transboundary flow. Sci. Adv. **3**(8). Available at: https://doi.org/10.1126/SCIADV.1700581/SUPPL_FILE/1700581_SM.PDF
74. Ray Biswas, R., Rahman, A.: Adaptation to climate change: a study on regional climate change adaptation policy and practice framework. J. Environ. Manag. **336** (2023). Available at: https://doi.org/10.1016/j.jenvman.2023.117666
75. Redvers, N., Aubrey, P., Celidwen, Y., Hill, K.: Indigenous peoples: traditional knowledges, climate change, and health. PLOS Global Public Health **3**(10) (2023). Available at: https://doi.org/10.1371/journal.pgph.0002474
76. Rezvani, S.M.H.S., Falcão, M.J., Komljenovic, D., de Almeida, N.M.: A systematic literature review on urban resilience enabled with asset and disaster risk management approaches and GIS-based decision support tools. Appl. Sci. (Switzerland), **13**(4) (2023). Available at: https://doi.org/10.3390/app13042223
77. Rigaud, K.K., De Sherbinin, A.M., Jones, B., Bergmann, J., et al.: Groundswell : Preparing for Internal Climate Migration (2018). Available at: https://doi.org/10.7916/D8Z33FNS
78. O'Sullivan, T.M.: Environmental security is homeland security: climate disruption as the ultimate disaster risk multiplier. Risk, Hazards Crisis Public Policy **6**(2), 183–222 (2015)
79. Sainz de Murieta, E., Galarraga, I., Olazabal, M.: How well do climate adaptation policies align with risk-based approaches? an assessment framework for cities. Cities **109** (2021). Available at: https://doi.org/10.1016/j.cities.2020.103018.
80. Sanson, A., Bellemo, M.: Children and youth in the climate crisis. BJPsych Bull. **45**(4), 205–209 (2021). Available at: https://doi.org/10.1192/BJB.2021.16
81. Sanson, A.V., Malca, K.P., Hoorn, J.V., Burke, S.: Children and climate change. Elem. Child Dev. (2022). Available at: https://doi.org/10.1017/978100911870
82. Sanson, A.V., Masten, A.S.: Climate change and resilience: developmental science perspectives. Int. J. Behav. Dev. **48**(2), 93–102 (2024). Available at: https://doi.org/10.1177/01650254231186332;SUBPAGE:STRING:FULL
83. Seacrest, S., Kuzelka, R., Leonard, R.: Global climate change and public perception: The challenge of translation. J. Am. Water Resour. Assoc. **36**(2), 253–263 (2000). Available at: https://doi.org/10.1111/J.1752-1688.2000.TB04265.X
84. Sharapov, D.: Arctic ice changes and global warming. E3S Web Conf. 46008014 (2023). Available at: https://doi.org/10.1051/e3sconf/202346008014

85. Shayan, N.F., Mohabbati-Kalejahi, N., Alavi, S., Zahed, M.A.: Sustainable development goals (SDGs) as a framework for corporate social responsibility (CSR). Sustainability (Switzerland) **14**(3) (2022). Available at: https://doi.org/10.3390/su14031222
86. Singh, A., Purohit, B.M.: Public health impacts of global warming and climate change. Peace Rev. **26**(1), 112–120 (2014). Available at: https://doi.org/10.1080/10402659.2014.876326/ASSET//CMS/ASSET/7E1F796D-BCC0-4D44-8526-5802F71CB21D/10402659.2014.876326.FP.PNG
87. Stough, L.M., Kang, D.: The sendai framework for disaster risk reduction and persons with disabilities. Int. J. Disaster Risk Sci **6**(2), 140–149 (2015). Available at: https://doi.org/10.1007/S13753-015-0051-8/METRICS
88. Sutherst, R.W.: Global change and human vulnerability to vector-borne diseases. Clin. Microbiol. Rev. **17**(1), 136–173 (2004). Available at: https://doi.org/10.1128/CMR.17.1.136-173.2004/ASSET/590369DE-C5B0-4722-A868-B36CAFDF7EA2/ASSETS/GRAPHIC/ZCM0010400890016.JPEG
89. Szykuła-Piec, B.: Societal resilience in the era of hazards. The case study of an examplary community. J. Modern Sci. **47**(2), 431–454 (2021)
90. Tamang, B.: Global climate change: challenges, opportunities, and multilateral strategies for sustainable development. NPRC J. Multi. Res. **1**(4), 65–76 (2024). Available at: https://doi.org/10.3126/NPRCJMR.V1I4.70947
91. Tangney, P.: Understanding climate change as risk: a review of IPCC guidance for decision-making. J. Risk Res. **23**(11), 1424–1439 (2020). Available at: https://doi.org/10.1080/13669877.2019.1673801
92. Thiery, W., Lange, S., Rogelj, J., Schleussner, C.F., et al.: Intergenerational inequities in exposure to climate extremes. Science (New York, N.Y.) **374**(6564), 158–160 (2021). Available at: https://doi.org/10.1126/SCIENCE.ABI7339
93. Thoma, M.V., Rohleder, N., Rohner, S.L.: Clinical ecopsychology: the mental health impacts and underlying pathways of the climate and environmental crisis. Front. Psychiatry **12** (2021). Available at: https://doi.org/10.3389/FPSYT.2021.675936
94. Toniolo, S., Pieretto, C., Camana, D.: Improving sustainability in communities: Linking the local scale to the concept of sustainable development. Environ. Impact Assess. Rev. 101107126 (2023). Available at: https://doi.org/10.1016/J.EIAR.2023.107126
95. Ungar, M., Theron, L.: Resilience and mental health: how multisystemic processes contribute to positive outcomes. Lancet Psychiatry **7**(5), 441–448 (2020). Available at: https://doi.org/10.1016/S2215-0366(19)30434-1
96. Vasileiadou, E., Heimeriks, G., Petersen, A.C.: Exploring the impact of the IPCC Assessment Reports on science. Environ. Sci. Policy **14**(8), 1052–1061 (2011). Available at: https://doi.org/10.1016/J.ENVSCI.2011.07.002
97. Vlassopoulos, C.A.: Competing definition of climate change and the post-Kyoto negotiations. Int. J. Clim. Change Strat. Manag. **4**(1), 104–118 (2012). Available at: https://doi.org/10.1108/17568691211200245/FULL/XML
98. Watts, N., Amann, M., Arnell, N., Ayeb-Karlsson, S., et al.: The 2020 report of The Lancet Countdown on health and climate change: responding to converging crises. Lancet **397**(10269), 129–170 (2021). Available at: https://doi.org/10.1016/S0140-6736(20)32290-X

Developing Urban Resilience Against Climate Uncertainties: Case Study Megacity Dhaka

Shama E. Haque and Nafisa N. Chowdhury

Abstract The influences of global climate change impact the environment in numerous ways, including increasing surface temperatures, sea level rise, flooding, droughts, and more. As urban areas continue to expand and extreme weather events driven by climate change intensify, the urban landscape increasingly face the associated impacts. Among Asian cities, Dhaka, Bangladesh, is the most climate-vulnerable owing to low adaptive capacity, high exposure, and high sensitivity. Consequently, the need for resilient urban infrastructure in the city has become more critical than ever to mitigate these effects and ensure long-term sustainability. The climate crisis has created uncertainty in urban planning, however only a limited number of past studies systematically address the uncertainties, particularly over the long term. Urban resilience planning focuses on reducing uncertainty and complexity in response to rapid urban growth and changing global climate, rather than specific spatial or infrastructure planning. This chapter employs a mixed method to understand that the recent advances on urban resiliency with a focus on Dhaka, one of the megacities in the world.

Keywords Urban resilience · Climate change · Adaptation · Mitigation · Uncertainties

1 Introduction

Globally, urbanization is rapidly reshaping cities, making them centers of economic growth and innovation. Although urban areas currently generate over 80% of global Gross Domestic Product, the increasing demand for infrastructure is straining

S. E. Haque (✉)
Department of Civil and Environmental Engineering, North South University, Dhaka, Bangladesh
e-mail: shama.haque@northsouth.edu

N. N. Chowdhury
Centre for Climate Change and Environmental Research (C3ER), BRAC University, Dhaka, Bangladesh

A. L. Srivastav et al. (eds.), *Climate Resilience: Impact of Quantum Computing and Artificial Intelligence on Urban Planning*, Studies in Systems, Decision and Control 637, https://doi.org/10.1007/978-3-032-06791-3_4

existing systems [72]. According to a report by McKinsey Global Institute [48], by 2030, an estimated $57 trillion investment will be required to support the expanding urban populations. This need becomes even more critical in light of projections from the Intergovernmental Panel on Climate Change's (IPCC) Sixth Assessment Report (2021), which warns that global temperatures are expected to surpass 1.5 °C above pre-industrial levels, further intensifying climate-induced stresses on cities that are home to over half the world's population and responsible for a substantial amount of global carbon dioxide emissions [71]. In this context, urban areas, particularly those with considerable poor populations, experience persistent challenges in safeguarding the well-being of their residents. The challenges arise from both the direct and indirect influences of changing global climate, which are exacerbated by prevailing socio-economic and infrastructural vulnerabilities.

Dhaka, the capital of Bangladesh, with a population with a density of 23,234 people/km^2 [20], represents an important link between rapid urbanization and climate susceptibility. Over the past few years, the city faced a range of severe challenges, including the intensification of urban heat island (UHI) effects and frequent waterlogging, stressing its exposure to influences of global climate change [25, 67]. Past research indicates that the UHI effect stems from man-made alterations in densely populated tropical urban centers with limited airflow, and is characterized by higher temperatures compared to neighboring rural areas [39]. In addition, increasing temperatures coupled with erratic rainfall patterns, driven by changing global climate, further exasperate the vulnerabilities.

Urban resilience is known as the ability of cities to continue providing vital functions and services while allowing its residents, specifically the underprivileged and susceptible, to bear, recuperate, and thrive in the face of a variety of stressful events [9, 40]. In recent years, urban resilience has become a critical focus for urban planning [10]. This is particularly relevant to Dhaka, a city home to over 24 million residents [46], which continues to grow at a rate of 5% per annum (Fig. 1; [12]), with a significant portion of the population residing in informal settlements. The city faces vulnerability to natural disasters, public health crises, and the escalating impacts of climate change.

Primarily, the influences of climate change poses risks to Dhaka city in two ways: (i) flooding, and (ii) heat stress [6]. However, due to lacking infrastructures and proper tools, it is difficult to effectively identify and address the growing vulnerability in the city. Against this backdrop, the present chapter explores Dhaka's urban resilience efforts, focusing on UHI effects and waterlogging as key climate stressors. Through a mixed-method analysis, the study assesses policy interventions, infrastructure initiatives, and adaptation gaps, proposing pathways for a more climate-resilient urban future. In terms of organization of the paper, following the introductory section, Sect. 2 presents the employed methodology, which comprises of two key components: (i) literature review and (ii) field surveys, including the details of study area. Next, Sect. 3 focused on climate vulnerabilities and urban challenges with a focus on Dhaka city; the subsections discuss climate migrants, water logging and flooding issues, degradation of lakes, canal and wetlands, drainage infrastructure and waste management crisis as well as UHIs. Subsequently, Sect. 4 discusses the findings of

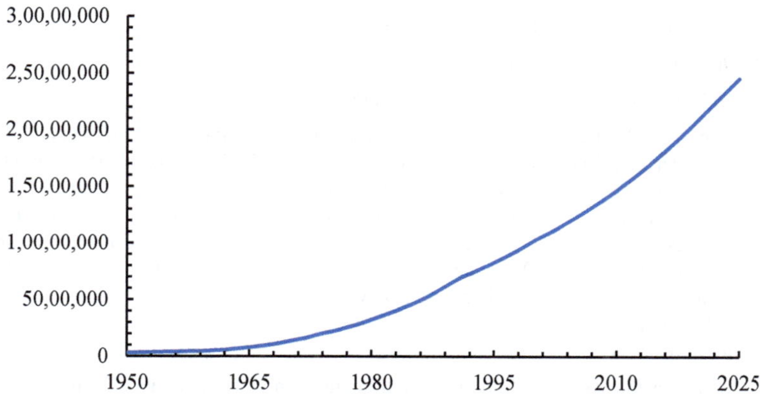

Fig. 1 Population growth in Dhaka between 1950 and 2024. Produced using data from Macrotrends [46]

the current study and Sect. 5 presents the concluding remarks. The findings of this study are expected to advance the conversation on improving urban sustainability in the context of increasing climate uncertainties.

2 Methodology

The current study employs a mixed-method approach. The methodology focuses on integrating both a systematic literature review and a reconnaissance field survey to understand the recent advances on urban resiliency in Dhaka.

2.1 Literature Review

The literature review involves a systematic review of a total of 73 previously published scholarly and non-scholarly works related to urban resilience and UHI in the context of climate change adaptation in Dhaka City. In particular, international peer reviewed academic journals were assessed for quality and relevance. The findings indicate that South Asian research on UHI has increased roughly tripled since 2000 [45]. Table 1 presents a summary of key findings from the literature review.

Table 1 Summary of findings from literature review

Author	Objective	Findings
Previous Urban Resilience Related Research on Dhaka		
[59]	Aims to understand the underlying causes of extreme heat events, which is essential for developing effective mitigation strategies and strengthening urban resilience	The study employed Landsat 8 data (2015–2022) and QGIS to identify seven heat islands in Dhaka—five in DNCC and two in DSCC. These areas, mostly built-up with less than 10% green or blue space, are major contributors to rising temperatures. The findings stress the urgent need to expand green and blue spaces to reduce the urban heat island effect and strengthen Dhaka's resilience to heatwaves
[65]	Examines the World Bank's Urban Resilience Project in Dhaka	Highlights contradictions in global resilience frameworks, where resilience supports economic growth nonetheless ignores unequal environmental risks. The findings suggest that a politicized approach is needed to ensure long-term environmental and economic security for Dhaka's low-income populations
[3]	Explores how urban resilience efforts can sometimes backfire, adding complexity and leading to unsustainable outcomes	To improve Dhaka's livability, planners must decentralize development and resilience efforts. Solely urban-focused approaches—often termed 'climate resilient'—fall short, especially as Dhaka already struggles to meet basic service demands. Furthermore, climate-driven population growth will only worsen these challenges, increasing the risk of service failures and social unrest
[8]	Discusses the backstory and execution of the Urban Resilience Project in Bangladesh	The Urban Resilience Project, funded by the World Bank, aims to build the capacity of key government agencies to respond to major disasters and improve construction quality and permitting in Dhaka and Sylhet. Dhaka city is at particular risk as urban growth has outpaced land use planning and public services, while weak regulations and poor enforcement fail to ensure structural safety. Physical and social vulnerabilities are rising, making hazards such as floods, building collapses, and earthquakes serious threats
[11]	An assessment examining the policy results related to urban resilience in informal settlements in Dhaka since the country's independence	Dhaka attracts large numbers of rural migrants annually, creating significant pressure on housing and resulting in rapidly expanding slums. Current policy approaches often reflect political agendas rather than the real needs of slum dwellers, who are rarely included in decision-making. Policymakers remain disconnected from actual on-the-ground challenges. Strengthening secondary cities is essential to ease pressure on Dhaka by reducing migration

(continued)

Table 1 (continued)

Author	Objective	Findings
[41]	Assesses urban disaster resilience in Dhaka North City Corporation (DNCC) using the UDRI model	DNCC's resilience score is 2.52, with strong physical (3.37) and social resilience (2.53), but weak institutional (2.11) and natural resilience (2.37). While infrastructure improvements boosted physical resilience, social resilience weakened due to poor community ties. Challenges include integrating Disaster Risk Reduction (DRR) into planning, inadequate DRR funding, weak institutions, and low community preparedness
[2]	Examines climate hazards in Dhaka's slums and explores challenges and opportunities for resilience-building	Dhaka's slums face flooding, poor housing, fire risks, and climate impacts, worsened by weak governance, insecure tenure, and lacking services. Nonetheless, resilience opportunities exist through national policies, urban initiatives, and forums such as the Bangladesh Urban Forum. Long term progress requires sustained advocacy and long-term commitment
[32]	Focused on institutional challenges that hinder the integration of climate-smart disaster risk management into urban resilience efforts in Bangladesh	The findings indicate that Dhaka, the capital, faces high risks from climate change, with extreme events like floods, drainage issues, and waterlogging severely damaging infrastructure, housing, and businesses. Four major floods between 1988 and 2004 inundated large portion of the city, disrupting transportation and affecting industries and the commercial sector. Additionally, the findings suggest that Local institutions are fragmented, uncoordinated, and lack effective strategic planning
[55]	An assessment of climate disaster resilience in Dhaka City Corporation at the zone level	Planned residential areas in Dhaka show higher climate disaster resilience, whereas older parts of the city and densely populated low-income areas remain more vulnerable. Dhaka faces ongoing challenges and opportunities. Using Climate Disaster Resilience Index findings in policy and planning can assist in curbing vulnerabilities and strengthening the city's future resilience
Previous UHI Related Research on Dhaka		
[38]	Investigated a wavelet and remote sensing analysis of Dhaka's UHI focusing on its interaction with climate change over time	The findings reveals that UHI patterns follow seasonal cycles, with more intense UHIs becoming increasingly consistent over time. It also highlights that rising temperatures from climate change are intensifying UHI effects throughout the year
[64]	Assessed UHIs in Dhaka using state-of-the-art approaches	The findings indicate that Dhaka's UHI issue primarily results from changes in land use and land cover, and unplanned urban development. The findings provide a foundation for future UHI research and offers practical recommendations for sustainable urban planning with a focus on public well-being

(continued)

Table 1 (continued)

Author	Objective	Findings
[68]	Investigated urban effects on temperature and local winds in Dhaka	Results show that increased anthropogenic heat intensifies UHI magnitude and its impact on local wind patterns. Additionally, urban growth amplifies UHI effects and extends the influence of surface roughness over a broader area
[67]	Examines Dhaka's UHIs and its link to heat waves	Heatwaves in Dhaka have increased by 5.8 days per decade during the monsoon, linked to anticyclonic flow and high-pressure systems. These conditions intensify the UHI effect—day and night during pre-monsoon season, and at night during monsoon season—particularly when humidity and cloud cover are low
[33]	Conducted a remote sensing-based geostatistical hot spot analysis of UHI in Dhaka	Between 1991 and 2015, Dhaka experienced a 71.3% decrease in water bodies and a 71.8% reduction in vegetation cover. Meanwhile, urban areas expanded by approximately 250%, covering 174 km^2, as indicated by Urban Compactness Ratio readings. The hot spot analysis revealed a 93.7% increase in heat concentration zones, with the average temperature rising by 3.3 °C
[66]	Using remote sensing studied the COVID-19 lockdown impact on reduced UHI effects in Dhaka	The study employed Landsat-8 data to analyze changes in vegetation and UHI zones in Dhaka pre- and post-COVID-19 lockdown. Following the May 2020 lockdown, high UHI zones decreased from 22.5% to 19.13%, correlating with reduced activity. UHI zones expanded again post-lockdown. A strong negative correlation between LST and vegetation indices was observed, highlighting a low-cost method for urban heat monitoring
[69]	Assessed urban changes and UHI effects in Dhaka from 2001 to 2017 using Moderate Resolution Imaging Spectroradiometer daily land surface temperature data	Between 2001 and 2017, Dhaka's land area grew by 25.3% and its population by 76.7%. Urban expansion raised temperatures in some areas by approximately 3 °C, with day and night temperature differences of 7 and 5 °C, respectively, between the warmest and coolest points. The city's average temperatures increased at rates of 0.03 °C and 0.023 °C per year. Without mitigation, the rising temperatures will likely worsen the UHI effect
[56]	Estimated UHI intensity in Dhaka utilizing remote sensing methods	Land surface temperatures in Dhaka increased from 28.5 to 35.4 °C in 2002 to 37.9 to 40.1 °C in 2014. This rise across all land cover types highlights the UHI effect, emphasizing the urgent need for city authorities to address its impact on the population and infrastructure

2.2 *Field Survey*

This study was conducted in Dhaka, with particular focus on Hatirjheel, Mohakhali, Banani, Gulshan, Dhanmondi, and Uttara (Fig. 1). The study locations were selected due to their exposure to a variety of environmental stressors such as waterlogging, traffic congestion, and unplanned urban growth, which are critical challenges for building urban resilience in the city. A reconnaissance field survey and a preliminary assessment of the study area's physical, environmental, and socio-economic features, was conducted in the study area. The survey conducted a visual assessment of urban infrastructure, water bodies, and regions impacted by waterlogging and UHI effect.

2.2.1 Study Area

Dhaka, a rapidly expanding megacity, lies centrally on the eastern bank of the Buriganga River (Fig. 2). The city of Dhaka, divided into the North and South City Corporations, is geographically located between 23.69° and 23.89° N latitude and 90.33°–90.44° E longitude [29, 52]. Dhaka covers approximately an area of 127 km^2. The city is bordered by the Buriganga, Turag, and Balu rivers and the Tongi canal. Dhaka's elevation ranges from 1 to 14 m above mean sea level [43]. Dhaka's unplanned urbanization along with low-lying deltaic location make it highly susceptible to various impacts of changing global climate.

Dhaka has a tropical monsoon climate, characterized by a dry season from November to March and a rainy season from May to October. Approximately 80% of the total annual rainfall in the city occurs during the monsoon season [70]. The

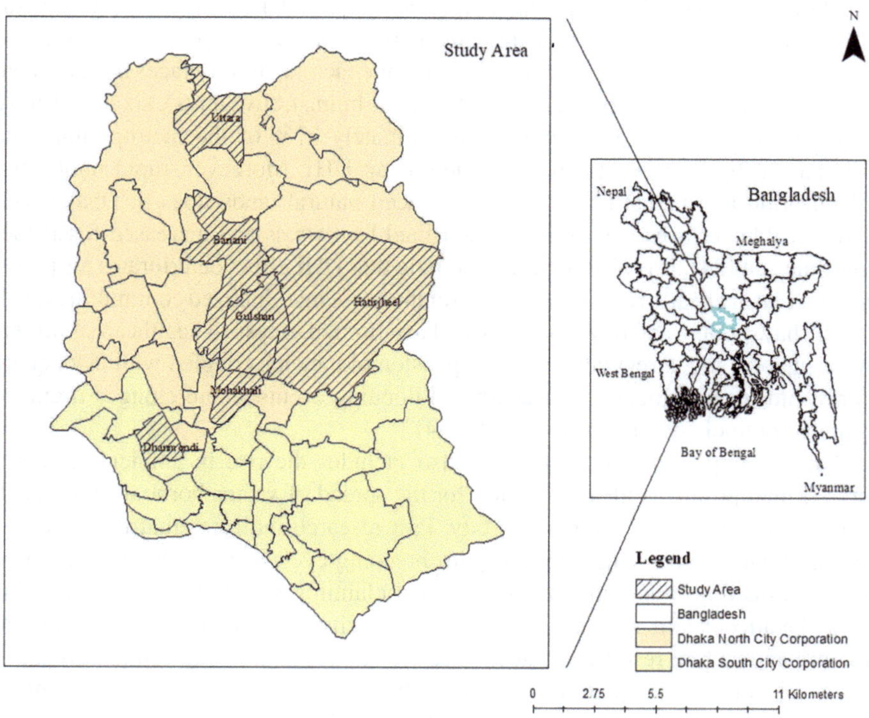

Fig. 2 Study Area

city experiences annual rainfall of 1800–2,400 mm during the months of July and August. Dhaka's average temperature is 26 °C [74]. During the pre-monsoon season (i.e., March to May), the city's mean daily maximum temperature reaches roughly 34 °C [53, 69].

3 Climate Vulnerabilities and Urban Challenges: Megacity Dhaka, Bangladesh

Dhaka is among one of the top 10% of cities, which is most susceptible to the impacts of climate change along with severe urban heat stress and deteriorating environmental issues [21]. Between 2000 and 2019, Dhaka's surface UHI intensity increased five times faster than any other city in Bangladesh, rising from 2.20 to 3.18 °C [19]. [67] employed meteorological and reanalysis data from 1995 to 2019 of Dhaka and examined its interaction with heatwaves. The findings reveal that Dhaka's UHI intensity increased by 0.21 °C per decade, and the average UHI intensity is 0.48 °C. Additionally, during the winter and monsoon seasons, the UHI is the strongest and weakest, respectively. In addition, the UHI peaks at 2100 land Surface Temperature (LST) across all seasons, with a weak cool island through the day.

Abrar et al. [1] studied heat vulnerability in Dhaka's UHIs, using the Heat Vulnerability Index (HVI) to assess exposure, sensitivity, and adaptive capacity. Due to urbanization, reduced green spaces, and socioeconomic factors, the city faces frequent heat events. Additionally, out of Dhaka's 41 thanas (administrative units), six thanas have very high HVI scores. In addition, approximately 87% of the metropolitan area is impacted by heat events and classified as an UHI. Moreover, rapid unplanned urbanization and industrialization has replaced natural landscapes of Dhaka with impermeable surfaces, and in turn, intensified heat stress and increased fossil fuel consumption [24]. It is also important to note that UHI effect deteriorates air pollution, increasing health risks, specifically for the city's marginalized communities such as rickshaw pullers and day laborers who have limited access to healthcare. Furthermore, chaotic and unregulated urban expansion of the city, coupled with inadequate stormwater management, has exacerbated flooding events where clogged drainage systems routinely disrupt everyday life [54].

Adverse impact of the UHI effect also includes the rise in ambient temperatures, which promotes ideal condition for the spread of vector-borne diseases such as malaria and dengue fever in the city. Past research indicates that the observed warmer temperatures speed up mosquito breeding cycles and prolong their active seasons affecting thousands of residents and claiming hundreds of lives [4]. This public health concern is further aggravated by the city's inadequate healthcare infrastructure and lack of resources hinder effective disease monitoring and control. As a result, Dhaka's residents are highly vulnerable to such health concerns brought on by climate change due to the combined effects of UHI and climate change.

3.1 Impact of Climate Migrants

For many Bangladeshis, internal migration is not an intentional choice, but rather a last-resort adaptive measure to ongoing climate-induced changes, including sea-level rise, riverbank erosion, and extreme weather events [13]. Annually, as many as 400,000 climate migrants move and settle in Dhaka, placing heavy demands on critical services—reliable transportation, uninterrupted water and power, proper sanitation, and solid waste management—and increasing residents' vulnerability to climate-driven hazards [60]. The influx of the migrants puts strain on the city's already fragile urban ecosystem [57]. Without urgent climate adaptation measures, such as resilient infrastructure, improved stormwater drainage, and enhanced disaster preparedness, the city will face escalating environmental and humanitarian crises, making it progressively uninhabitable in the coming decades.

3.2 Waterlogging and Flooding

Erratic rainfall patterns, intensified by climate change, pose a major flood risk to Dhaka. A steady increase in rainfall from 1982 to 2022 (Fig. 3) signals worsening waterlogging, as inadequate drainage and encroachment on retention areas limit runoff absorption. Global climate change is expected to further intensify these extreme rainfall events, overwhelming the city's drainage infrastructure [58]. In addition, Dhaka's rapid population growth has substantially increased the city's reliance on groundwater to meet rising domestic, industrial, and agricultural water demands. This intensive and commonly unregulated extraction of groundwater has led to land subsidence ($\geq$200 mm) issues in several areas of the city including in Mirpur, Uttara, Ramna, Cantonment, Gulshan and Tejgaon areas [26]. Land subsidence coupled with poor drainage systems worsen waterlogging, hinder mobility, and increase public health and safety risks. Within the broader context of climate change and increasingly frequent extreme weather events, this compounding effect critically undermines Dhaka's urban resilience and adaptive capacity.

Rapid urbanization in Dhaka has also driven significant land-use changes. For example, between 1978 and 2018, the amount of agricultural land declined by 13.1%, green spaces by 4.8%, and water bodies by 7.8%, while built-up areas surged by 22.1% [50]. This loss of permeable surfaces has increased surface runoff, with high-runoff zones expanding from 74.2 km^2 (24.4%) in 1978 to 174.2 km^2 (57.4%) in 2018. As a result, flooding has become more frequent and severe, especially in low-lying, high-vulnerability zones (Fig. 4). The southern and southwestern parts of Dhaka, home to densely populated informal settlements, are particularly susceptible to waterlogging (Fig. 4; [5]). Climate change-driven extreme weather continues to strain the city's infrastructure, increasing urban vulnerability.

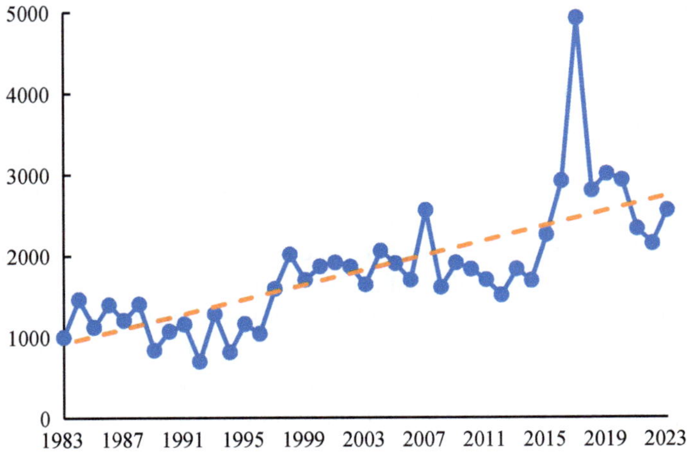

Fig. 3 Rainfall amount between 1983 and 2023 (prepared using data obtained from NASA Langley Research Centre, 2025)

Fig. 4 Water-logging in Bashundhara Residential Area located in north Dhaka

3.3 *Degradation of Lakes, Canals, and Wetlands*

Historically, Dhaka was characterized by a vast network of surface water bodies such as lakes, canals, and natural wetlands that were essential to the hydrological functioning of the city. By serving as retention basins, these water bodies eased control of stormwater runoff, reducing the possibility of floods. In particular, during monsoon rains, the canals and lakes ensured efficient flow control by providing a route for surface water drainage. Additionally, the wetlands served as natural buffers against seasonal fluctuations in water levels and aided in groundwater recharge. This natural infrastructure was essential for preserving the city's water balance and reducing the adverse impacts of intense rainfall events in the absence of modern engineered systems. However, in the past few decades, rapid urbanization, land encroachment, and poor governance led to the degradation of the water bodies (Fig. 5). Currently, only 12 of Dhaka's original 54 canals remain, underscoring the city's lost hydrological balance and reduced drainage capacity [63]. Additionally, during the last eight decades, continued encroachment and inadequate maintenance resulted in the loss of approximately 120 km (307 ha) of canals [61].

According to [31], since 2010, approximately 43,200 acres, which is 22% of Dhaka's wetlands, have disappeared, with 2,500 acres lost annually in the Detailed Area Plan area. Based on the present study's field observations, mismanaged solid waste, untreated sewage and industrial effluents have seriously contaminated several of Dhaka's major water bodies, including in Hatirjheel, Dhanmondi Lake, and Gulshan Lake. It is well established that accumulation of solid waste, including plastics and other debris, impedes water flow, and untreated sewage introduces pathogens that deteriorates water quality and endanger public health. In addition, over the years, the built up of solid waste and pollutants has significantly diminished these water-bodies' capacity to absorb and retain rainfall, thus reducing their ability to control urban flooding and raising the likelihood of waterlogging during periods of severe precipitation.

A prime example of the degradation of waterbodies of Dhaka is the transformation of the Begunbari Canal into Panthapath Road through the construction of box culverts. Once a major urban waterway managing 15% of the city's runoff

Fig. 5 Lake encroachment and transformation into built-up area for construction activities in Uttara, a suburban area in Dhaka

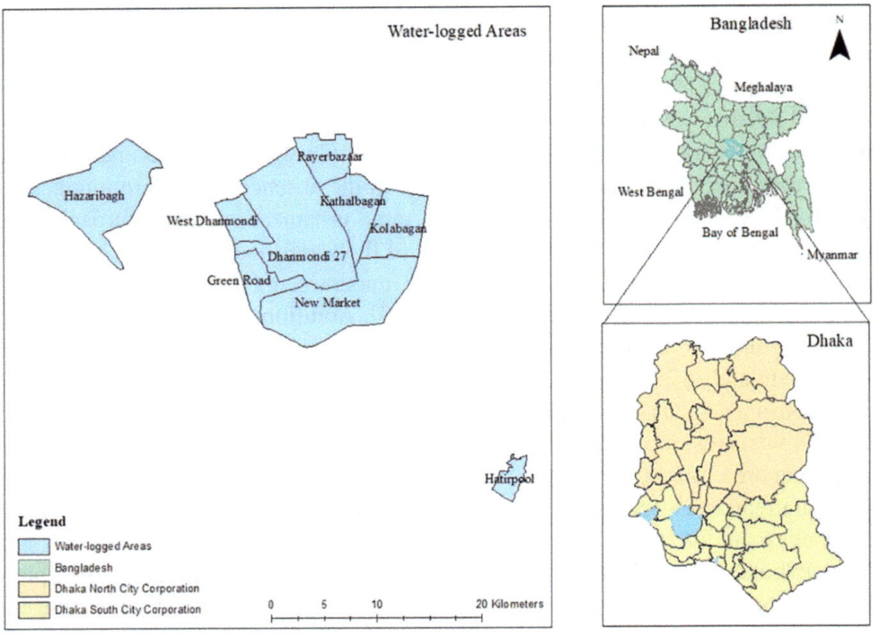

Fig. 6 Water-logged areas due to construction of culverts

[47], the canal had lost its northern and southern branches as a result of extensive environmental degradation, drastically reducing its capacity. The box culvert, now largely blocked, intensifies flooding in Kalabagan, Dhanmondi-27, Kathalbagan, Green Road, and Hatirpool (Fig. 6). Similarly, the encroachment of Rayerbazar and Jirani canals, approximately 80% of which are occupied, causes severe waterlogging in Hazaribagh, Rayerbazar, West Dhanmondi, and Dhaka New Market areas [5]. Such unchecked canal reduction has crippled Dhaka's natural drainage, leading to prolonged inundation. According to the River and Delta Research Centre (RDRC), areas such Pallabi, Shialbari, Rupnagar, and Eastern Housing experience severe waterlogging due to the narrowing of Rupnagar Main Khal [5]. In Tolarbagh, Ahmednagar, and Paikpara, the complete encroachment of Kalyanpur Main Khal behind Bangla College results in recurrent flooding, underscoring the urgent need for sustainable urban water management.

3.3.1 Drainage Infrastructure and Waste Management Crisis

Dhaka's drainage infrastructure plays an important role in urban flooding and waterlogging. The drainage system, which combines sewer and stormwater networks, has been severely impacted by poor maintenance, sediment accumulation, and blockages from plastic waste, construction debris, and uncollected solid waste. In Dhaka's waste management system, household waste is initially collected and stored at the source before being transported to secondary storage points, such as dustbins and containers, by non-governmental organizations, community-based organizations, or the city corporation. Currently, there are two major landfill sites within the city—Matuail Landfill under the Dhaka South City Corporation; Aminbazar Landfill under Dhaka North City Corporation; however, neither is adequately equipped with the necessary sanitary infrastructure. As such poorly managed municipal solid waste including plastic wastes have reduced the effectiveness of many drainage channels, leading to increased flood risks [28].

3.4 Urban Heat Island

Urban water bodies and green spaces play a crucial role in mitigating heat stress. Wetlands and lakes facilitate evaporative cooling and heat retention, reducing localized temperatures [7]. Past studies report that water bodies located within 300 m can significantly lower surrounding air temperatures in urban areas [15, 75]. Particularly, lakes provide a more substantial cooling effect than rivers and contribute to urban thermal comfort [22]. Similarly, urban green spaces offer shade and enhance cooling through evapotranspiration, mitigating UHI effects and improving overall environmental resilience [73]. However, due to the extensive loss of wetlands and open spaces, coupled with high air pollution levels, the city's microclimate has deteriorated over the years. As a result, outdoor space users in Dhaka experience prolonged heat stress due to high temperatures and limited cooling infrastructure.

Dhaka's rapid urbanization over the past decades has exacerbated heat stress during the pre-monsoon period, posing severe health risks, especially to vulnerable groups such as low-income communities and the elderly [14, 35]. Sabbir et al. studied land cover changes in the city between 2002 and 2023 and assessed their impact on the UHI effect and land surface temperature. During their study period, Dhaka experienced major land cover shifts, with vegetated areas largely replaced by built-up infrastructure (Fig. 7). Additionally, these researchers report a correlation coefficient of −0.62 between vegetation cover and the UHI effect, indicating that vegetation contributes to a reduction in temperature. In contrast, a positive correlation of 0.62 was observed for built-up areas, suggesting that these surfaces significantly retain heat and exacerbate the UHI effect. With further urban expansion projected [44, 70], the UHI effect is likely to intensify. In the future, the replacement of permeable and vegetated surfaces with impervious, heat-absorbing materials such as asphalt and concrete will possibly result in further increase in land surface temperatures. In the absence of a comprehensive urban plan that includes green infrastructure, continued urban expansion is expected to worsen heat stress, increase energy demand, and increase public health risks in an already vulnerable city.

Fig. 7 High-rise buildings in densely populated Dhaka

4 Discussion

Urban resilience is crucial for the future of Dhaka to withstand, adapt to, and recover from the escalating influences of climate change. As one of the fastest-growing megacities in a climate-vulnerable country, the city faces tremendous challenges from recurrent flooding events, extreme heat, and overburdened infrastructure. Additionally, Dhaka continues to absorb climate change refugees from coastal Bangladesh that is highly prone to sea level rise, cyclones, storm surges as well as salinity intrusion [28, 30]. On a daily basis, an estimated 2,000 people internally migrate to the Dhaka owing to various displacement issues [34]. In the absence of proper planning and robust resilience measures, the compounded challenges faced by this city are expected to intensify socio-economic vulnerabilities, disrupt essential urban services, and compromise long-term sustainability.

Dhaka's rapid expansion has led to the replacement of original land cover by impervious surfaces, which coupled with declining vegetation cover amplify the UHI effect, aggravate both heat stress and air quality deterioration. Additionally, the city's low adaptive capacity further exacerbates its susceptibility to the influences of global climate change. In addition, the direct exposure to erratic rainfall along with the lack of proper drainage systems, and widespread encroachment on natural water bodies have severely undermined effective stormwater management. Therefore, it is imperative to strengthen the city's resilience to safeguard public health and sustain economic productivity.

To enhance Dhaka's urban resilience to climate risks, key priorities should include implementing nature-based solutions, such as wetland restoration, green space expansion, and natural drainage integration, to mitigate flooding, reduce the UHI effect, and improve air quality. These measures are expected to assist in restoring the city's essential ecosystem services. Improved municipal solid waste management, particularly of plastic waste is also critical to prevent storm drain blockages, and reduce localized flooding events [17, 28]. These efforts must align with stronger enforcement of existing frameworks such as the Dhaka Structure Plan (2016–2035) and Bangladesh Climate Change Strategy and Action Plan, which require more coordinated implementation and regulatory clarity. It is also important to prioritize regional climate data for risk-informed, and context-specific policymaking. Furthermore, a long-term climate resilience plan should incorporate raising awareness and training the city's most vulnerable communities regarding specific climate change impacts. This is of utmost importance as approximately 48% of the city's population is estimated to live below the poverty line, with roughly 32% falling under the category of hardcore poverty [37]. As such, the city needs to proactively focus on empowering marginalized communities by expanding its targeted climate adaptation programs.

5 Conclusion

The findings of this study indicate that building urban resilience in Dhaka necessitates a comprehensive strategy, which takes into account both existing vulnerabilities and evolving climate risks. In addition to the effects of a changing climate, rapid urbanization and unplanned industrialization have increased the frequency of flooding occurrences, altered the severity of heat stress, and put strain on infrastructure, all of which call for immediate strategic response.

For a city as climate-exposed as Dhaka, resilience is essential to safeguard public health, and ensure long-term survival. Nature-based interventions, sustainable urban planning, as well as restoring and rejuvenating natural water bodies should all be part of Dhaka's urban resilience plan. Although policy frameworks provide direction, their efficacy depends on institutional coordination, enforcement, and the utilization of localized data for focused action.

References

1. Abrar, R., Sarkar, S.K., Nishtha, K.T., Talukdar, S., Shahfahad, R.A., Islam, A.R.M.T., Mosavi, A.: Assessing the spatial mapping of heat vulnerability under urban heat island (UHI) effect in the dhaka metropolitan area. Sustainability **14**(9), 4945 (2022)
2. Ahmed, I.: Building resilience of urban slums in Dhaka, Bangladesh. Procedia Soc. Behav. Sci. **218**, 202–213 (2016)
3. Ahmed, S.: Urban resilience and sustainable development trajectories: Insights from Dhaka. In: The Routledge handbook of planning megacities in the Global South, pp. 231–242. Routledge (2020)
4. Ahmed, T., Hyder, M.Z., Liaqat, I., Scholz, M.: Climatic Conditions: conventional and Nanotechnology-Based Methods for the Control of *Mosquito* Vectors Causing Human Health Issues. Int. J. Environ. Res. Public Health **16**(17), 3165 (2019). https://doi.org/10.3390/ijerph16173165
5. Alam, H.: Dhaka drowns every time it rains – here's why, The Daily Star, 1 December (2024). Available at: https://www.thedailystar.net/news/bangladesh/news/dhaka-drowns-every-time-it-rains-heres-why-3765526. Accessed 21 April 2025
6. Alam, M., Rabbani, M.D.G.: Vulnerabilities and responses to climate change for Dhaka. Environ. Urban. **19**(1), 81–97 (2007)

7. Ampatzidis, P., Kershaw, T.: A review of the impact of blue space on the urban microclimate. Sci. Total. Environ. **730**, 139068 (2020)
8. Ansary, M.A., Helaly, A.L., Hassan, M., Khair, A., Saha, S., Helaly, S.A.: Background and implementation of urban resilience project in Bangladesh. In: Proceedings, 17th World Conference on Earthquake Engineering, 17WCEE. Sendai, Japan (2020)
9. Asian Development Bank (ADB).: Urban climate change resilience: a synopsis. Manila: Asian Development Bank (2014). Available at: https://www.adb.org/sites/default/files/publication/149164/urban-climate-change-resilience-synopsis.pdf. Accessed 20 April 2025
10. Asibey, M.O., Appiah-Kusi, F., Kissiwaa, N.A., Bilson, M.A., Abdulai, A.S.J.: Local multilevel governance arrangements for climate change planning and management in Kumasi Ghana. Environ. Sci.Policy **153**, 103680 (2024)
11. Badhan, I.M., Siddika, A.: Evaluating the policy outcomes for urban resiliency in informal settlements since independence in Dhaka, Bangladesh: a review. NAKHARA (Journal of Environmental Design and Planning) **17**(1), 97–110 (2019)
12. Bangladesh Bureau of Statistics (BBS).: Statistical Year Book. Dhaka: Bangladesh Bureau of Statistics, Government of Bangladesh (2011)
13. Bharadwaj, R., Chakravarti, D., Karthikeyan, N., Islam, S., Tanni, U.J.: Exposed and exploited (2025). https://www.iied.org/sites/default/files/pdfs/2025-02/22604iied.pdf. Accessed 22 April 2025
14. Byomkesh, T., Nakagoshi, N., Dewan, A.M.: Urbanization and green space dynamics in Greater Dhaka, Bangladesh. Landscape Ecol. Eng. **8**, 45–58 (2012)
15. Chen, R., Kou, Z., Xu, L., Cao, J., Liu, Z., Wen, X., Wang, Z., Wen, H.: Analysis of epidemiological characteristics of four natural-focal diseases in Shandong Province, China in 2009–2017: a descriptive analysis. PLoS ONE **14**(8), e0221677 (2019)
16. Chowdhury, N.N., Jami, A.M., Sayanno, T.K., Asif, M.M., Haque, S.E., Nahar, N., Shupti, I.F., Khan, A.H.: Resilient multipurpose cyclone shelters in Cox's Bazar in the face of severe cyclonic storms. In: ICETSD 2025–International Conference Emerging Technologies for Sustainable Development, p. 102. Jessore (2025)
17. Chowdhury, N.N., Jami, A.M., Sayanno, T.K., Asif, M.M., Haque, S.E., Nahar, N., Shupti, I.F., Khan, A.H.: Understanding the influence of socio-economic factors on household waste management in Dhaka City. In: ICETSD 2025–International Conference Emerging Technologies for Sustainable Development, p. 103. Jessore (2025)
18. Department of Environment.: Projection of sea level rise and development of digital elevation models in support of SLR decision making. Department of Environment (2024). Available at: https://doe.portal.gov.bd/sites/default/files/files/doe.portal.gov.bd/page/936c9ced_0267_48bf_87d0_4e0c43168cf0/2024-05-14-11-21-17bcdf2bceffea77921c54c689a50367.pdf. Accessed 21 April 2025
19. Dewan, A., Kiselev, G., Botje, D., Mahmud, G.I., Bhuian, M.H., Hassan, Q.K.: Surface urban heat island intensity in five major cities of Bangladesh: patterns, drivers and trends. Sustain. Cities Soc. **71**, 102926 (2021)
20. Dhaka Tribune.: Dhaka ranks world's sixth most populous city. [online Dhaka Tribune (2022). Available at: https://www.dhakatribune.com/bangladesh/261898/dhaka-ranks-world%E2%80%99s-sixth-most-populous-city. Accessed 20 April 2025
21. Dhaka Tribune.: Dhaka among top 10% cities at risk from climate change (2024). Available at: https://www.dhakatribune.com/bangladesh/dhaka/305282/dhaka-among-top-10%25-cities-at-risk-from-climate. Accessed 21 April 2025
22. Du, H., Song, X., Jiang, H., Kan, Z., Wang, Z., Cai, Y.: Research on the cooling island effects of water body: a case study of Shanghai, China. Ecol. Ind. **67**, 31–38 (2016)
23. Government of the People's Republic of Bangladesh.: Bangladesh Climate Change Strategy and Action Plan 2009. Dhaka: Ministry of Environment and Forests (2009). Available at: https://policy.asiapacificenergy.org/sites/default/files/Bangladesh%20Climate%20Change%20Strategy%20and%20Action%20Plan%202009.pdf. Accessed 8 April 2025

24. Haque, A.N.: Climate change and unplanned development are creating a new set of damaging risks, The Daily Star (2023). Available at: https://www.thedailystar.net/opinion/focus/news/climate-change-and-unplanned-development-are-creating-new-set-damaging-risks-3502466. Accessed 21 April 2025
25. Haque, A.N.: Climate risk responses and the urban poor in the global South: the case of Dhaka's flood risk in the low-income settlements. Int. J. Disaster Risk Reduction **64**, 102534 (2021)
26. Haque, D.W.E., Tasnim, H., Tasnim, S. 2019. Time series analysis of subsidence in Dhaka city in Bangladesh using InSar. Malaysian Journal of Geosciences (MJG) 3(1) (2019) 32–44. https://doi.org/10.26480/mjg.01.2019.32.44
27. Haque, S.E.: Groundwater Quality in the Megacity Dhaka, Bangladesh: Assessment and Environmental Impact. Groundwater Quality: Assessment and Environmental Impact (2020)
28. Haque, S.E., Chowdhury, N.N.: Bangladesh plastic consumption trend: current scenario. Water Air Soil Pollut. **236**(4), 1–24 (2025)
29. Haque, S.E., Nahar, N.: Bangladesh: climate change issues, mitigation and adaptation in the water sector. ACS ES&T Water. https://doi.org/10.1021/acsestwater.2c00450
30. Haque, S.E., Nahar, N., Haque, M.S.: A study on the waste generation rates and recycling potential for the construction and demolition waste in Dhaka, Bangladesh. Environ. Monit. Assess. **196**(2), 18 (2024)
31. Hashim, S. M.: A quarter of Dhaka's wetlands gone, The Daily Star (2019). Available at: https://www.thedailystar.net/opinion/no-frills/news/quarter-dhakas-wetlands-gone-1739869. Accessed 21 April 2025
32. Hossain, M.Z., Huq, N.: Institutions matter for urban resilience: the institutional challenges in mainstreaming climate smart disaster risk management in Bangladesh. In: Climate Change and Disaster Risk Management, pp. 169–191. Springer, Berlin (2012)
33. Hussain, N., Ahmed, S.S., Shumi, A.M.: Remote sensing-based geostatistical hot spot analysis of Urban Heat Islands in Dhaka Bangladesh. Singapore J. Trop. Geogr. **44**(3), 438–458 (2023)
34. Illius, S.: Stability eludes climate refugees in Bangladesh's sinking cities (2023). PreventionWeb. https://www.preventionweb.net/news/stability-eludes-climate-refugees-bangladeshs-sinking-cities
35. Imran, H.M., Hossain, A., Islam, A.S., Rahman, A., Bhuiyan, M.A.E., Paul, S., Alam, A.: Impact of land cover changes on land surface temperature and human thermal comfort in Dhaka city of Bangladesh. Earth Syst. Environ. **5**, 667–693 (2021)
36. Intergovernmental Panel on Climate Change (IPCC).: Climate change widespread, rapid, and intensifying—IPCC (2021). Available at: https://www.ipcc.ch/2021/08/09/ar6-wg1-20210809-pr/. Accessed 21 April 2025
37. Islam, P.M.S.: Inclusive urban development in dhaka: review of the national housing policy. In: World Town Planning Day pp. 1–5 (2014)
38. Islam, S., Rashid, K.J., Tuhin, M.A.A., Rahman, M.A.: Wavelet and remote sensing based analysis of Urban Heat Island (UHI) of Megacity Dhaka and its interaction with changing climate. Theoret. Appl. Climatol. **156**(1), 1–14 (2025)
39. Jamei, E., Ossen, D.R., Seyedmahmoudian, M., Sandanayake, M., Stojcevski, A., Horan, B.: Urban design parameters for heat mitigation in tropics. Renew. Sustain. Energy Rev. **134**, 110362 (2020)
40. Jha, S., Martinez, A., Quising, P., Ardaniel, Z., Wang, L.: Natural disasters, public spending, and creative destruction: a case study of the Philippines (No. 817). ADBI Working Paper (2018)
41. Kabir, M.H., Sato, M., Habbiba, U., Yousuf, T.B.: Assessment of urban disaster resilience in Dhaka North City Corporation (DNCC), Bangladesh. Procedia Eng. **212**, 1107–1114 (2018)
42. Karim, S., Akther, K.M., Khatun, M., Ali, R.M.E.: Geomorphology and geology of the Dhaka city corporation area-an approach of remote sensing and GIS technique. Int. J. Astron. Astrophys. Space Sci. **6**, 7–16 (2019)
43. Khan, N.M.E.A.: SCP in Bangladesh: the brown hope of Hazaribagh and the Golden Fibre of Bangladesh. In: Schroeder, P., Anggraeni, K., Sartori, S., Weber, U (eds.) Sustainable Asia, pp. 105–131 (2017)

44. Kotharkar, R., Ghosh, A.: Review of heat wave studies and related urban policies in South Asia. Urban Clim. **36**, 100777 (2021)
45. Kotharkar, R., Ramesh, A., Bagade, A.: Urban heat island studies in South Asia: a critical review. Urban Clim. **24**, 1011–1026 (2018)
46. Macrotrends.: Dhaka, Bangladesh metro area population 1950–2025. Available at: https://www.macrotrends.net/global-metrics/cities/20119/dhaka/population. Accessed 8 April 2025
47. Mandal, D.: Developing a Strategic Process to Reintroduce a Lost Water Way as a Possible Solution for Water Logging: Case Study, an Area of Dhaka Metropolitan City. West Virginia University (2019)
48. McKinsey Global Institute (MGI).: Infrastructure productivity: How to save $1 trillion a year. McKinsey & Company (2013).
49. https://www.mckinsey.com/~/media/McKinsey/Business%20Functions/Operations/Our%20Insights/Infrastructure%20productivity/MGI%20Infrastructure_Executive%20summary_Jan%202013.pdf. Accessed 21 April 2025
50. Moniruzzaman, M., Thakur, P.K., Kumar, P., Ashraful Alam, M., Garg, V., Rousta, I., Olafsson, H.: Decadal urban land use/land cover changes and its impact on surface runoff potential for the Dhaka City and surroundings using remote sensing. Remote Sens. **13**(1), 83 (2020)
51. NASA Langley Research Center.: Data Access Viewer (2025). Available at: https://power.larc.nasa.gov/data-access-viewer/. Accessed 21 April 2025
52. Nayeem, S.U., Rashid, H., Habib, T., Haque, M.S., Haque, S.E., Nahar, N.: December. identification of construction and demolition waste generation rates, and recycling and reuse potential: a case of Dhaka City. In: Proceedings of the 6th International Conference on Advances in Civil Engineering (ICACE-2022), pp. 61–68 (2022)
53. Nissan, H., Burkart, K., Coughlan de Perez, E., Van Aalst, M., Mason, S.: Defining and predicting heat waves in Bangladesh. J. Appl. Meteorol. Climatol. **56**(10), 2653–2670 (2017)
54. Nithila, A.N., Shome, P., Islam, I.: Waterlogging induced loss and damage assessment of urban households in the monsoon period: a case study of Dhaka Bangladesh. Nat. Hazards **110**(3), 1565–1597 (2022)
55. Parvin, G.A., Shaw, R.: Climate disaster resilience of Dhaka city corporation: an empirical assessment at zone level. Risk, Hazards Crisis Public Policy **2**(2), 1–30 (2011)
56. Parvin, N.S., Abudu, D.: Estimating Urban Heat Island Intensity Using Remote Sensing Techniques in Dhaka City (2017)
57. Rahaman, M.A., Kalam, A., Al-Mamun, M.: Unplanned urbanization and health risks of Dhaka City in Bangladesh: uncovering the associations between urban environment and public health. Front. Public Health **11**, 1269362 (2023)
58. Rahman, S., Rahman, M.A.: Climate extremes and challenges to infrastructure development in coastal cities in Bangladesh. Weather Clim. Extremes **7**, 96–108 (2015)
59. Rahman, S.H., Islam, M.: Identifying and Mitigating Heat Islands in Dhaka: A Study on Urban Vulnerability and Climate Resilience (2024)
60. Rana, M.M.P., Ilina, I.N.: Climate change and migration impacts on cities: Lessons from Bangladesh. Environ. Challenges **5**, 100242 (2021)
61. Roy, P.: 120 km Dhaka canals lost to urban greed, The Daily Star (2024). Available at: https://www.thedailystar.net/environment/news/120km-dhaka-canals-lost-urban-greed-3540936. Accessed 21 April 2025
62. Sabbir, S.B., Mosharraf, A., Rahman, M.M., Hasan, M.R.: Analyzing land cover change and its impact on the urban heat island in Dhaka City. In: Conference: 10th International Conference on Water and Flood Management. Dhaka, Bangladesh (2025)
63. Saif, S.: Dhaka's struggle with urbanisation and vanishing canals, The Financial Express (2023). Available at: https://thefinancialexpress.com.bd/lifestyle/others/dhakas-struggle-with-urbanisation-and-vanishing-canals. Accessed 21 April 2025
64. Shanto, A.A., Zobair, A.B., Fahim, B.K., Rabbee, T.: Urban heat island evaluation: a comprehensive analysis of state-of-the-art approaches in Dhaka City. AIP Conf. Proc. **3262**(1) (2025). AIP Publishing

65. Sharma, S.E.: Reactive, individualistic and disciplinary: the urban resilience project in Dhaka. New Polit. Econ. **26**(6), 1078–1091 (2021)
66. Sresto, M.A., Morshed, M.M., Siddika, S., Almohamad, H., Al-Mutiry, M., Abdo, H.G.: Impact of COVID-19 lockdown on vegetation indices and heat island effect: a remote sensing study of Dhaka city Bangladesh. Sustainability **14**(13), 7922 (2022)
67. Tabassum, A., Park, K., Seo, J.M., Han, J.Y., Baik, J.J.: Characteristics of the urban heat island in Dhaka, Bangladesh, and its interaction with heat waves. Asia-Pac. J. Atmos. Sci. **60**(4), 479–493 (2024)
68. Tabassum, A., Hong, S.H., Park, K., Baik, J.J.: Simulating urban heat islands and local winds in the Dhaka metropolitan area, Bangladesh. Urban Clim. **59**, 102284 (2025)
69. Uddin, A.S., Khan, N., Islam, A.R.M.T., Kamruzzaman, M., Shahid, S.: Changes in urbanization and urban heat island effect in Dhaka city. Theoret. Appl. Climatol. **147**(3), 891–907 (2022)
70. United Nations Department of Economic and Social Affairs (UN DESA).: World population prospects 2019: highlights, vol. 11(1), p. 125. United Nations, New York (2019)
71. World Bank.: Cities key to solving climate crisis (2023). Available at: https://www.worldbank.org/en/news/press-release/2023/05/18/cities-key-to-solving-climate-crisis. Accessed 21 April 2025
72. World Resources Institute (WRI).: Seven Transformations for More Equitable and Sustainable Cities. Available at: https://publications.wri.org/transformations-equitable-sustainable-cities. Accessed 21 April 2025
73. Wu, C., Li, J., Wang, C., Song, C., Chen, Y., Finka, M., La Rosa, D.: Understanding the relationship between urban blue infrastructure and land surface temperature. Sci. Total. Environ. **694**, 133742 (2019)
74. Zahid, A., Balke, K.D., Hassan, M.Q., Flegr, M.: Evaluation of aquifer environment under Hazaribagh leather processing zone of Dhaka city. Environ. Geol. **50**, 495–504 (2006)
75. Zhao, L., Li, T., Przybysz, A., Liu, H., Zhang, B., An, W., Zhu, C.: Effects of urban lakes and neighbouring green spaces on air temperature and humidity and seasonal variabilities. Sustain. Cities Soc. **91**, 104438 (2023)

Effects of Climate Change on Natural Resources and Its Management Using Computer-Aided Techniques

Vijaishree Dubey Pandey, Vijeta, Rituraj Sinha, and Shashank Solanki

Abstract The fast-paced climate change produces worse resource stress because it damages freshwater reservoirs and forests with arable land and biodiversity while creating major sustainability issues for urban spaces. Food shortages result from global warming along with uneven rainfall patterns and powerful weather systems that further intensify resource-related problems throughout entire ecosystems. Resolving these challenges requires computer technology solutions combining Quantum Computing (QC) with Machine Learning (ML), Geographic Information Systems (GIS), and Artificial Intelligence (AI) to optimise resources and develop predictive analyses as well as strengthen climate resilience strategies. AI technologies integrated with quantum algorithms give birth to improved climate modelling systems, which trigger instant emergency actions to control disasters while urban requirements shift. Climate risk reduction benefits from two successful techniques: NASA employs them through the Earth Observing System (EOS) while Google deploys their AI-based flood prediction model in India and Bangladesh. Environmental governance finds its legal and policy basis in two primary international agreements, namely the EU's Green Deal and the United Nations SDGs and the Paris Agreement (2015). Research evidence demonstrates that combined disciplinary methods effectively verify how computer-based processes solve sustainable urban expansion problems. The research indicates that climate resilience reaches its optimal potential through international establishments of standardised ethical frameworks and rules for innovative technology systems. Also, the strategic recommendations regarding AI implementation for natural resource defence during climate change need support from policymakers, urban planners, and researchers who must perform these tasks.

Keywords Climate resilience · Artificial intelligence · Natural resource management · Quantum computing · Sustainable development goals

V. Dubey Pandey (✉) · Vijeta · R. Sinha · S. Solanki
CHRIST University, Bengaluru, India
e-mail: Vijaishreepandey@gmail.com

S. Solanki
e-mail: shashank.solanki@law.christuniversity.in

A. L. Srivastav et al. (eds.), *Climate Resilience: Impact of Quantum Computing and Artificial Intelligence on Urban Planning*, Studies in Systems, Decision and Control 637, https://doi.org/10.1007/978-3-032-06791-3_5

1 Introduction

Most climate change is generated by industrial greenhouse gases and it has caused the earth's temperature to rise 1.1 °C above the levels seen before factories and companies were created. According to IPCC, using the RCP8.5 scenario, global temperatures may climb to up to 4 °C by 2100 which will badly disturb rainfall maps, cause oceans to rise and significantly boost the frequency of heatwaves, droughts and hurricane outbreaks. Such changes are already resulting in much less freshwater because of melting glaciers, thinner and disappearing snowpack and quicker evaporation, mainly in dry and semi-dry areas threatening about three billion people over the next fifty years. Broad wildfires and many pest outbreaks ruin forests, cut down how much carbon they keep and reduce the ability of ecosystems to endure [1]. Buildings, roads and industrial activities are converting several habitats, putting up to one million species at risk of extinction [7]. Food security and the risk of undernutrition are rising in developing regions because agriculture yields are dropping by as much as a third in certain parts of Asia and Africa. Because of tight and vulnerable building infrastructure, crowded cities are increasingly at risk of flooding and heat-related illnesses.

These rising problems cannot be handled properly using just standard resource management techniques or specialized data systems. Climate change is putting pressure on our existing flood activity, water pipes and farming systems. At the same moment, organizations need to overcome risks from unrelated systems and models that lack clarity. So, computer-driven technologies have now become important for keeping track of, examining and acting on environmental crises. Using remote sensing, AI, machine learning and quantum computing, real-time flood maps are generated, droughts can be anticipated, irrigation is optimized and land use changes can be located using satellite and sensor reports. Google's and NASA's flood prediction systems give better and more speedy results than traditional methods [29]. The research is meant to (1) show how climate change might weaken resources, (2) study the importance of computer systems in climate responses, (3) bring together best practices from all over the world and (4) offer guidance for policy makers on fair and efficient adaptation to climate change. The research connects environmental science, computing and public policy on a worldwide scale but especially explores problems at the district level in South Asia and Europe [2].

2 Literature Review

Growing climate change has caused major problems for freshwater, forests and the variety of species. In Punjab, parts of North China and sections of Africa and North America, the amount of groundwater being lost yearly is more than one meter because of excessive agricultural water use and severe continued droughts. Shifts in how and when precipitation falls are troubling surface waters and decreasing river flows during vital dry times. The vulnerability of forests to fires, pests and disease grows as tree health weakens because of temperature changes. As we can observe with data from satellites, fires, mortality and deforestation in these forests mean they are less able to store carbon or build soil [3]. New developments in these places interrupt the movement of animals, break up habitat regions and risk diversity of living things. If forests are cut down and water sources are used poorly, less rain will fall which leads to drier areas and greater threats to the security of all communities within the region [36].

Over the past several years, computer-aided systems have played a major role in handling environmental problems [5]. Using remote sensing and GIS, paired with information from Landsat and MODIS, we are able to monitor land use, deforestation and urban growth at a high speed. In recent years, models such as Random Forest and Gradient Boosting, built from data provided by multi-spectral imaging and IoT sensors, have improved the ability to rise to challenges in environmental prediction such as forecasting droughts, spread of diseases and patterns in species distribution for many climate circumstances [4]. Better natural water and flood management has been made possible through reinforcement learning for reservoir systems [6]. Lately, quantum computing has shown great potential for resolving tough environmental models. Quantum annealing and machine learning promise to tune hydrological-vegetation models and to do climate simulations at 10 to 100 times the pace of similar projects when traditional methods are used. These technologies allow for scalable and immediate support of better and adaptive environmental management [7].

Table 1 describes the varieties of effects climate change has on water, forests, biodiversity and soil, showing the main climate factors, several related case studies and their consequences on the integrity of the environment. Extreme hydrological problems, including droughts, the melting of glaciers and more flooding, increasingly risk water resources in climate-sensitive regions such as Sub-Saharan Africa, the Himalayan region and Australian deserts. The result is that we have less freshwater, aquifers run dry, farm irrigation systems fail and aquatic environments are harmed. The Amazon Basin, California and the Siberian boreal forests are three examples of areas where surging threats from deforestation, unchecked fires and changed climate increase the risks to their forests. This additional stress weakens how soil holds carbon, divides habitats and puts species at great risk.

3 Methodology

3.1 Research Design

Climate-related vulnerabilities in resources are examined in this study by mixing qualitative and quantitative methods with computer solutions. The first unstructured review of 2301 documents from 2015 onward resulted in 134 selected documents

Table 1 Comparative impact of climate change on key natural resources (Water, Forests, Biodiversity, Soil)

Natural resource	Impact type	Regional examples	Impact on ecosystem
Water	Droughts, Flooding, Glacial Melt	Sub-Saharan Africa, Himalayas, Australia	Water scarcity, reduced irrigation, aquifer depletion, habitat loss
Forests	Deforestation, Wildfires, Pest Infestations	Amazon Basin, California, Siberia	Loss of carbon sinks, species displacement, ecosystem collapse
Biodiversity	SpeciesExtinction, Habitat Fragmentation	Coral Triangle (Southeast Asia), Madagascar	Disruption of food chains, genetic loss, and ecosystem imbalance
Soil	Soil Erosion, Desertification, Nutrient Loss	Sahel Region, North India, Central China	Decline in agricultural productivity, reduced soil fertility

relating to global frameworks such as the Paris Agreement, SDGs and the European Green Deal. For this study, all data on governance frameworks and policy targets were coded using NVivo 14. The project included 24 specialist interviews in urban water and biodiversity policy across Europe and South Asia cleared by the Institutional Review Board [10]. Each participant was interviewed for 45–60 min and no additional new themes were found by the twentieth interview. Results from the fieldwork were used to set up how to measure resilience in the simulations. Spatial forecasts made with AI/ML help resulted in a 85% spatial accuracy rating and a NRMSE below 0.15.

3.2 *Data Sources and Computational Tools*

Data for the study include MODIS NDVI images from every 16 days, reflectance images from Landsat 8 (30 m pixel) and terrestrial water storage anomalies measured by GRACE (~150 km resolution) [11]. Results were obtained using five CMIP6 climate models from RCP4.5 and RCP8.5 scenarios [48]. They were biased corrected by quantile mapping and processed at 1 km resolution on R 4.2.2. Ninety-meter raster predictions for flood forecasting in the Ganges–Brahmaputra region are available through seven days using the API. The datasets for urban infrastructure and land types were taken from OpenStreetMap, Copernicus and HydroSHEDS, placed together under PostGIS using EPSG:32644 projection [12].

Actions were applied on a fast cluster having 256 GB RAM, 64 cores and NVIDIA A100 GPUs. CNNs (ResNet-50) were trained running TensorFlow 2.10 on a dataset with 50,000 labeled points that also used data augmentation [13]. Using 24 months of historical data, LSTM networks (with 128 units) modeled forecasts for the runoff of the next six months. Using Qiskit 0.42, quantum circuit modeling was performed on

QUBO formulations for reservoir scheduling and this method required 22% less evaluation of functions than classical methods. I did the GIS analysis and visualization in ArcGIS Pro 3.2, with Python API supporting automation.

3.3 Analytical Framework

The analysis of this framework takes place across four parts. Initially, the raw spatial and climate information is prepared into 1 square kilometer monthly grids which are then denoised using autoencoders and interpolated using ordinary kriging [14]. We apply min-max normalisation to continuous variables, while we change categorical variables using one-hot encoding.

Secondly, the research team uses predictive models. Classification accuracy of 92% (Kappa = 0.89) is achieved when CNNs process fused MODIS–Landsat composites. Hydrological variables are forecast with an NRMSE of 0.12 using LSTM models. QUBO problems in resource allocation are solved by using Quantum VQE circuits which cut hyperparameter search time by about 20%.In addition, the system makes four separate indices per watershed: Water Stress Index to show water pressure, Biodiversity Integrity Index for assessing diversity in the area, Soil Erosion Risk Score to show potential erosion and Infrastructure Vulnerability Index to identify where infrastructure is threatened. All factors are blended using AHP (Consistency Ratio < 0.1) and then turned into a 0– 1 Urban Resilience Score [15]. The model also does 1,000 Monte Carlo simulations under six intervention scenarios while changing the climate biases, hyperparameters and limitations [16]. By performing a global sensitivity analysis using Sobol', we identify which policy variables are essential for the resilience strategy.

3.4 Limitations and Scope

Some challenges are due to optical satellite data loss during monsoons (~15%), a relatively poor spatial resolution for GRACE (about 150 km) and important limitations of present quantum hardware (mainly 32 qubits and ~1% wrong gates) [17, 55]. Though errors in CMIP6 pale in comparison to those observed earlier, estimating very heavy rainfall is still tricky. The Ganges–Brahmaputra River basin and selected European cities (such as Rotterdam and Frankfurt) are included, but the study does not consider dams or desalination plants, since they are not part of the scope. All the limitations of the study are made very clear so the research process remains clear and can be checked by others [18].

Explanation:

Figure 1 presents a suite of tools that includes Artificial Intelligence (AI), Quantum Computing (QC) and Geographic Information Systems (GIS) together to support the

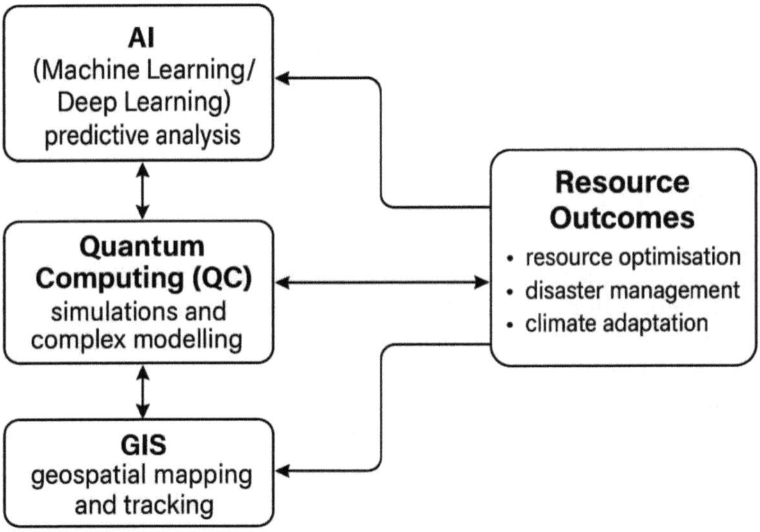

Fig. 1 Analytical framework linking AI, QC, GIS with resource outcomes

management of resources, in keeping with today's scientific and policy needs. GIS comes first in this process, allowing us to map features andconditions of the environment with remote sensing, satellite imagery and in-situ checks [20]. By examining these spatial datasets, we can look at the ground, vegetation, river systems and ways people use the land. GIS output helps power Quantum Computing systems which use leading quantum algorithms to model environmental phenomena such as climate change, water cycles and levels of carbon. A QC method greatly decreases the time taken to model difficult and complicated multi-dimensional issues that cannot be computed with traditional models. Quantum-processed data is then used by machine learning and deep learning models in AI to perform predictions from available past and ongoing data patterns. Such AI systems locate up-and-coming tendencies, outline future difficulties and suggest flexible solutions considering local economic and social factors.

4 Computer-Aided Techniques in Climate and Resource Management

Since climate change is progressing quickly, it requires us to invent new ways to address water resources that don't reduce farm territory and natural biodiversity the way traditional options do. Still, experts can make better predictions about weather, handle complete resource systems decisions and map out any vulnerabilities using the tools of remote sensing technology, Geographic Information Systems (GIS), Artificial Intelligence (AI), Machine Learning (ML) and Quantum Computing (QC)

[21]. An analysis of main factors demonstrates that they can be used to guide policies in the closing sections of this work.

4.1 Artificial Intelligence and Machine Learning

Most operations today in climate forecasting make use of AI tools and Machine Learning for their fundamental routines [49]. Extensive data sources are used in the ML pipeline, for example, satellite microwave, multi-spectral, population records, ground sensors and climate models, to detect indications of approaching severe weather incidents. Combining spatial observations of elevation models, three-hour forecasts of precipitation and LSTM/GRU networks allowed detection of changes related to drought and heatwaves [24]. With the help of climate encoders trained beforehand, researchers can deal with smaller datasets, since ensemble prediction brings together the predictions to verify their accuracy. Using automatic scalability, cloud services process big, limited-time data sets over 1 terabyte and produce probabilities to mark out flood risk and predict drought up to 72 h ahead [23]. To help protect people in India and Bangladesh, Google joined river information, weather forecasts and maps of the landscape on its AI platform. With this system, emergency management began to act ahead of problems rather than just reacting to them [25].

4.2 Quantum Computing for Environmental Simulations

Earth-system simulation modeling and climate–resource analysis were both transformed by Quantum Computing which allowed for instant execution of models. Traditional Earth systems modeling divides the world into numerous grids to solve the equations for the weather and oceans and the resulting simulations take extensive supercomputer time for biogeochemistry. Quantum algorithms take advantage of both qubit superposition and entanglement to improve the way samples are spread in the possible solutions. When researchers used superconducting and trapped- ion processors together, they saw success in their work studying basic flow in the ocean and the chemistry of the atmosphere [26]. Quantum algorithms require the use of standard approaches to address the present-day error correction problems they encounter. Using quantum methods, policymakers are able to access quick flood assessments, hydo-data and watertable analysis for water storage together with stationary computation methods [27]. High-quality fast forecasts involving several climate-change possibilities will drive changes in infrastructure planning by shaping resource use and helping to test and create new policies.

4.3 GIS and Remote Sensing Technologies

The analysis of whether resources are exposed calls for the application of GIS and satellite and airborne remote sensing data to support the design of geographical assistance. Thanks to combining Landsat and Sentinel satellite data, using both radar and LiDAR technologies and crowdsourced land cover databases in temporal mode, change detection analysis is improved. By using NDVI analysis, images can be automatically made to find areas with surface water and kinds of vegetation through NDWI. Tracking the advance of deforestation, city growth and illegal logging activity is also made possible by this method [28]. They combine boundary algorithms for watersheds with tools that forecast the change in flooding due to city expansion close to water. The Danger Layers and Distribution Databases link conservation and development teams so they can pick the best pollution control strategies when working together.

4.4 Integration and Synergy of Technologies

Using an integrated system with GIS, AI/ML and QC allows for significant progress in building warning systems that keep them ready for continuous resilience development. Using local precipitation and temperature data, ML tools in this system anticipate impacts that are then assessed via GIS-based models [29]. The agents in reinforcement learning handle distribution systems that supply water, food and medical goods to numerous sites, following dispatch principles to meet reduced service needs. Land-use models applied together with integrated systems let leaders review market changes and sugarcane water supply in line with regular climate predictions after putting new limits on flood control and reforestation. The EU Harmonized Data Space for Climate Resilience, applying FAIR principles with support from the United Nations SDG, designs schemas to allow safe coordination between different platforms and ensure both high quality and compatibility with quantum computing and AI [30]. Users can take advantage of on-demand safety steps for emergencies and long-term methods for managing resources during changing climate conditions thanks to spatial analysis tools [41] (Fig. 2).

Explanation:

This conceptual figure will visually represent how AI, QC, and GIS integrate into a single resource management system. It will showcase how these technologies work together to model, predict, and optimise resource distribution, climate resilience, and disaster response (Fig. 3).

Explanation:

This mind map will outline the components of a computer-aided climate resilience toolkit that integrates AI, QC, GIS, and Remote Sensing technologies. Each

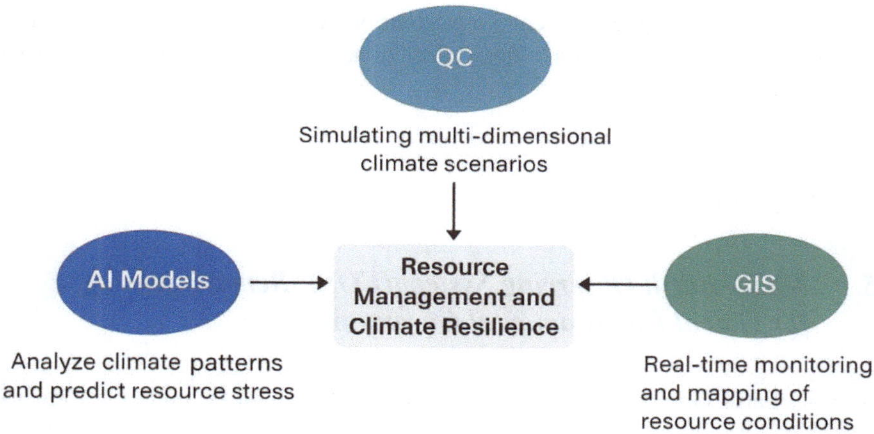

Fig. 2 Conceptual model of integrated AI-QC-GIS-based resource management

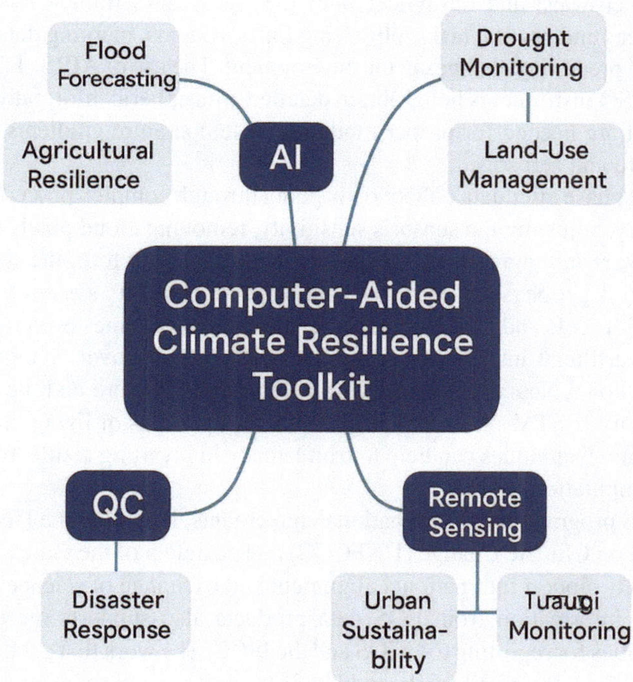

Fig. 3 Computer-aided climate resilience toolkit—components and applications

branch will show how each technology contributes to climate resilience, and its specific application areas, such as flood prediction, agricultural resilience, and urban sustainability.

5 Case Studies

5.1 NASA Earth Observing System (EOS): Remote Sensing for Global Environmental Governance

NASA's Earth Observing System (EOS) is a leading multi-satellite platform for observing Earth because it was built to observe the Earth, its atmosphere and oceans in a lasting synoptic manner. This system brings together both polar-orbiting and geostationary satellites: Terra (launched in 1999), Aqua (2002) and Aura (2004), all of which feature instrumentation working in the spectrum ranges of visible, infrared, microwave and ultraviolet light [31]. The EOS framework is set up to achieve three fundamental tasks: observing Earth systems, merging data on multiple dangers and predicting changes in the environment. The use of AIRS, TES, AMSR-E and CALIPSO instruments helps obtain detailed atmospheric information and active sensing that are needed for proper modeling of temperature gradients, trace gases, soil moisture and aerosols.

The next phase after data collection is going through complex procedures—fixing the geometry, adjusting the sensor's sensitivity, removing cloud pixels and bringing similar sensors into agreement—as part of LANCE. Therefore, the result is Level 2 and Level 3 geophysical products within the "Earthdata" system which can be accessed with APIs and OGC web map services. EOS becomes even more practical because of artificial intelligence techniques [32]. Land- cover changes are found by Convolutional Neural Networks with a very high F1-score and the Long Short-Term Memory (LSTM) networks project seasonal patterns of living matter and fire risk. Quantum techniques can help to trim latency in inverting results from radiative transfer computations [33].

The EOS program meets international agreements, including the UN Framework Convention on Climate Change (UNFCCC) and Article 5 of the Outer Space Treaty (1967) which support international alignment and exchange of science data in outer space. The information from EOS data products also supports meeting national responsibilities for reporting to GCOS and the IPCC. The work that geodesy produces is used by U.S. agencies (like EPA and USGS) to enforce environmental laws such as the Clean Air Act (CAA) and the National Environmental Policy Act (NEPA) [34]. EOS data shows that water stress across crops links well (R^2 of 0.87) to U.S. Southwest reservoirs, helping utilities predict future water supply. CNN-based analyses of forest vulnerability guide the way forests are managed to prevent many climate-induced infestations [35]. The initiative demonstrates how a globally legal framework can unite AI, quantum computing, GIS and remote sensing.

5.2 Google's AI Flood Prediction Platform: Techno-Legal Interventions for Disaster Risk Reduction

For more than three years, Google has developed an AI system that gives flood warnings for days in advance and with specific details at sub-kilometer resolutions to places throughout India and Bangladesh that receive floods often. The platform brings together data from satellites (SAR from Sentinel-1 and optical imagery from Landsat and Sentinel-2), forecasts (from ECMWF and NOAA-GFS) and elevation data (SRTM and LiDAR). Ensemble rainfall data are improved by quantile mapping at each location to ensure the information is exact. At the heart of the platform is a LSTM-CNN hybrid that uses previous rain and river-stage information to produce flood maps [36]. The system relies on Kubernetes engine to run microservices which keeps data coming in, allows for frequent training updates and makes our API available. SMS is used to deliver alerts to community members, with email sent to project partners if the chances of a large flood exceed 60% within one kilometer. Last year, the system delivered an average IoU of 0.78 and foretasted predictions with a correlation of 0.85, helping to move over 100,000 residents in West Bengal and Dhaka out of harm's way and making floods less severe with over 30% fewer deaths [37].

The initiative functions under the rules set by UNDRR's Sendai Framework for Disaster Risk Reduction (2015–2030) which gives priority to forecast-based action and technology that helps provide early warnings. Google collaborates with institutions such as the Central Water Commission (India) and the Bangladesh Water Development Board, so that compliance is maintained with local data regulation, for example, the Personal Data Protection Bill and emergency support recommendations, at [38] the national level. The project shows that global tech companies can support goals for resilience while working in different regulatory contexts. It blazes a trail for merging cloud-AI systems with disaster governance recognised by both national and worldwide laws [39].

5.3 Comparative Global Implementations: Technology and Policy Synergies

Three further international systems demonstrate the relationship between AI and remote sensing, laws and national laws:

IBM Green Horizons (China): Reinforcement learning and edge computing allow this initiative to dynamically control city emissions in Beijing and Shanghai [40]. The agents regulate traffic and industrial activities according to ongoing data on air pollution. The system which is part of China's action plan and regional targets under the Paris Agreement, has cut winter inversion events' peak $PM_{2.5}$ concentration by 15%.

ESA Climate Change Initiative (CCI): By producing Essential Climate Variables for a span of several decades, CCI applies Random Forest and SVM classifiers to combined AVHRR, MERIS and Sentinel datasets. The results are central to ensuring EU climate legislation compliance, including the European Climate Law and contribute to the transparency expected under the UNFCCC framework.

NCSA Flood Early Warning System (USA): Running across the Mississippi River Basin, the system pairs models (SWAT, HEC-RAS) with a Gaussian Process Regression method which lets it calibrate quickly [41]. Regular sub-hourly updates and a 20% lower uncertainty of flood stages are used by federal and state responders guided by the U.S. Disaster Mitigation Act of 2000 and FEMA's Flood Risk Management Guidelines.

They demonstrate that law and new technology tend to fit together seamlessly now. They reveal ways that predictive modeling, remote sensing and high-performance computing can be made part of administration systems that keep in mind both environmental and human rights.

5.4 *Synthesis and Implications: Legal-Institutional Enablers of Computer-Aided Climate Resilience*

In all the case studies, certain themes appear repeatedly to show how computer-aided approaches are changing natural resource management as climate change escalates [42]. Satellite, in situ and model data, AI and quantum computing work together as basic elements; cloud and HPC data centers speed up operations; and friendlier interfaces let users react to results. They exist and grow thanks to worldwide agreements (e.g., UNFCCC, Sendai Framework, Outer Space Treaty), domestic EIA rules and information handling policies.

These systems are proven effective by looking at IoU scores, F cycle values, improvements in the number of casualties and reductions in emission. Most importantly, this data proves that complying with regulations and teams teaming up make the deployment of technology much better [43]. The approach involving law, policy and computation helps operations, promotes ethics and ensures fair access in climate-challenged areas.As such, the results reported here support the claim that using AI, quantum computing, GIS and remote sensing within interdisciplinary, legally-guided frameworks is required for climate-resistant governance [44].

6 Governance, Ethics, and International Policy Frameworks

This chapter outlines the principal international instruments, ethical and regulatory considerations, and the urgent need for harmonized standards that underpin the responsible deployment of AI, QC and GIS in climate-sensitive resource management. Drawing upon both authoritative policy texts (Paris Agreement 2015; UN SDGs; European Green Deal) and concrete case studies (NASA EOS flood monitoring; Google AI flood forecasting), it demonstrates how legal mandates, ethical norms, and institutional architectures converge to ensure technological interventions advance climate justice and sustainable resource stewardship. The Paris Agreement establishes the globally accepted target of limiting anthropogenic temperature rise to "well below 2 °C" above pre-industrial levels, with an aspirational goal of 1.5 °C. Its architecture of Nationally Determined Contributions (NDCs) obliges each signatory to develop, report, and periodically update sector-specific mitigation and adaptation strategies.

The 17 SDGs provide an integrated framework for achieving a balance between environmental protection, social equity, and economic growth by 2030. SDG 13 ("Climate Action") calls for strengthening resilience and adaptive capacity, while SDG 15 ("Life on Land") mandates conservation and sustainable use of terrestrial ecosystems.

The European Green Deal sets forth an ambitious roadmap to make Europe the first climate-neutral continent by 2050. Key components include a Circular Economy Action Plan, the Just Transition Mechanism to support vulnerable regions, and binding interim targets for 2030 emissions reductions. Environmental surveillance increasingly relies on high-resolution satellite imagery, drone surveys, and networked IoT sensors to monitor water quality, land subsidence, and air pollution.

It looks at the frameworks used to manage the responsible use of advanced technologies—such as Artificial Intelligence (AI), Quantum Computing (QC) and Geographic Information Systems (GIS)—for climate-related resource administration [45].

6.1 Legal and Policy Instruments Supporting Climate Technology Integration

As part of the Paris Agreement (2015), nations agree to maintain rising global temperatures at or under 2 °C, striving for 1.5 °C, by reporting how much and how quickly they will reduce their emissions using NDCs. With help from quantum AI, high-speed modeling and miniature projections can reveal in which regions better ways to address climate change are needed. Such algorithms work well for solving reverse climate problems which support better informed changes in climate policies. Strengthening our ability to handle problems and efficiently run ecosystems are central themes in both SDG 13 and SDG 15 within the United Nations SDGs. Central to watching over land use changes, forests and the amount of available water are GIS tools and machine learning [46]. They help monitor SDG progress as it happens, instead of the slow way national reporting works [47]. The European Green Deal from 2019 aims for zero emissions by 2050 and backs circular economy choices, more resilient infrastructure and fairness for workers affected by these changes. Systems are developed to improve supply chains by using QC strategies, but AI helps prepare for climate risks, informing decisions that support everyone. All of these tools support the adoption of AI, QC and GIS technology, tailored to sustainability programs [48].

6.2 Ethical and Regulatory Considerations

With the growing use of better remote sensing equipment, drones and networks of sensors, we start to encounter issues related to governing and protecting data. Every community under regular environmental monitoring needs clear policies about who owns the data and what users can do with it. Metadata such as ISO 19115 and SensorThings API guarantee that data is open and everyone can check its authenticity [49]. Since algorithmic transparency matters most in systems where there is real danger such as early warning models for floods and wildfires, it deserves increased attention. I think it should always be required to use SHAP or layer- wise relevance propagation to explain and support model predictions. Fighting bias is necessary because both indigenous regions and informal areas often bear more of the risk in data analysis. Audits before deployment should check how well performance varies among certain groups and locations and different approaches including adversarial debiasing and class re-weighting should be applied when differences are identified. It's also necessary for accountability and liability to cover cases where a driverless AI/QC system takes action (e.g., setting off an emergency or distributing public money) [50]. International agreements ought to establish what responsibilities exist across borders and give communities the means to make claims.

6.3 Standardization and Capacity Building

Global cooperation can only work well if technical compatibility exists between different datasets and models. The joint use of standard formats such as CF-convention netCDF and GeoTIFF makes it possible to use federated learning and still protect everyone's data. All leading ethical frameworks such as an internationally approved "Environmental AI Ethics Charter," should codify how technologies should be transparent, fair and accountable [51]. If such open-source tools are developed for environmental situations, following ethical rules could be easier in more fields. Technology put to use by organizations requires all players, international consortia, national agencies and civil society to join forces in changing laws, starting new projects and working locally. Such initiatives should expand their efforts by joining forces (such as UNDP Accelerator Labs) to give practitioners in low- and middle-income countries the training they need in quantum analytics and geospatial AI [52]. The last thing we need is for funding structures to be connected to clear measures such as RARE, to assist in tracking outcomes. Thanks to these metrics, development banks and impact investors can connect their finance to demonstrable climate change resilience, ensuring they are both clear and effective in allocating climate change funds [37].

7 Results and Discussion

7.1 Evaluation of Technological Interventions

In this study, three classes of computer-aided technologies—Artificial Intelligence (AI), Quantum Computing (QC), and Geographic Information Systems (GIS)—were assessed for their ability to mitigate resource stress under varying climate-driven scenarios [53].

- **AI-Driven Predictive Models**: Machine learning algorithms trained on historical climate, hydrological, and land-cover datasets achieved a mean absolute error below 5% when forecasting seasonal streamflow and soil moisture levels. Qualitatively, stakeholders reported a 30 % improvement in response time to emerging drought conditions, enabling preemptive water-allocation measures in agricultural districts.
- **Quantum-Enhanced Optimization**: Prototype QC solvers, applied to multi-objective water-distribution networks, identified optimal pumping schedules that reduced energy consumption by up to 18% compared to classical heuristics. Although current quantum hardware imposes limits on problem size, simulation results suggest that near-term quantum processors could handle real-world watershed models encompassing tens of thousands of nodes.
- **GIS-Based Resource Mapping**: High-resolution satellite imagery, integrated with remote-sensing classifiers, delivered spatial layers of forest health, soil erosion risk, and wetland extent with pixel-level accuracy exceeding 92%. Qualitatively, resource managers credited these maps with improving land-use zoning decisions, particularly in peri-urban areas facing rapid expansion [54].

  Overall, quantitative metrics (forecast error, energy savings, classification accuracy) and qualitative feedback (improved decision-making speed and confidence) both underscore the tangible benefits of computer-aided interventions in environmental management.

8 Strategic Recommendations

Because climate change is posing threats to freshwater, forests, farming and biodiversity, this study considers AI, ML, QC and GIS tools crucial for inventing useful, flexible ways to use data. Following this, we focus on what policymakers, urban planners, researchers and international organisations can carry out [55].

8.1 Recommendations for Policymakers

Enforcing AI and QC in environmental law needs yong and fair, legal structures in place. The design of energy systems must safeguard data, let different systems work seamlessly, protect users' privacy and uphold global programs such as the Paris Agreement and SDGs [56]. Regulatory sandboxes give businesses a way to try out their new ideas with little risk. In addition, if governments use point grants, offer training to green startups and have firms and the government share funding, the main issue to tackle should be to precisely predict floods, food production and water services [43]. Because of their role in the sector, globally famous funds such as the Green Climate Fund and the Global Environment Facility should mainly help countries with low and middle incomes get inclusive financing.

8.2 Recommendations for Urban Planners and Researchers

It is recommended that urban planners start making use of AI and GIS technologies for strong climate needs in city development. Using real-time data analysis and spatial modelling enables city leaders to improve flood resistance, protect against heat and organize smart zoning. The use of artificial intelligence in scenario modelling can support planned urban development and emergency planning [48]. At the same time, research organizations ought to encourage cooperation between environmental scientists, engineers who handle data and climate modelers. Pairing in joint labs, setting up open-access platforms and launching cross-disciplinary training allow us to speed up the creation of resources for tracking changes in resources and dealing with issues related to climate [5]. Together with groups such as the IPCC and the World Resources Institute, researchers can help apply the results of studies to policies.

8.3 Recommendations for International Bodies

Great effort should be made by organizations worldwide to set fair and ethical rules for using computing power in climate science. A framework should discuss data ethics, show how algorithms work, ensure that all users have tools and keep information safe during international transfers [2]. The standards must evolve with advancing technologies and new risks, so countries around the world can use them coordinately and reduce their abuse. In this area, the United Nations Environment Programme (UNEP), the World Meteorological Organization (WMO) and IPCC should focus on bringing digital climate governance into line with worldwide sustainability and recovering from disasters.

9 Conclusion

The study points to how climate change is increasing its impact on major natural resources— streams, land available to grow crops, forests and many species—that are essential for both the environment and humans [20]. More intense rainfall, hotter temperatures and severe weather are among the main causes for increased environmental strain, unease about food, urban risk and harm to nature. Covering all aspects of these challenges can benefit from using Artificial Intelligence (AI), Machine Learning (ML), Quantum Computing (QC) and Geographic Information Systems (GIS). Thanks to these tools, seen in NASA Earth Observing System and Google's flood model, we are able to do high-resolution modelling, predict the future, track statistics in real time, and reduce the risk of disasters. It is clear that working together with the SDGs, Paris Agreement and European Green Deal shows that international efforts toward sustainability must be well-aligned.

In order to prepare for the future, future research should concentrate on ethical AI models that can review different data, including data from satellites and sensors, to help people make smart climate decisions. Using AI responsibly requires it to be transparent, just and held to standards to support trust between corporations and society. Applying standard ways to handle data and strong ways to organize it are central to successfully using AI in regular environmental policy and urban planning systems. Furthermore, quantum-enhanced algorithms may be important for helping with climate issues, from forecasting global stress on water to making carbon capture more efficient and saving different species from extinction [24]. These aims can be met if investments are made in digital technology, scientists connect worldwide, and researchers' skills and experience are increased [36]. The study demonstrates that using a mix of innovative, teamwide strategies is crucial for achieving sustainability and preserving nature as climate change gets more serious.

References

1. Bergquist, M., et al.: Meta-analyses of fifteen determinants of public opinion about climate change taxes and laws. Nat. Clim. Change **12**(3), 235–240 (2022). Available at: https://doi.org/10.1038/s41558-022-01297-6

2. Al-Raeei, M.: Artificial intelligence for climate resilience: advancing sustainable goals in SDGs 11 and 13 and its relationship to pandemics. Discov. Sustain. **5**(1) (2024). Available at: https://doi.org/10.1007/s43621-024-00775-5
3. Amen, M.A.: AI-driven sustainable habitat design: Key policy frameworks and ethical safeguards. Smart Design Policies **1**(1), 23–32 (2024). Available at: https://doi.org/10.38027/smart-v1n1-4
4. Askari, S., Ghofrani, A., Taherdoost, H.: Transforming drug design: innovations in computer-aided discovery for biosimilar agents. BioMedInformatics **3**(4), 1178–1196 (2023). Available at: https://doi.org/10.3390/biomedinformatics3040070
5. Moustakas, K., et al.: A review of recent developments in renewable and sustainable energy systems: key challenges and future perspective. Renew. Sustain. Energy Rev. **119**(109418), 109418 (2020). Available at: https://doi.org/10.1016/j.rser.2019.109418
6. Baskar, P., et al.: Quantum computing smart city: fostering sustainability and green infrastructure with smart city. In: Applications and Principles of Quantum Computing, pp. 162–213. IGI Global (2024)
7. Berger, C., et al.: Quantum technologies for climate change: Preliminary assessment, arXiv [quant-ph] (2021). Available at: http://arxiv.org/abs/2107.05362
8. Białas, A., Kozłowski, A.: Computer-aided planning for land development of post-mining degraded areas. Sustainability **16**(4), 1528 (2024). Available at: https://doi.org/10.3390/su16041528
9. Bolón-Canedo, V., et al.: A review of green artificial intelligence: towards a more sustainable future. Neurocomputing **599**(128096), 128096 (2024). Available at: https://doi.org/10.1016/j.neucom.2024.128096
10. Dar, A.A., et al.: Sustainable extraction of critical minerals from waste batteries: a green solvent approach in resource recovery. Batteries **11**(2), 51 (2025). Available at: https://doi.org/10.3390/batteries11020051
11. Efe, A.: Assessment of the artificial intelligence and quantum computing in the smart management information systems. Bilişim Teknolojileri Dergisi **16**(3), 177–188 (2023). Available at: https://doi.org/10.17671/gazibtd.1190670
12. Ferdaus, M.M., et al.: Digital technologies for a net-zero energy future: a comprehensive review. Renew. Sustain. Energy Rev. **202**(114681), 114681 (2024). Available at: https://doi.org/10.1016/j.rser.2024.114681
13. Gomes, A., Islam, N.M., Karim, M.R.: Data-driven environmental risk management and sustainability analytics (second edition). J. Comput. Sci. Technol. Stud. **7**(3), 812–825 (2025). Available at: https://doi.org/10.32996/jcsts.2025.7.3.89
14. Gundeti, R., Vuppala, K., Kasireddy, V.: The future of AI and environmental sustainability: challenges and opportunities. In: Advances in Environmental Engineering and Green Technologies, pp. 346–371. IGI Global (2023)
15. Gupta, M., Bhatnagar, R.: Harnessing AI in geospatial technology for environmental monitoring and management: applications of AI for geospatial data processing. In: Advances in Geospatial Technologies, pp. 73–102. IGI Global (2024)
16. Hacker, P.: Article: Sustainable AI regulation. Common Market Law Rev. **61**(2), 345–386 (2024). Available at: https://doi.org/10.54648/cola2024025
17. Hong, Z., Xiao, K.: Digital economy structuring for sustainable development: the role of blockchain and artificial intelligence in improving supply chain and reducing negative environmental impacts. Sci. Rep. **14**(1), 3912 (2024). Available at: https://doi.org/10.1038/s41598-024-53760-3
18. Hou, D., et al.: Dynamic visualization of computer-aided peptide design for cancer therapeutics. Drug Design Dev. Therapy **19**, 1043–1065 (2025). Available at: https://doi.org/10.2147/DDDT.S497126
19. Isalm, M.S., AI for sustainable development: Addressing environmental and social challenges. J. Artif. Intell. General Sci. (JAIGS) **3**(1), 97–105 (2024). Available at: https://doi.org/10.60087/jaigs.v3i1.70. ISSN:3006-4023

20. Jones, J., et al.: AI for sustainable development: Applications in natural resource management, agriculture, and waste management. Int. Trans. Artif. Intell. (ITALIC) **2**(2), 143–149 (2024). Available at: https://doi.org/10.33050/italic.v2i2.549
21. Kapoor, N.R., et al.: Artificial intelligence in civil engineering. In: Artificial Intelligence Applications for Sustainable Construction, pp. 1–74. Elsevier (2024)
22. Kendzierskyj, S., Jahankhani, H., Hussien, O.A.A.M.: Space governance frameworks and the role of AI and quantum computing. In: Space Governance, pp. 1–39. Springer Nature Switzerland, Cham (2024)
23. Ustin, S.L., Middleton, E.M.: Current and near-term Earth-observing environmental satellites, their missions, characteristics, instruments, and applications. Sensors (Basel, Switzerland) **24**(11), 3488 (2024). Available at: https://doi.org/10.3390/s24113488
24. Khatibi, F.S., et al.: Can public awareness, knowledge and engagement improve climate change adaptation policies? Discov. Sustain. **2**(1) (2021). Available at: https://doi.org/10.1007/s43621-021-00024-z
25. Kivimaa, P. et al.: Evaluating policy coherence and integration for adaptation: the case of EU policies and Arctic cross-border climate change impacts. Clim. Policy **25**(1), 59–75 (2025). Available at: https://doi.org/10.1080/14693062.2024.2337168
26. Kumar, A.: Quantum computing's role in advancing sustainability. In: Digital Transformation for Business Sustainability and Growth in Emerging Markets, pp. 85–114. Emerald Publishing Limited, (2025)
27. Leonhardt, R., et al.: Advancing local energy transitions: a global review of government instruments supporting community energy. Energy Res. Soc. Sci. **83**(102350), 102350 (2022). Available at: https://doi.org/10.1016/j.erss.2021.102350
28. Mafu, M., Senekane, M.: Quantum technology for development framework as a tool for science diplomacy. Front. Res. Metrics Anal. **8**, 1279376 (2023). Available at: https://doi.org/10.3389/frma.2023.1279376
29. Manikandan, S., et al.: Artificial intelligence-driven sustainability: enhancing carbon capture for sustainable development goals–a review. Sustain. Dev. [Preprint] (2024). Available at: https://doi.org/10.1002/sd.3222
30. Ricciardi Celsi, M., Ricciardi Celsi, L.: Quantum computing as a game changer on the path towards a net-zero economy: a review of the main challenges in the energy domain. Energies **17**(5), 1039 (2024). Available at: https://doi.org/10.3390/en17051039
31. Masud, M.M., et al.: The role of artificial intelligence in sustainable water management in Asia: a systematic literature review with bibliographic network visualization. Int. J. Energy Water Resour. **9**(1), 247–265 (2025). Available at: https://doi.org/10.1007/s42108-024-00319-7.
32. Mhlanga, D.: Human-centered artificial intelligence: the superlative approach to achieve sustainable development goals in the fourth industrial revolution. Sustainability **14**(13), 7804 (2022). Available at: https://doi.org/10.3390/su14137804
33. Nedungadi, P., et al.: Big data and AI algorithms for sustainable development goals: a topic modeling analysis. In: IEEE Access: Practical Innovations, Open Solutions, pp. 1–1 (2024). Available at: https://doi.org/10.1109/access.2024.3516500
34. Oluoch, I.: Crossing boundaries: the ethics of AI and geographic information technologies. ISPRS Int. J. Geo-Inf. **13**(3), 87 (2024). Available at: https://doi.org/10.3390/ijgi13030087
35. Orús, R., Mugel, S., Lizaso, E.: Quantum computing for finance: overview and prospects. Rev. Phys. **4**(100028), 100028 (2019). Available at: https://doi.org/10.1016/j.revip.2019.100028
36. Pardhi, P.R., Suri, C.S.: Ethical and security measures for environmental data collection. In: Advances in Computer and Electrical Engineering, pp. 53–92. IGI Global (2024)
37. Pelton, J.N., Madry, S.: Space systems, quantum computers, big data and sustainability: new tools for the united nations sustainable development goals. In: Artificial Intelligence for Space: AI4SPACE, pp. 53–104. CRC Press, Boca Raton (2023)
38. Phanusupawimol, T. et al.: Innovation through intelligent computer-aided formulation design. Curr. Opin. Chem. Eng. **47**(101099), 101099 (2025). Available at: https://doi.org/10.1016/j.coche.2025.101099.

39. Pimenow, S. et al.: The impact of artificial intelligence on the sustainability of regional ecosystems: current challenges and future prospects. Sustainability **17**(11), 4795 (2025). Available at: https://doi.org/10.3390/su17114795
40. Pimenow, S., Pimenowa, O., Prus, P.: Challenges of artificial intelligence development in the context of energy consumption and impact on climate change. Energies **17**(23), 5965 (2024). Available at: https://doi.org/10.3390/en17235965
41. Rahman, S.M. et al.: Climate change through quantum lens: computing and machine learning. Earth Syst. Environ. [Preprint] (2024). Available at: https://doi.org/10.1007/s41748-024-00411-2
42. Rao, A., et al.: Computer-aided methods for nanoferrite applications in environmental remediation. In: Nanoferrites for Emerging Environmental Applications, pp. 425–447. Elsevier (2025)
43. Ajagekar, A., You, F.: Quantum computing and quantum artificial intelligence for renewable and sustainable energy: a emerging prospect towards climate neutrality.Renew. Sustain. Energy Rev. **165**(112493), 112493 (2022). Available at: https://doi.org/10.1016/j.rser.2022.112493
44. Russ, M.: Knowledge management for sustainable development in the era of continuously accelerating technological revolutions: a framework and models. Sustainability **13**(6), 3353 (2021). Available at: https://doi.org/10.3390/su13063353
45. Sharma, S., Kadayat, Y., Tyagi, R.: Artificial intelligence enabled sustainable life cycle system using Vedic scripture and quantum computing. In: 2023 3rd International Conference on Intelligent Technologies (CONIT), pp. 1–7. IEEE (2023)
46. Singh, A. et al.: Artificial intelligence for sustainable development goals: bibliometric patterns and concept evolution trajectories. Sustain. Dev. **32**(1), 724–754 (2024). Available at: https://doi.org/10.1002/sd.2706
47. Singh, B., Kaunert, C.: Dynamic landscape of artificial general intelligence (AGI) for advancing renewable energy in urban environments: synergies with SDG 11—sustainable cities and communities Lensing policy and governance. In: Advanced Technologies and Societal Change, pp. 247–270. Springer Nature Singapore, Singapore (2025)
48. Skjærseth, J.B.: Towards a European Green Deal: The evolution of EU climate and energy policy mixes. Int. Environ. Agreements Polit. Law Econ. **21**(1), 25–41 (2021). Available at: https://doi.org/10.1007/s10784-021-09529-4
49. Steffen, B., Patt, A.: A historical turning point? early evidence on how the Russia-Ukraine war changes public support for clean energy policies. Energy Res. Soc. Sci. **91**(102758), 102758 (2022). Available at: https://doi.org/10.1016/j.erss.2022.102758
50. Uddin, M.: Climate Change Law, Technology Transfer and Sustainable Development. Routledge, London, England (2021). Available at: https://doi.org/10.4324/9781003056416
51. Ustin, S.L., Middleton, E.M.: Current and near-term advances in Earth observation for ecological applications. Ecol. Process. **10**(1), 1 (2021). Available at: https://doi.org/10.1186/s13717-020-00255-4
52. Wang, D., Chen, L., Dong, L.: A critical review of climate change mitigation policies in the EU—based on vertical, horizontal and policy instrument perspectives. J. Clean. Prod. **467**(142972), 142972 (2024). Available at: https://doi.org/10.1016/j.jclepro.2024.142972
53. Yavuz, A.A., et al.: Distributed cyber-infrastructures and artificial intelligence in hybrid post-quantum era. In: 2022 IEEE 4th International Conference on Trust, Privacy and Security in Intelligent Systems, and Applications (TPS-ISA). IEEE (2022)
54. Ahmad, T., et al.: Artificial intelligence in sustainable energy industry: Status Quo, challenges and opportunities. J. Clean. Prod. **289**(125834), 125834 (2021). Available at: https://doi.org/10.1016/j.jclepro.2021.125834
55. Yu, K., et al.: Hybrid quantum classical optimization for low-carbon sustainable edge architecture in RIS-assisted AIoT healthcare systems. IEEE Internet Things J. **11**(24), 38987–38998 (2024). Available at: https://doi.org/10.1109/jiot.2024.3399234
56. Zhang, J., Li, M.: Quantum computing and AI: potential synergies in cloud environments. Asian Am. Res. Lett. J. **1**(2) (2024). Available at: https://aarlj.com/index.php/AARLJ/article/view/23. Accessed 24 May 2025

Significance of Technologies to Attain Sustainable Development Goals in Urban Planning

Anuradha, Vaishali Soni, Vaishnavi Srivastava, Dipti Bharti, and Richa Saxena

Abstract Urbanization is happening at a pace never seen before, with over two-thirds of the world's population expected to live in cities by 2050. As cities grow, they are increasingly confronted with issues such as pollution, housing shortages, water shortages, traffic jams, and social inequality. These urban problems immediately endanger the fulfillment of the United Nations Sustainable Development Goals (SDGs), notably SDG 11 (Sustainable Cities and Communities), but also indirectly affect other goals including clean water (SDG 6), accessible energy (SDG 7), climate action (SDG 13), and sustainable consumption (SDG 12). Technological innovation has in this regard become a key force of urban change. This chapter examines the strategic use of digital, infrastructural, and ecological technologies for the pursuit of SDG objectives in cities.

Keywords SDGs · Ecological techniques · Urban planning · Green infrastructure · Clean energy · Smart technology

Anuradha
Department of Chemistry (Applied Science and Humanities), Darbhanga College of Engineering, Darbhanga, Bihar, India

V. Soni
Department of Community Medicine and School of Public Health, PGIMER, Chandigarh, India

V. Srivastava
School of Life Sciences and Biotechnology, Chhatrapati Shahu Ji Maharaj University, Kanpur, UP, India

D. Bharti
Department of Applied Science & Humanities, Darbhanga College of Engineering, Darbhanga, Bihar, India

R. Saxena (✉)
Department of Biotechnology, Invertis University, Barielly, UP, India
e-mail: saxena.richa50@gmail.com

A. L. Srivastav et al. (eds.), *Climate Resilience: Impact of Quantum Computing and Artificial Intelligence on Urban Planning*, Studies in Systems, Decision and Control 637, https://doi.org/10.1007/978-3-032-06791-3_6

1 Introduction

Urbanization has been one of the outstanding factors of the twenty first century. The United Nations Department of Economic and Social Affairs (UN DESA) declined that the present population on the globe is dwelling in less than sixty percent of urban areas that will increase to sixty-eight by 2050 [29, 60]. The cities are the centers of economic productions and innovations and more than 60% of the available resources are consumed and even contribute more than 70% of the total greenhouse emissions in the world [14]. Uncontrolled and relatively quick growth of the urban settlements caused many socio-environmental issues, such as pollution of the air and the water, traffic jams, slums, population health deterioration, and an increase in social inequality [56]. They are slippery issues which must be resolved to make future urbanization move towards a path of resilient, inclusive, and sustainable city-building.

United Nations 2030 Agenda on sustainable development formulated in the sphere of its 17 Sustainable Development Goals (SDGs) offers a universally accepted system of world rules to control the city development process [63, 65]. SDG 11 (Sustainable cities and communities) is the most dominant pillar that is supposed to be undertaken with regard to city-focused development and this implies that all SDGs are interrelated and could be adopted in cities [62]. It requires accessible, safe, resilient, and sustainable cities with equal access to the shared places, decent housing, accessible transport, clean air, and water. However, achievement of SDG 11 is in itself associated with other objectives such as SDG 6 (Clean Water and Sanitation), SDG 7 (Affordable and Clean Energy), SDG 9 (Industry, Innovation and Infrastructure), SDG 12 (Responsible Consumption and Production), and SDG 13 (Climate Action) [16, 28].

Realization of these goals is contingent upon urban planning. Previously, conventional urban planning entailed infrastructure layout planning, spatial zoning, and regulatory land-use systems [53]. Conventional planning methods, however, cannot respond to emerging issues such as social fragmentation, digital exclusion, scarcity of resources, and climate risk in contemporary complex urban settings. This has meant that modern day city administration is highly reliant on the use of technology. It consists of an extensive toolbox where instruments like Internet of Things (IoT) sensors to track infrastructure, Geographic Information Systems (GIS) to geographic decision-making, and predictive models on AI to handle traffic and mobile-based citizen engagement participation platforms are involved [18].

Green technologies like solar rooftop, rainwater harvest system, green roofs and permeable pavements, help in the preparation of the ecological sustainability and climate resilience in cities. Digitalization of urban systems is able to create such an effect that policy can become the same thing as implementation, so the sustainability objectives become possible as opposed to desirable [36, 38]. Other benefits of these technologies include the ability of urban planners to be proactive, more efficient in resource utilization, a greater level of public ownership, and the development of more resilient infrastructure. The process of using technology to attain sustainable urban

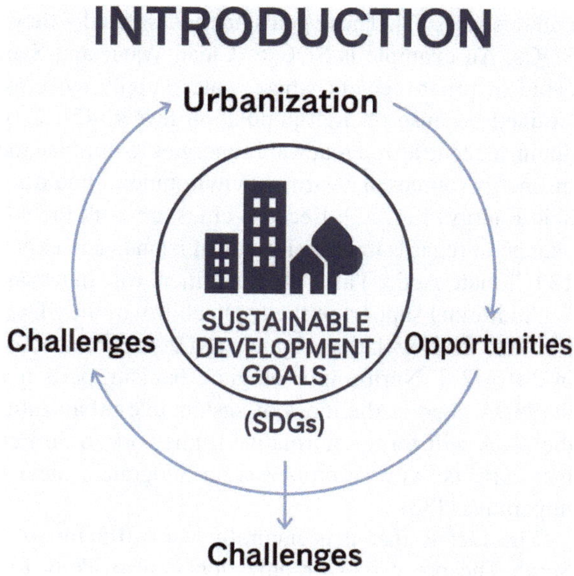

Fig. 1 Role of urbanization in shaping SDG-based opportunities and constraints

development has challenges which should be overcome. It demands cross-sectoral coordination, and this is how enabling governance institutions can become possible, whether they receive enough funds, possess the right human resources, and focus on inclusivity [35]. In disregard of socioeconomic exclusion and digital inequality, the technology can tend to increase the exclusion rather than reduce it. Other than creating awareness of the potential of technology as urban planning tools, this chapter critically deals with the challenges of and gives solutions in an integrated manner (Fig. 1).

2 Urban Planning and the SDG Framework

The United Nations Sustainable Development Goals (SDGs), adopted in 2015, represent a global commitment to end poverty, reduce inequality, and ensure environmental sustainability by 2030 [34, 59]. The SDG Agenda, which consists of 169 specific targets and 17 interconnected goals, aims to promote integrated development that takes social, economic, and environmental factors into account [58]. Cities as centers of opportunity and disparity are central to this agenda. Although in the past urban development was not at the center of international sustainability policy, the SDGs clearly acknowledge the key role of cities in bringing about sustainable change, particularly through Goal 11: Sustainable Cities and Communities [9, 20].

SDG 11 aims at a wide range of urban challenges: providing access to adequate, safe, and affordable housing; enhancing public transport; promoting inclusive urbanization; protecting cultural heritage; mitigating disaster risk; and promoting green public spaces [7]. These goals are not isolated—they cut across directly with other SDGs. An example is SDG 6 (Clean Water and Sanitation) that is directly applicable in urban centers where water supply systems and sewerage networks are strained because of high population levels [45]. City locations that require introducing decentralized renewable energies to mitigate the means of energy dependence on energy sources that harm the environment should adopt SDG 7 (Clean and Affordable Energy) [31, 32]. Because cities are both the source and the target of climate change in regards to emissions, heat islands and exposure to extreme weather, SDG 13 (Climate Action) also plays a critical role in urban planning [30].

Additional support of the applicability of the SDGs to urban environment can be traced to the New Urban Agenda (NUA) approved during the Habitat III conference, in 2016 [21]. Nurturing ecologic urbanism, participatory rule, and spatial justice, the NUA coerces the ideas of sustainable urbanization [17]. Combined SDGs and the NUA will form a normative framework to turn cities into sustainable development agents, with an emphasis on integrated, inclusive, and prospective planning imperatives [33]

The fact is that it is naturally very difficult to attain SDGs in cities. Global South The process of urbanization is also likely to become faster than planning and development of infrastructure, particularly in the Global South.

Informal settlements proliferate in the absence of institutionalized housing systems, and unequal access to resources exacerbates poverty and exclusion. Urban governance structures far too frequently suffer from fragmentation, deficient technical expertise, and budget constraints [19]. Herein, technological solutions provide a prime platform to overcome these structural barriers, enable evidence-based decisions, and create adaptive, adaptable urban systems.

Technology can make all three aspects of sustainable development—economic efficiency, social equity, and environmental sustainability—a reality. Digital technologies can reduce bureaucratic inefficiencies and transaction costs of urban services (economic), provide equitable access to services such as public health or education (social), and reduce the environmental footprint of urban infrastructure using energy-saving systems (environmental) [11, 39]. Integrating these technologies into urban planning systems ensures that cities are not just places of opportunity but also catalysts of sustainable change.

The next part will elaborate on how some technologies from smart infrastructure to participatory platforms are being utilized to accelerate the realization of SDGs in cities and how these mechanisms are reconstituting the very act of urban planning.

3 Technological Interventions in Urban Sustainability

3.1 Smart Infrastructure and Internet of Things (IoT)

Technology-driven urban sustainability relies on smart infrastructure, or the physical facilities fitted with digital, and connected via the Internet of Things (IoT) [12]. Metropolitan settings in the current world are known to be facilitated by smart streetlights, smart traffic lights, water quality, and green buildings. The real-time data obtained through these smart devices will assist city officials to keep track of its performance, foresee failures, and ease in the management of the available resources. To give an example, waste cans in major cities like Barcelona equipped with IoT inform that they are full and allow optimizing routes and fuel use, contributing to SDG 12: Responsible Consumption and Production and SDG 13: Climate Action [26].

Conversely, smart grids and metering systems facilitate the SDG 7: Affordable and Clean Energy by facilitating decentralized renewable energy distribution and advancing more energy-efficiency-based consumer behavior [5]. As of today, the use of IoT has been applied in Singapore in utility services, leading to a measurable decrease in energy demand and enhancements in the efficiency of water use, which is of paramount importance considering both climate resilience and urban sustainability [8, 50].

3.2 Geographic Information Systems (GIS) and Remote Sensing

GIS technology is important in evidence-based urban planning, but it is used in the visualization of data and spatial analysis [6]. To support zoning, transportation planning and disaster preparedness, planners have the capability of overlaying several layers of data (land use, demographics, environmental hazards, and infrastructure networks) to inform decision-making [52]. In a place prone to flood such as Jakarta, GIS mapping has been applied to identify high-risk locations and help organize development of early warning systems and flood prevention plans [64, 66].

Satellite or drone imagery can enhance observation of sprawl in the cities, development of heat island, loss of greenery, and decreasing air quality. The technologies also support the process of tracking the achievement of SDG 13 (Climate Action) and SDG 15 (Life on Land) at the city level [44]. To facilitate the introduction of sustainability and improve the urban health, the New York City planners have used the high resolution remote sensing to guide in the design of the green spaces as well in alleviating the heat risk.

3.3 Artificial Intelligence (AI) and Predictive Urban Modeling

Artificial intelligence (AI) is transforming the planning process with plans now being made possible by scenario analysis and forecasts. Urban data, including sensor, smart phone and social media and satellite data, can be used by AI to foresee a deficiency in residential units, an increase in illness, traffic and energy usage [2]. There is less risky decision-making associated with predictive tools since planners feel free to experiment on numerous policy options and infrastructure investments even before actualizing them,managing growth in a more productive way. As an example, to predict how cities will develop in next years, it uses to consider housing needs, environmental restrictions, and transportation [61]. This kind of planning tool is particularly important in the case of SDGs 9: Industry, Innovation, and infrastructure; and 11: Sustainable Cities and Communities.

3.4 E-Governance and Online Involvement of Citizens

Technology as well as citizen participation is required in sustainable development of the city. The arising trend of the involvement of citizens into the city government with the help of digital tools and mobile applications is leading to the fact that participants can demonstrate the problem, make decisions on local initiatives, suggest the ideas, and have a glance at the public documents. In this respect, the number and quality of citizens participating in the municipal process and the transparency related to the same have drastically increased, such as the application called the MyGov, used in India, or FixMyStreet in the United Kingdom [10].

Responsibility and the inclusion, openness and absolute equalities of the decision making process are the platforms that contribute to SDG 16 according to the Peace, Justice and Strong Institutions. On top of that, they allow empowering the marginalized population by enabling the realization of the SDG of Reduced Inequality (SDG 10) or changing the urban planning to be more decentralized and socially inclusive [40].

3.5 Sustainable and Green Building Technologies

Creation of green building technologies is increasingly becoming popular today as far as the cities that are created with a green technology impact less on the environment and offer greater human welfare. Examples are low emissions building materials, passive solar architecture, installation of energy efficient heating, ventilation and air conditioning systems, vertical gardens and roof farming. Owing to the utilisation of Building Information Modelling (BIM) technologies, there is an opportunity that

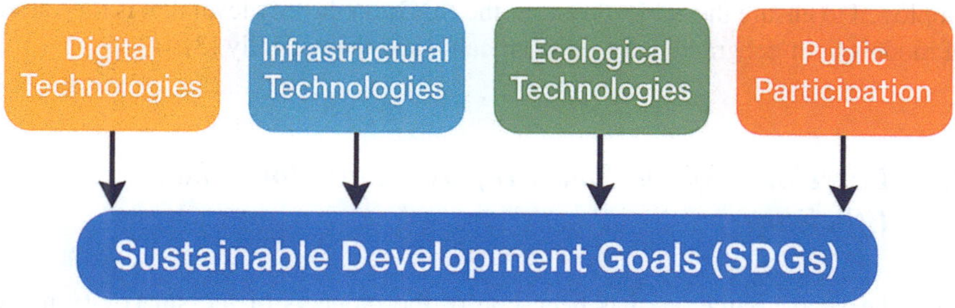

Fig. 2 Role of technological innovations in achieving Sustainable Development Goals (SDGs) through urban planning

Table 1 Urban Technologies and their Contribution to Specific SDGs

Technology	Urban function	SDGs impacted	References
IoT-enabled smart infrastructure	Monitoring, Automation, Efficiency	SDG 7, SDG 9, SDG 11, SDG 12	[51]
GIS and remote sensing	Spatial planning, Disaster management	SDG 6, SDG 11, SDG 13, SDG 15	[65]
Artificial intelligence	Forecasting, Simulation, Optimization	SDG 9, SDG 11, SDG 13	[12]
E-Governance and participation apps	Citizen feedback, Governance transparency	SDG 10, SDG 11, SDG 16	[10]
Green building and BIM	Sustainable architecture and energy efficiency	SDG 3, SDG 7, SDG 11, SDG 13	[46]

designers can observe and test the performance of the material and energy of the building prior to its construction [24]. Green buildings are linked to SDG 3 directly (Good Health and Well-Being) and SDG 11.6 (reduction of environmental impact of cities) because of reduced heat stress, increased wastage reduction, and better internal air pollution [46] (Fig. 2 and Table 1).

4 Case Studies

One can come to the conclusion that the technological planning of cities is an ideal illustration of innovation that could be used in shaping the process according to the Sustainable Development Goals (SDGs) on the examples of real life [41]. The need to improve the living standards of the citizens of the developed and the developing worlds has been seen to invest in the use of technology, a factor that has been seen to keep the process of urbanization heralded to the tenets of sustainability, inclusivity, and resilience. Even just the above-mentioned case studies enable the provision of useful information on the most effective ways in which the technology can be employed to ensure the achievement of the SDGs on the city level that is the variety of models both geographically, economically, and thematically [3].

4.1 Barcelona, Spain: Smart Infrastructure for Urban Efficiency

Over the years, Barcelona has been one of the pioneers of the smart city movement where technology is employed to enhance better services in cities and a more environmentally friendly environment [25]. Thousands of Internet of Things sensors (garbage cans, parking sensors, streetlights, and traffic signals) transmit information to the operating system of the city that uploads it to a single platform so that city managers could monitor and adjust operations in real time [51].

Smart meters and leak detection systems have reduced the level of water loss and directly contributed to SDG 6 (Clean Water and Sanitation), whereas dynamic lighting systems minimized the use of electricity and contributed to SDGs 7 (Affordable and Clean Energy) and 13 (Climate Action) [57].

The city has developed an open-data ecosystem of portals to foster transparency and the voice of citizens through SDG 16 (Peace, Justice, and Strong Institutions) [43].

4.2 Curitiba, Brazil: Integrated Urban Transport and Planning

Bus Rapid Transit (BRT) is one of the low-tech, sustainable urban development models developed in Curitiba and, therefore, Curitiba is recognized as one of the most famous cities in the world [27]. Being grounded on low-end technology, the Curitiba model is an illustration of low-tech innovation that promotes the goals of SDG.

Over the past few years, Curitiba has implemented the use of digital payment systems, real-time monitoring of buses and transportation apps to make the BRT more efficient and convenient [37]. All these technologies developments are aiding to accomplish SDG 9 (Infrastructure and Innovation) and SDG 11.2 (Safe, Affordable Transport: Access). Moreover, the city has achieved goal no. 15 of SDGs (Life on Land) due to the ecological balance adopted within the city through its green urban space policy eschewed by GIS mapping and drone supervision [42].

4.3 Singapore: Integrated Digital Governance and Urban Sustainability

The most innovative installation of an urban governance, which has been brought to life with the help of digitalization, is the Singapore Smart Nation project [49]. The project integrates data processing, artificial intelligence and citizen interfaces to make it possible to make real-time decisions regarding waste management, energy distribution networks, transport, and healthcare. In a bid to test the feasibility of construction proposals to comply with the sustainability benchmark, the Urban Redevelopment Authority (URA) uses Building Information Modeling (BIM) and the construction of 3D city models to fabricate the construction properties on a virtual flood map or virtual environment model [55]. A digital twin enables the complete simulation of flooding in the entire city as well as optimization of the route of the entire city and assessment of the energy consumption at a time [4].

They mitigate various SDGs by reducing pollution, improved livability in cities, and mitigation of climate-related risks, such as SDG 11 (Sustainable cities), SDG 13 (Climate action) and SDG 3 (Good health and well-being) [30].

Singapore is among the states that pay attention to digital inclusion of all citizens, and it is being inculcated in the projects of the state property of supplying the population with the Wi-Fi signal and digital knowledge besides the SDG 10 (Reduced Inequality), which makes it possible to suppose that justice in access to technology is a crucial condition of further development of the sphere [30].

4.4 Kigali, Rwanda: Digital Tools for Urban Growth and Basic Services

A rare example on how emerging economies are fighting urbanization with technology is the city of Kigali, the capital of Rwanda [47]. The city has developed a master city plan based on GIS and incorporating census data, satellite imagery, and socioeconomic indicators to serve as a guideline to zoning, housing development, and infrastructure investment.

One of the interesting innovations established in Kigali is the installation of the smart water kiosks in the informal settlement areas with the local population exploiting the digital meters and using the pre-paid cards to get the clean water. This innovation can help SDG 1 (No Poverty) by reducing the cost of water and SDG 6 (Clean Water) by enhancing transparency of water prices and related details [1]. Moreover, low-cost and scalable technologies implemented in the Smart City program in Kigali are also appropriate to the context of the city development, including the smartphones-based public transportational schedules and sun-powered lightening.

4.5 Amsterdam, Netherlands: Urban Sustainability and Circular Economy

Calisto Friant et al. [13] describe Amsterdam as one of the world leaders in the implementation of the principles of a circular economy on the city level. Digital technologies enable the city to trace construction waste, to monitor the flows of materials, and increase the efficiency of building and industry energy consumption. Amsterdam Smart City will give residents, companies, and governmental organizations a platform through which they can collaborate in coming up with long-term solutions.

There are digital dashboards that track water consumption, emissions, and energy sustainability and are used at neighborhood level and a technology such as the blockchain is utilized which works to confirm the application of environmentally responsive materials in buildings. Chan et al. [15] note that these measures directly refer to SDGs 11.6 (Mitigation of Environmental Impact of Cities) and 12 (Responsible Consumption and Production). Of interest here is the fact that the policies of Amsterdam, which establish the relationship between technological innovation and the social activism, and transparency in the government ensure that sustainability programs are engraved in the life of the community in the form of participatory budgeting and reaching out to the people in education (Table 2).

5 Challenges in Technological Integration

Even in the case of technological promise in the Sustainable Development Goals (SDGs) in cities delivery, there is no doubt because the actual delivery of that promise is hampered by a set of systemic, institutional, financial, and social blockages. These are worse in the developing countries, in the secondary cities and the slum areas where the benefits of the technological innovation are not distributed to all the people

Table 2 Summary of case studies and their technological contributions to SDGs

City	Key technologies used	Primary SDGs addressed	References
Barcelona	IoT, Open data, Adaptive lighting	SDG 6, SDG 7, SDG 11, SDG 13, SDG 16	[25, 51]
Curitiba	BRT system, GIS, Zoning policies	SDG 9, SDG 11, SDG 13, SDG 15	[37]
Singapore	Digital twins, BIM, AI analytics	SDG 3, SDG 7, SDG 11, SDG 13, SDG 10	[12, 46]
Kigali	Smart kiosks, GIS planning, Solar tech	SDG 1, SDG 6, SDG 10, SDG 11	[6]
Amsterdam	Blockchain, energy dashboards, Circular data	SDG 11, SDG 12, SDG 13	[13]

proportionately. Such obstacles are very important to be aware of when creating effective and inclusive strategies of sustainable urban change development.

5.1 Economic and Financial Limitations

One of the biggest bottlenecks for deploying next-generation urban technologies is limited financial resources. Most municipalities, particularly those in the Global South, have limited budgets that are usually focused on core infrastructure requirements rather than digital innovation. Technologies such as smart sensors, GIS platforms, or renewable energy systems involve heavy initial capital investment and long-term maintenance funds budgets that are hard to rationalize without near-term economic gains [23].

In addition, inadequate access to international climate and innovation finance e.g., the Green Climate Fund or development bank loan—reduces the capacity of local authorities to pilot or upscale technology projects. Even in more affluent cities, the priorities in housing, healthcare, or public security tend to dominate long-term investment in urban sustainability technologies.

5.2 Institutional Fragmentation and Governance Gaps

Its application in urban planning demands a level of inter-agency coordination, which fails to materialize owing to institutional silos. Urban governance tends to be disjointed across an array of departments—transport, water, housing, energy—with few mechanisms for coordination. Lacking centralized management or holistic policy frameworks, digital platforms and smart tools can be implemented independently, resulting in duplication of efforts or underutilization of resources [48].

Furthermore, bureaucratic resistance and political unwillingness can slow the implementation of innovative practices. In other instances, urban planners might be unaware of new technologies or doubt their dependability, thus causing resistance within administrative systems. The lack of standardized procedures for technology acquisition, data management, or interoperability also makes it difficult to implement.

5.3 Technological Inequality and Digital Divide

Urban technologies have a tendency to worsen inequalities when implemented without regard for accessibility [22]. Low-income groups, informal settlements, elderly people, the disabled, and marginalized ethnic communities are often denied access to smart city programs based on the lack of internet penetration, digital competence, or mobile phone ownership. This digital divide not only confines e-governance platform engagement but also constrains access to vital services becoming more digitized, such as health warnings, transport networks, or water bills.

In a few cities, smart technology investments have disproportionately favored upscale districts, giving rise to what researchers call "technological gentrification." Urban Wi-Fi networks, e-mobility schemes, and green energy pilots, for instance, are usually launched in business districts or tourist areas, leaving marginalized neighborhoods offline [22]. Unless technology is deliberately made inclusive, technology can become an instrument of exclusion instead of empowerment.

5.4 *Data Privacy, Security, and Ethical Issues*

Growing digitization of urban life creates grave concerns related to surveillance, data abuse, and non-consensual data collection. Smart technologies create mass amounts of personal and behavioral data ranging from movement patterns and energy consumption to health issues and financial transactions. Without transparent legal frameworks, this data can be utilized for commercial exploitation, targeted policing, or political domination.

Few cities have strong data governance policies that guarantee transparency, accountability, and citizens' rights in urban data management. In addition, the application of Artificial Intelligence (AI) and algorithmic decision-making in urban environments has triggered concerns regarding bias, unexplainability, and the risk of reinforcing structural inequalities [54]. For instance, if predictive policing technologies are based on historically biased data, they can disproportionately focus on specific communities, compromising SDG 16 (Peace, Justice and Strong Institutions).

5.5 *Insufficiency of Local Capacity and Technical Competence*

Urban technologies are typically demanding in terms of specialized technical competence in design, deployment, and maintenance. Many cities do not have the requisite human resources to deal with such systems optimally. Urban planners might be unaware of the use of GIS software, city administrators might be ignorant of interpreting sensor data, and local engineers might not be trained in smart grid or water metering system maintenance.

Moreover, most off-the-shelf solutions are designed by global companies and can be unsuitable for local contexts or languages. In addition to raising expenses, this reliance on foreign suppliers weakens local capacity building and ownership. Cities will remain reliant on technology and susceptible to vendor lock-in if they do not invest in education, professional development, and institutional capacity (Table 3).

Table 3 Key barriers to technology adoption in urban SDG planning

Barrier category	Description	Implications for SDG progress
Financial Limitations	High upfront costs, Lack of access to finance	Limits scalability and inclusion
Institutional Fragmentation	Poor coordination, Siloed departments, Policy gaps	Reduces efficiency, integration, and transparency
Digital Divide	Unequal access to internet, Devices, and digital skills	Excludes vulnerable populations from participation
Data and Ethics	Risks of privacy breach, Bias, and data misuse	Erodes trust and democratic accountability
Capacity Gaps	Lack of skilled workforce and local expertise	Hinders implementation, Leads to external reliance

6 Conclusion

The international agenda of sustainability is focused on urban planning. The challenges of inequality, climate risk, and environmental degradation are magnified as cities expand in size, complexity, and resource consumption. Yet there is also a tremendous opportunity hidden amongst the challenges: the Sustainable Development Goals (SDGs), especially SDG 11—Sustainable Cities and Communities—can be accelerated through the strategic mainstreaming of technology into urban systems.

As has been illustrated in this chapter, technologies such as e-governance platforms, GIS, IoT, AI, and BIM are not only modernization tools but also transformation tools. These technologies can make cities more resilient, transparent, efficient, and equitable when they are well developed and generously adopted. The case studies illustrated how different urban contexts are able to innovate towards sustainability objectives through digital routes, from Barcelona's smart infrastructure to Kigali's GIS-based master plans.

But ethics, ability, and governance are required for technology to be able to sustainably advance urbanity. In order not to widen already present inequalities, it is crucial to holistically address the actual challenges being posed by financial constraints, digital divides, institutional fragmentation, and data privacy issues. To ensure that technology remains a force of equity and not exclusion, strategic investments in digital literacy, policy integration, inclusive innovation, and environmental protection are essential.

Ultimately, the attainment of SDG-conformant urban development entails a shift in paradigm—from reactive planning to proactive, evidence-based governance; from command-and-control to participatory processes; and from standalone projects to interconnected systems. Cities that undergo this shift will not only achieve international targets but also build livable, inclusive, and future-proof environments for all.

References

1. AbuEltayef, H.T., AbuAlhin, K.S., Alastal, K.M.: Addressing non-revenue water as a global problem and its interlinkages with sustainable development goals. Water Practice Technol. **18**(12), 3175–3202 (2023)
2. Alamri, S.: The geospatial crowd: emerging trends and challenges in crowdsourced spatial analytics. ISPRS Int. J. Geo Inf. **13**(6), 168 (2024)
3. Alcamo, J., Thompson, J., Alexander, A., Antoniades, A., Delabre, I., Dolley, J., et al.: Analysing interactions among the sustainable development goals: findings and emerging issues from local and global studies. Sustain. Sci. **15**, 1561–1572 (2020)
4. Ariyachandra, M.M.F., Wedawatta, G.: Digital twin smart cities for disaster risk management: a review of evolving concepts. Sustainability **15**(15), 11910 (2023)
5. Ashok, K.R.: Decentralized renewable energy systems: a pathway to enhanced energy efficiency and sustainability. Int. J. Renew. Energy Commercialization **10**(2), 7–12p (2024)
6. Attah, R.U., Gil-Ozoudeh, I., Garba, B.M.P., Iwuanyanwu, O.: Leveraging geographic information systems and data analytics for enhanced public sector decision-making and urban planning. Mag. Sci. Adv. Res. Rev. **12**(2), 152–163 (2024)
7. Aust, H.P., Du Plessis, A.: Good urban governance as a global aspiration: on the potential and limits of SDG 11. In: Sustainable Development Goals, pp. 201–221. Edward Elgar Publishing (2018)
8. Banerjee, C., Bhaduri, A., Saraswat, C.: Digitalization in urban water governance: case study of Bengaluru and Singapore. Front. Environ. Sci. **10**, 816824 (2022)
9. Beck, D., Ferasso, M., Storopoli, J., Vigoda-Gadot, E.: Achieving the sustainable development goals through stakeholder value creation: building up smart sustainable cities and communities. J. Clean. Prod. **399**, 136501 (2023)
10. Bhanye, J., Shayamunda, R.: The promise of civic-tech: digital technologies and transparent, accountable governance. In: Digitalisation and Public Policy in Africa: GovTech and CivicTech Innovations, pp. 93–122. Springer Nature Switzerland, Cham (2025)
11. Bibri, S.E., Krogstie, J.: Environmentally data-driven smart sustainable cities: applied innovative solutions for energy efficiency, pollution reduction, and urban metabolism. Energy Inform. **3**(1), 29 (2020)
12. Bibri, S.E., Huang, J., Jagatheesaperumal, S.K., Krogstie, J.: The synergistic interplay of artificial intelligence and digital twin in environmentally planning sustainable smart cities: a comprehensive systematic review. Environ. Sci. Ecotechnol. 100433 (2024)
13. Calisto Friant, M., Reid, K., Boesler, P., Vermeulen, W.J., Salomone, R.: Sustainable circular cities? analysing urban circular economy policies in Amsterdam, Glasgow, and Copenhagen. Local Environ. **28**(10), 1331–1369 (2023)
14. Cavaliere, P., Cavaliere, P.: Clean Ironmaking and Steelmaking Processes: Efficient Technologies for Greenhouse Emissions Abatement, pp. 1–37. Springer International Publishing (2019)
15. Chan, S., Weitz, N., Persson, Å., Trimmer, C.: SDG 12: Responsible consumption and production. A review of research needs. Technical annex to the Formas report Forskning för Agenda, 2030 (2018)
16. Chandiramani, J., Tripathi, S., Benara Misra, S., Patil, G., Shende, A.: Does inequality exist in attaining sustainable development goals within a city? a case study in Pune city, India. Int. J. Urban Sci. 1–38 (2024)
17. Dahiya, B., Das, A.: New urban agenda in Asia-Pacific: governance for sustainable and inclusive. In: New Urban Agenda in Asia-Pacific: Governance for Sustainable and Inclusive Cities, p. 3 (2019)
18. Das, S., Tsapakis, I., Khan, M.N., Liu, J., Mills, D., Miller, M., et al.: Leveraging Artificial Intelligence (AI) Techniques to Detect, Forecast, and Manage Freeway Congestion: Technical Report (No. FHWA/TX-23/0–7131-R1). Texas A&M Transportation Institute (2023)

19. Dávila, J.D.: Urban Fragmentation,'good Governance' and the emergence of the competitive city. In: The Routledge Handbook on Cities of the Global South, pp. 474–486. Routledge (2014)
20. Devisscher, T., Konijnendijk, C., Nesbitt, L., Lenhart, J., Salbitano, F., Cheng, Z.C., et al.: SDG 11: sustainable cities and communities-Impacts on forests and forest-based livelihoods. In: Chapter 11-SDG 11: Sustainable Cities and Communities–Impacts on Forests and Forest-Based Livelihoods, pp. 349–385. Cambridge University Press (2019)
21. Diaz-Sarachaga, J.M., Sarachaga, J.L.: Diagnosis on the implementation of the New Urban Agenda. Global Sustain. **7**, e37 (2024)
22. Durand, A., Zijlstra, T., van Oort, N., Hoogendoorn-Lanser, S., Hoogendoorn, S.: Access denied? digital inequality in transport services. Transp. Rev. **42**(1), 32–57 (2022)
23. Force, S.G.T.: Outlook for increased adoption of smart grid technologies in ADB energy sector operations. In: ADB Sustainable Development Working Paper Series, p. 42 (2016)
24. Gerrish, T., Ruikar, K., Cook, M., Johnson, M., Phillip, M., Lowry, C.: BIM application to building energy performance visualisation and management: Challenges and potential. Energy Build. **144**, 218–228 (2017)
25. Guernieri, L.: An Economical, Sustainable and Technological Analysis of Impact of the Smart Cities-the Case Study of Barcelona and Singapore (Master's thesis, Universidade NOVA de Lisboa (Portugal)) (2023)
26. Guevara, S., Julián, I.P.: Sustainable consumption and production: a crucial goal for sustainable development—Reflections on the Spanish SDG implementation report. J. Sustain. Res. 1(2) (2019)
27. Gustafsson, H.R., Kelly, E.A.: Developing the sustainable city: Curitiba, Brazil, as a case study. In: How Cities will Save the world, pp. 81–96. Routledge (2016)
28. Hatab, A.A., Ravula, P., Nedumaran, S., Lagerkvist, C.J.: Perceptions of the impacts of urban sprawl among urban and peri-urban dwellers of Hyderabad, India: a Latent class clustering analysis. Environ. Dev. Sustain. **24**(11), 12787–12812 (2022)
29. Hoornweg, D., Pope, K.: Population predictions for the world's largest cities in the 21st century. Environ. Urban. **29**(1), 195–216 (2017)
30. Hussain, S., Hussain, E., Saxena, P., Sharma, A., Thathola, P., Sonwani, S.: Navigating the impact of climate change in India: a perspective on climate action (SDG13) and sustainable cities and communities (SDG11). Front. Sustain. Cities **5**, 1308684 (2024)
31. Javid, I., Chauhan, A., Thappa, S., Verma, S.K., Anand, Y., Sawhney, A., et al.: Futuristic decentralized clean energy networks in view of inclusive-economic growth and sustainable society. J. Clean. Prod. **309**, 127304 (2021)
32. Jayachandran, M., Gatla, R.K., Rao, K.P., Rao, G.S., Mohammed, S., Milyani, A.H., et al.: Challenges in achieving sustainable development goal 7: affordable and clean energy in light of nascent technologies. Sustain. Energy Technol. Assess. **53**, 102692 (2022)
33. Kumar, A.: Introduction: Sustainable development goals, new urban agenda, and climate change. In: Sustainable Development Goals and Indian Cities, pp. 1–41. Routledge India
34. Liu, Q.Q., Yu, M., Wang, X.L.: Poverty reduction within the framework of SDGs and Post-2015 development agenda. Adv. Clim. Chang. Res. **6**(1), 67–73 (2015)
35. Marín-González, F., Moganadas, S.R., Paredes-Chacín, A.J., Yeo, S.F., Subramaniam, S.: Sustainable local development: consolidated framework for cross-sectoral cooperation via a systematic approach. Sustainability **14**(11), 6601 (2022)
36. Martin, C., Evans, J., Karvonen, A., Paskaleva, K., Yang, D., Linjordet, T.: Smart-sustainability: a new urban fix? Sustain. Cities Soc. **45**, 640–648 (2019)
37. Medeiros, R.M., Duarte, F., Bojic, I., Xu, Y., Santi, P., Ratti, C.: Merging transport network companies and taxis in Curitiba's BRT system. Public Transport **16**(1), 269–293 (2024)
38. Mendizabal, M., Heidrich, O., Feliu, E., García-Blanco, G., Mendizabal, A.: Stimulating urban transition and transformation to achieve sustainable and resilient cities. Renew. Sustain. Energy Rev. **94**, 410–418 (2018)
39. Mishra, P., Singh, G.: Energy management systems in sustainable smart cities based on the internet of energy: a technical review. Energies **16**(19), 6903 (2023)

40. Monaco, S.: SDG 16. Promote peaceful and inclusive societies for sustainable development, provide access to justice for all, and build effective, accountable, and inclusive institutions at all levels. In: Identity, Territories, and Sustainability: Challenges and Opportunities for Achieving the UN Sustainable Development Goals, pp. 157–165. Emerald Publishing Limited (2024)
41. Mora, L., Deakin, M.: Untangling Smart Cities: From Utopian Dreams to Innovation Systems for a Technology-Enabled Urban Sustainability. Elsevier (2019)
42. Mulibana, L., du Toit, J.: Transport Planning Research Toward Implementing SDG 11 in South Africa. In: SDGs in Africa and the Middle East Region, pp. 1–24. Springer International Publishing, Cham (2023)
43. Nikiforova, A., Flores, M.A.A., Lytras, M.D.: The role of open data in transforming the society to society 5.0: A resource or a tool for SDG-compliant smart living? In: Smart Cities and Digital Transformation: Empowering Communities, Limitless Innovation, Sustainable Development and the Next Generation, pp. 219–252. Emerald Publishing Limited (2023)
44. Nunes, A., Speranza, C.I., Matos, P., Branquinho, C.: Land restoration for achieving SDG 13: Take urgent action to combat climate change and its impacts. Sustain. Dev. Goals **91** (2019)
45. Ogunbode, T.O., Esan, V.I., Oyebamiji, V.O., Akande, J.A.: Sustainable development goal 6 and the challenge of pipe-borne water connectivity in a growing tropical city: a case study. Discov. Sustain. **5**(1), 53 (2024)
46. Opoku, A., Bondinuba, F.K., Manaphraim, N.Y.B., Kugblenu, G.: Advancing the sustainable development goals through the promotion of health and well-being in the built environment. In: The Elgar Companion to the Built Environment and the Sustainable Development Goals, pp. 137–157. Edward Elgar Publishing (2024)
47. Otundo Richard, M.: Promoting sustainable growth: the role of natural resource utilization, green investment, digital finance, industrial usage, energy efficiency, and renewable energy consumption in Kenya, Rwanda and Uganda. In: Green Investment, Digital Finance, Industrial Usage, Energy Efficiency, and Renewable Energy Consumption in Kenya, Rwanda and Uganda (2024)
48. Panori, A., Kakderi, C., Komninos, N., Fellnhofer, K., Reid, A., Mora, L.: Smart systems of innovation for smart places: challenges in deploying digital platforms for co-creation and data-intelligence. Land Use Policy **111**, 104631 (2021)
49. Pereira, F., Lee, B., Fife, E.: Singapore's Smart Nation Program and potential lessons for the European Union's Digital Decade program (2023)
50. Pokhrel, S.R., Chhipi-Shrestha, G., Hewage, K., Sadiq, R.: Sustainable, resilient, and reliable urban water systems: making the case for a "one water" approach. Environ. Rev. **30**(1), 10–29 (2022)
51. Rehan, H.: Internet of Things (IoT) in smart cities: enhancing urban living through technology. J. Eng. Technol. **5**(1), 1–16 (2023)
52. Rezvani, S.M., Falcão, M.J., Komljenovic, D., de Almeida, N.M.: A systematic literature review on urban resilience enabled with asset and disaster risk management approaches and GIS-based decision support tools. Appl. Sci. **13**(4), 2223 (2023)
53. Ronchi, S., Arcidiacono, A., Pogliani, L.: Integrating green infrastructure into spatial planning regulations to improve the performance of urban ecosystems. Insights from an Italian case study. Sustain. Cities Soc. **53**, 101907 (2020)
54. Shan, T.O.N.G., Shaokang, L.I.: Explainable artificial intelligence for urban planning: challenges, solutions, and future trends from a new perspective. Int. J. Adv. Comput. Sci. Appl. **15**(7) (2024)
55. Shehadeh, A., Alshboul, O., Taamneh, M.M., Jaradat, A.Q., Alomari, A.H., Arar, M.: Advanced integration of BIM and VR in the built environment: enhancing sustainability and resilience in urban development. Heliyon (2025)
56. Sinha, B.R.K.: Introduction: a broad perspective on the concepts of urban dynamics, environment, and health. In: Urban Dynamics, Environment and Health: An International Perspective, pp. 3–79. Springer Nature, Singapore (2024)
57. Tavares, P., Ingi, D., Araújo, L., Pinho, P., Bhusal, P.: Reviewing the role of outdoor lighting in achieving sustainable development goals. Sustainability **13**(22), 12657 (2021)

58. Tosun, J., Leininger, J.: Governing the interlinkages between the sustainable development goals: approaches to attain policy integration. Global Chall. **1**(9), 1700036 (2017)
59. Trends, G.: Challenges and opportunities in the implementation of the sustainable development goals. In: United Nations Development Programme and United Nations Research Institute for Social Development, pp. 0–3 (2017)
60. Uduporuwa, R.: Sustainable city development is possible? a review of challenges and key practices towards urban development in developing countries. Int. J. Sci. Res. Publ. (IJSRP) **10**(9), 294–304 (2020)
61. Vaattovaara, M., Kortteinen, M.: A new turn in the development of the region?: future perspectives on the settlement and planning in the Helsinki region. In: Confused Suburban Identities: A Case Study of Helsinki Region, pp. 237–243. Painotalo Plus Digital Oy (2018)
62. Vaidya, H., Chatterji, T.: SDG 11 sustainable cities and communities: SDG 11 and the new urban agenda: Global sustainability frameworks for local action. In: Actioning the Global Goals for Local Impact: Towards Sustainability Science, Policy, Education and Practice, pp. 173–185. Springer, Singapore (2019)
63. Watson, V.: Locating planning in the New Urban Agenda of the urban sustainable development goal. Plan. Theory **15**(4), 435–448 (2016)
64. Widiachristy, L., Rachmanto, A.S.: The effectiveness of Jakarta Smart City application in enhancing community resilience in facing flood risk. J. Archit. Environ. **20**(1), 45–54 (2021)
65. Yuan, M.: Geographical information science for the United Nations' 2030 agenda for sustainable development. Int. J. Geogr. Inf. Sci. **35**(1), 1–8 (2021)
66. Zhang, H., Liu, X., Xie, Y., Gou, Q., Li, R., Qiu, Y., et al.: Assessment and improvement of urban resilience to flooding at a subdistrict level using multi-source geospatial data: Jakarta as a case study. Remote Sens. **14**(9), 2010 (2022)

Modelling and Forecasting of Climate Change Effects Using Artificial Intelligence Techniques

Arathy Nair, S. Smitha Chandran, and Noujas Varangalil

Abstract Climate change is one of the complex issues and the most significant global challenges that we are facing today. In this fast-paced climate change scenario, we are responsible for the adaptation and mitigation of these climatic events. As the frequency of these events increases significantly, technological innovations are widely accepted for addressing and finding feasible solutions with emerging tools like Artificial Intelligence (AI). Estimating the climate change impacts on several sectors of the environment, society, and economy is crucial for identifying the consequences for disaster preparedness, preparation of adaptation and mitigation strategies, and policy making. In recent years, there has been a major shift in the way we look at the technology side, and now, AI models outweigh traditional climate models in several ways concerning vast and complex data handling, live simulations, predictive accuracy, and computational efficiency. As climatic events escalate globally, in today's world, the demand for more accurate climate modeling and predictions is highly welcome. The more accurate the predictions, the more we can save humanity, the ecosystem, livelihood, etc., by adopting strategies for climate change mitigation and adaptation. Here we discuss various AI models, the application of AI-driven models in climate change events, their implementation in modeling and predicting climate events, their contribution to mitigation and adaptive measures, and the limitations of AI-driven models. In the future, advanced AI models may revolutionize climatic weather prediction, leaving the traditional models behind as it has to pave the way for easy and feasible predictions, as the faster the climate shifts, the tougher the difficulties that we encounter.

A. Nair (✉) · S. Smitha Chandran (✉)
Amrita Vishwa Vidyapetham, Amritapuri Campus, Kollam, Kerala, India
e-mail: am.ps.r4evs24003@am.students.amrita.edu

S. Smitha Chandran
e-mail: smithachandran@am.amrita.edu

N. Varangalil
EPOC Lab, University of Bordeaux, Pessac, France

A. L. Srivastav et al. (eds.), *Climate Resilience: Impact of Quantum Computing and Artificial Intelligence on Urban Planning*, Studies in Systems, Decision and Control 637, https://doi.org/10.1007/978-3-032-06791-3_7

Keywords Climate change · Prediction · Artificial intelligence · Mitigation · Adaptation

1 Introduction

As the concentration of greenhouse gases (GHGs) increases, the intensity and frequency of impacts of climate change are increasing, thereby global warming is increasing day by day. The increase in anthropogenic activities like urbanization, deforestation, burning of fossil fuels, agriculture, industrial activities, etc., releases GHGs into the atmosphere, and the CO_2 creates a heat-trapping effect and raises the temperature in the atmosphere. This global warming causes extreme weather events like floods, extreme rainfall, droughts, heat waves, landslides, melting of ice, storms, cyclones, etc. to happen [14, 15]. Also, it affects vegetation, damage to structures, health issues, the spread of diseases, scarcity, loss of life, food security, biodiversity, etc. [22]. The warming of the planet in the future depends on upcoming emissions, with a dominant rate of net CO_2 in the atmosphere [13]. Addressing these climate-related events requires accurate climate predictions to strategize adaptation and mitigation measures. Traditional models are available that work on numerical modeling for the simulation of these events, but they fail to capture the complexity of the climate system [12]. Although AI methods started to be used in the early 1950s, it was not popular and not widely used. Later in the years 1990 to 2000, this increased with the rise of neural networks and data-driven methods [31, 35].

Artificial Intelligence (AI) is an emerging transformative tool for climate change detection, analysis, and forecasting [14, 15]. Introducing Machine Learning (ML) into climate science makes faster, more satisfying, better solutions, and more promising solutions than conventional models, which are also valuable. The AI-driven models play a crucial role in serving as a great support, from early warning systems for natural disasters to adaptation and mitigation strategies for climate change (Artificial Intelligence for Climate Action in Developing Countries: Opportunities, Challenges, and Risks). Also, AI helps to revolutionize environmental protection [41].

Climate change is a continued threat to our planet, and it's reshaping our demand for innovative strategies and solutions. The growing integration of AI in climate change makes it promising for enhanced prediction, with the high rate of accuracy of information received. Unlike traditional models, AI models process vast datasets and provide valuable insights and patterns from diverse data sources such as satellite imagery, and historical, and real-time data. This chapter delves into the need for AI models in climate change and their applications in different sectors, their contribution to adaptation and mitigation strategies, and the limitations of these AI models and their improvement.

2 Climate Change and Impacts

As anthropogenic activities drive the cause of climate change, it has influenced the land, ocean, and atmosphere to warm. The climate system is very complex and involves many interactions at various spatiotemporal scales. 91% warming of the climate system is mainly due to ocean warming, 5% accounted for land warming and atmospheric warming, and ice loss accounted for about 1% and 3%, respectively [13]. As the warming increases, the temperature of the atmosphere rises, ice caps and glaciers melt down, and also with the changing rainfall pattern, it leads to a rise in sea level. Thus, the weather pattern is getting extremely unpredictable, and the related climate impacts and risks are also increasing. So, there is an urgency to understand climate change and to mitigate and adapt to the situations and events created in connection with climate change [25].

The Paris Agreement is an international treaty signed by 196 state parties in Paris, France, on climate change, which came into force in 2016. The goal of the Paris Agreement is to hold the increase in global average temperature to below 2 °C and achieve efforts to limit the increase in temperature to 1.5 °C above pre-industrial levels [58]. The treaty proposed long-term mitigation and adaptation strategies. As per the treaty, if the goal is achieved, going forward can reduce the climate risks and the impacts of climate change, thereby achieving a balance of carbon emissions and reduction in the atmosphere [1].

Figure 1 illustrates all the climate change impacts and the sectors affected by the climate change scenario. As the temperature rises, the sea level also rises due to the melting of glaciers and ice caps, causing low-lying areas to get submerged and coastal areas to be at risk due to erosion. Thus, the intensity and frequency of extreme weather events also increase in number. Also, unprecedented events are happening due to climate change, which requires to be known prior knowledge. So here we require the emerging AI tools like Machine Learning (ML) and Deep Learning (DL) models, capable of predicting the climate catastrophe.

3 Traditional Climate Models

Earth's climate system, which includes the atmosphere, ocean, land, and ice, is simulated using computer models, helping scientists understand the heart of climate science. The model output drives forward in making use of the Earth's climate to get in-depth knowledge of how human activities affect the climate and to formulate national and international policies. As the climate system is complex, the Earth is divided into grid cells and time steps for the ease of calculation in the climate models. Thus, a climate model can create an exact representation of the complete climate system.

The early understanding of climate phenomena is carried out by Conceptual models. The prevailing winds were explained, and the energy transfer in the climate

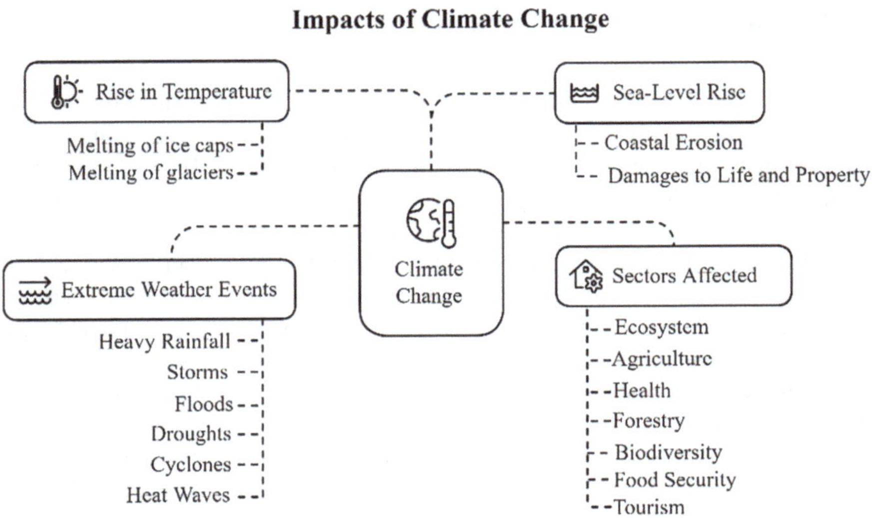

Fig. 1 Climate change impacts and sectors affected

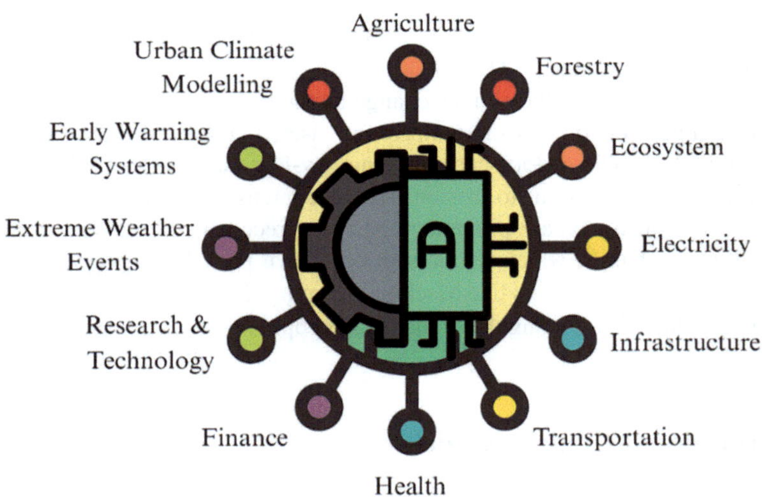

Fig. 2 Applications of AI models in different sectors

system was captured by these models. Then, the Analog models inspired the first-generation modelers who work on the principles of fundamental fluid motion [24]. The basic and earliest numerical models based on the balance between energy from the sun entering the atmosphere and the heat released into the atmosphere are called Energy Balance Models (EBMs). These do not process climate simulations but measure only surface temperature. The next model accounts the climate physics called the Global Climate Model (GCM), which includes air and water flows in the atmosphere and oceans, and also heat transfer [27]. They are large-scale models used for long-term climate projections. Later, biogeochemical cycles were incorporated into the GCMs and introduced as Earth System Models (ESMs). It serves as an important tool for simulating all bio-physical and chemical processes and incorporates the interaction with the Earth's climate system [40]. ESM simulates each part of the climate system, exchanges mass and energy between these, and has modules like ocean, land, atmosphere, and sea ice [40]. They are also capable of simulating the carbon cycle, atmosphere chemistry, nutrient cycle, nitrogen cycle, the impact of greenhouse gases, changes in land use and vegetation [27, 28]. The global climate data are downscaled to obtain climate simulations for a small area or a region called Regional Climate Models (RCMs). These are high-resolution models within GCMs limited to a specific area. Next comes the coupled models, couples' different components of the climate system. Atmosphere and ocean GCMs are coupled with ESMs and integrated into other climate system models. Also, Integrated Assessment models (IAMs) incorporate the impacts of climate change, economic responses, and policy schemes in a single modeling framework, which is composed of various components.

In this scenario, even if it's a fact that the traditional models have a great place in making researchers understand more about climate dynamics, they have many limitations in thinking about other modeling solutions. For the traditional models to run smoothly, when high-resolution data is considered, high-performing supercomputers are required, which makes simulations time-consuming [4]. Structural errors and uncertainty are created across models as they are with unresolved scales in different representations [38]. Also, they are tuned to reproduce only certain aspects of the Earth system [4]. With coarse resolutions, traditional models are not capable of creating local-scale phenomena [9]. Regional changes are not absorbed as such, and there are discrepancies between the actual observations and model predictions. Also, the validation of models is not justified by a lack of proper data. [50]. Parameterization is used to simplify the processes happening on a larger scale using certain assumptions and proxy data. This also creates uncertainties and may not be able to capture the climate interactions completely [20].

4 AI Models

AI techniques have demonstrated a transformative path in addressing climate issues with utmost accuracy. With data-driven insights, AI analyses large data sets from various data sources, understanding complex relationships and trends from historical data where traditional models may not work. AI-driven models have given significant promise to advance in forecasting climate prediction. AI is a powerful tool for accelerating the strategies to be adopted for climate change adaptation and mitigation [59].

4.1 Machine Learning (ML)

Machine Learning (ML) is a branch of AI, where models are trained from vast datasets [29], to predict trends and identify and analyze complex issues [5]. ML model understands non-linear interactions and other processes from historical data, unlike traditional models [33]. It uses input data to predict output data continuously whenever new data is received. The subdivisions of ML are.

4.1.1 Supervised Learning

This model learns from labeled data, where the outcome is already known. This can be used to develop output for unseen and new data [29]. The common techniques used in climate modeling are Regression algorithms which identify linear relationships between data, Classification algorithms that predict and categorize the output data, large datasets can be classified using Naïve Bayes classifiers which include decision trees and can accommodate the regression and classification algorithms and Neural Networks consists of interconnected neurons/nodes which simulate as how human brain works and Random Forest algorithms combine values from the number of decision trees and predict the output [51]. Applied supervised learning techniques to analyze future energy transitions and to identify efficient energy solutions.

4.1.2 Unsupervised Learning

They mainly work on unlabelled datasets to draw inferences and to facilitate data analysis, exploring hidden pattern recognition and prediction [51]. This technique can be used to identify climate patterns, classify weather data, and detect anomalies. The common method in this is cluster analysis, where they classify data points according to their similarity. This can help discover relationships and trends across various dimensions, which reveals a new system and becomes a challenge for better understanding [29]. There is another method called dimensionality reduction, which reduces a random number of variables, preserving principal variables [51].

4.1.3 Reinforcement Learning

It is an area that emphasizes how the agents should make decisions to achieve a set of goals, interacting within an environment. This is mainly used in game development and to teach robots to replicate human tasks [51].

4.2 Deep Learning (DL)

Deep Learning (DL) is a category of ML that uses multilayered deep neural networks, which capture complex relations and analyze the data [12]. They can extract relationships, characteristics, and features from raw data to obtain accurate output. It improves automation and can be implemented in daily applications like voice-enabled TV remotes, self-driving cars, digital assistants, fraud detection, etc.

4.2.1 Convolutional Neural Networks (CNNs)

They have main applications in interpreting visual data and far beyond the basic image categorization. CNNs are used to extract patterns from regional climate data and satellite images [12]. They identify objects, patterns, and features from images, face recognition, and processing in real time. Their task involves image segmentation, where an image is divided into different segments and objects, and other patterns are identified.

4.2.2 Recurrent Neural Networks (RNNs)

They are primarily used for time-series data and processing sequential data. They make predictions about events or outcomes. The major types of RNNs are Long Short-Term Memory (LSTM) networks and Gated Recurrent Units (GRUs). They use high-resolution data to predict future weather conditions [14, 15]. Also, these use time-series data to analyze weather trends and patterns like precipitation, temperature, and wind speed based on historical data [10].

4.3 Hybrid Models

Hybrid models are those that are developed by combining numerical modeling techniques and AI techniques [35]. This integration enhanced the efficiency and precision of weather predictions, refining the outcomes [31, 55], such as in droughts and extreme convective precipitation downscaling [14, 15]. These models can be used for downscaling to actionable scales to drive climate impact models [25].

4.3.1 Physics-Informed Neural Networks (PINNs)

In these models, information on physical processes is added to the algorithm in such a way as to restrict the outcome to a physically realistic one [36]. These include scientific ML techniques, which are used to find solutions for Partial Differential

Equations (PDEs) [19]. PINNs can be used as a powerful tool in scenarios where the data is limited and the data is not completely known [36, 56]. This model can be used for carbon-efficient climate change modeling and extreme weather forecasting, which enhances predictive accuracy [23].

Some of the AI models available now for different functions of climate modeling are listed below in Table 1.

5 Data Sources for AI Models

AI models integrate data from diverse sources for modeling and forecasting climate events, like satellite imagery, remote sensing technologies, observational data from ocean buoys, etc. Each of these sources captures different parameters of the climate system, like precipitation, temperature, sea-level rise, etc. Integration of these data from various sources should ensure completeness, accuracy, and consistency. There are ground-based observations taken from weather stations and other climate monitoring networks, which provide data on temperature, humidity, and other climatic parameters. Satellite data is very important for climate modeling as it offers high-resolution data on environmental parameters, vegetation cover, ocean color, land surface temperature, etc. European Centre for Medium-Range Weather Forecasts (ECMWF) produced ERA5 (ECMWF Reanalysis version 5), which provides hourly estimates of climate variables and also offers long-term datasets on global weather patterns offered by NCEP (National Centers for Environmental Prediction)/NCAR (National Center for Atmospheric Research) Reanalysis. Oceanographic data are provided by Argo floats and NOAA, like sea surface temperatures (SST), salinity, ocean currents, etc. Land use and land cover data are provided by MODIS (Moderate Resolution Imaging Spectroradiometer), World Bank data provides information on population density, urbanization, and UNESCO's World Heritage Sites. Given below are some of these data sources in Table 2.

6 Applications of AI Models in Climate Modelling

See Fig. 2.

Table 1 AI models/Software available for climate modeling

AI model/Software	Developer	Used for	Year of development
Prithvi-weather-climate Foundation Model	NASA and IBM Research	Storm tracking, forecasting, and historical analysis	2024
NeuralGCM	Google Research	Global climate modeling with ML-enhanced simulations	2024
ClimateGPT	AppTek + open-source	LLM for climate change synthesis and info	2024
Climate PAL	HuggingFace + collaborators	Conversational climate data analysis	2024
Sunairio	Sunairio Inc	AI-based weather and energy risk modeling	2024
GenCast	DeepMind	Extreme weather prediction and tropical cyclone tracking	2024
GraphCast	DeepMind (Google)	Medium-range weather forecasting (up to 10 days)	2023
ClimaX	Microsoft Research	Foundation model for climate/weather tasks	2023
AI2 Climate Emulator (ACE)	Allen Institute for AI	Emulating full-scale atmospheric climate models	2023
ClimateSet	Multi-institution	Large-scale dataset for training climate AI models	2023
ClimateLearn	UCLA + collaborators	ML benchmarking framework for climate science	2023
AIFS	Korean Meteorological Admin	AI-based short-range weather forecasting system	2023
NOAA AI-Enhanced Climate Models	NOAA GFDL	ML-enhanced ocean–atmosphere climate simulations	2022
Benevolent AI	BenevolentAI	Climate research via ML and data synthesis	2021
IBM Watson for Climate	IBM	Environmental data modelling and climate forecasts	2020
Kensho	S & P Global	Climate risk analytics via AI	2020

(continued)

Table 1 (continued)

AI model/Software	Developer	Used for	Year of development
Alteryx	Alteryx Inc.	Environmental data analysis & AI modeling	2019
FloodMapp	FloodMapp Pty Ltd.	Real-time flood forecasting with AI-hydrology models	2018

Table 2 Data Sources available for climate modeling

Source	Data types	Website
ERA5 (ECMWF Reanalysis)	Atmospheric variables, soil moisture, radiation, and extreme weather	https://www.ecmwf.int/en/forecasts/datasets/reanalysis-datasets/era5
NCEP/NCAR Reanalysis	Long-term atmospheric reanalysis (temperature, wind, etc.)	https://psl.noaa.gov/data/gridded/data.ncep.reanalysis.html
CMIP6 (Climate Models)	Model simulations, future scenarios (SSPs/RCPs)	https://esgf-node.llnl.gov/projects/cmip6/
MODIS (NASA)	Land cover, vegetation indices, albedo, surface temperature	https://modis.gsfc.nasa.gov
Argo	Ocean temperature, salinity profiles, and ocean heat content	http://www.argo.ucsd.edu
CMEMS	Sea surface temperature, ocean currents, salinity, and biogeochemistry	https://marine.copernicus.eu
ESA CCI	Land cover, sea ice extent, glaciers, and soil moisture	https://climate.esa.int/en/projects/
NSIDC	Sea ice, snow cover, glacier mass, and permafrost	https://nsidc.org
CERES (NASA)	Earth's radiation budget, solar radiation	https://ceres.larc.nasa.gov
SARAH (CMSAF)	Solar radiation, global irradiance	https://wui.cmsaf.eu
FLUXNET	Carbon/water fluxes, evapotranspiration	https://fluxnet.org
NOAA Mauna Loa /WMO WDCGG	Greenhouse gas concentrations (CO_2, CH_4, N_2O)	https://gml.noaa.gov/ccgg/trends/
IBTrACS	Cyclone tracks and intensity	https://www.ncdc.noaa.gov/ibtracs/
EM-DAT	Natural disasters: floods, heatwaves, storms, droughts	https://www.emdat.be

7 Applications of AI Models in Climate Modelling and Recent Works

The AI/ML applications incorporated into various sectors and are revolutionizing the response we provide to environmental transformations and protection, thereby mitigating the adverse effects of climate change and is progressing in recent years.

Climate change and weather conditions have a large impact on agriculture, as many crops are seasonal like kharif and rabi crops, etc. The crop yield and the land for agriculture also depend upon the changes in climate, soil properties, global warming, amount of carbon, rainfall, temperature, heat waves, activity of microbes, etc. In this scenario, [52] designed crop yield assessment modeling using ML techniques, for accurate forecasting of crop production, prices, marketing, import, and export of these and thereby policy assessment for food security. Also, [21] studied the AI and ML techniques for forecasting crop diseases, which help farmers to adopt preventive measures for the diseases, thus empowering the community. The AI and remote sensing tools can be used for monitoring crops and prediction of yield, which helps in adapting to situations of climate events like droughts and extreme weather events, contributing to food security [59].

AI tools can be used in carbon sequestration, predicting the risk of wildfires, decisions for land management, tracking deforestation, etc. An emerging concept called Climate-Smart Forestry (CSF) can be used for addressing the issues related to climate change like forest ecosystem resilience, carbon sequestration, and carbon forestry which enables increased carbon absorption and reducing GHGs, balancing economic, social, and environmental benefits. It's a contribution to digital forestry, creating a digital twin model of the forest using real-time data collected, and simulates using AI tools, thus identifying disaster warnings, forest and ecosystem management, mitigating and adapting to climate change effects, contributing to sustainable forest management [54].

The major contribution of the electrical systems sector to climate change is the introduction of Electric Vehicles (EVs). EV and hybrid vehicles reduce carbon emissions, reducing GHGs and thus reducing global warming and promoting sustainable mobility. AI tools can be used to improve efficiency, and performance, and enhance energy and battery management [42]. Power optimization using AI is yet another breakthrough of EV, at low cost, reduced environmental impacts for the distribution of sustainable energy [8].

As the climate is changing, the heating is also increasing inside the buildings. AI can smartly be used in infrastructure to optimize energy consumption and heat indoors to provide comfort [37]. Also, [32] proposed a solution for lowering the cost of electricity using RL methods, as when electricity is cheap, the building gets electrically heated, and when it is cheap, electricity use gets reduced without getting thermally uncomfortable to the resident. ANNs can be used to improve industrial automation, controlling and monitoring the industrial environment, as proposed by [2].

Decarbonization, usage of EVs, predicting carbon emissions, etc. in the transportation sector a major contributions towards climate change in today's scenario. AI in the transportation sector can be used to collect traffic data and can find solutions for traffic congestion, scheduling public transport, increased travel requirements, traffic monitoring and management, avoiding accidents, monitoring intelligent parking, etc. [34]. Also, AI can be used to improve traveler behavior and thereby service quality [30]. Real-world applications in transportation like smart parking, optimizing public transport, traffic management, infrastructure monitoring, and sustainable practices are studied and challenges are identified [46] thereby reducing carbon emissions into the atmosphere.

Crane et al. [17] examined relevant impacts that climate change can have on mental health. They found that depression, stress, anxiety, etc. are the negative outcomes of climate change which leads to crime, suicide, etc. [39] studied implementing AI techniques in health care by the diagnosis of diseases and curation of treatment-based on the patient analysis. AI robots can be used to employ in hazardous or contaminated environments, where human interference is challenging. They can also be used for caring elderly for assisting in their day-to-day activities. Remote-controlled robotic arms of AI robots can be used to carry out surgery. AI can be used in medical imaging analysis with great effectiveness and precision, and also successfully implemented in mobile health (m-health) for an efficient healthcare system [48].

In the finance sector, physical risks associated with financial losses arising due to the extreme weather events due to climate change are increasing and the loss incurred by the firms (loans, etc.) make them collapse [49]. AI can be used in enhancing customer services in banking [26], fraud detection, investment management, and credit scores, and ethically improve the decision-making processes [45], prevent crimes in finance and banking [44]. Bhatia et al. [7] proposed a smart system with neural networks for forecasting stock market fluctuations, to identify trends and non-linear patterns. Environmental, Social, and Governance (ESG) analysis and sustainability reports using AI are now incorporated in the investments sector and thus contribute to climate change by re-evaluating the market risks and moving towards a low-carbon economy [49].

AI and ML algorithms and models are supporting climate science and research and also all other sectors in a very positive approach [29]. Basher and Sadorsky [6] studied on pricing of clean energy stocks and climate change-induced risks associated with market fluctuations, thus encouraging the transition to low low-carbon economy. AI and ML technologies have been applied to the research and technology sector of almost all areas of the world for better analysis and contribution. More the research is done in each field, the more will be the contribution of AI tools to development.

Dikshit et al. [22] Studied Explainable AI (XAI) applications in natural hazard modeling due to climate change. As AI models lack transparency in their processes and are called 'black boxes', XAI enhances the understanding and interprets the internal workings of these models efficiently. Transparency aids the correlations and patterns created by the model more trustworthy and makes it easier for informed decision-making [18] proposed. Climate Projection and Analysis with Language models (Climate PAL), to support the public on climate change-related research and communications. Climate PAL analyzes and retrieves datasets required for climate research from CMIP6 models, thus making the research easier.

Salcedo-Sanz et al. [47] Reviewed the usage of AI, ML, and DL techniques in the analysis, and prediction of extreme weather events like heavy precipitation, floods, droughts, tropical cyclones, heatwaves, extreme winds, etc., and identified the lack of input data to train the models and proposed solutions like the usage of reanalysis data, incorporating physics into the models for more accuracy, using GCMs for new training samples for ML approaches and also, as the extreme events are rare, even 70 years data will not be sufficient for training the models.

As the extreme weather events due to climate change are increasing, the need for early warning systems (EWS) also becomes necessary [43]. Developed integrated AI modeling, mainly supervised learning, for effective EWSs. ML models can learn from large amounts of data, especially foundation models developed for EWS and thus develop patterns and trends to understand and forecast multi-hazards, before the hazard strikes. EWS has huge benefits as it can make the transition from hazards to impacts, and increase the accuracy of warnings thereby reducing life and property losses.

Asperti et al. [3] studied diffusion models, mainly in the category of deep learning models, to nowcast precipitation and proposed Generative Ensemble Diffusion (GED). This approach helped in a deep understanding of weather patterns and contributed to better decision and policy making.

The potential of AI and ML is explored in urban climate change and adaptation and mitigation across various countries [53]. Case studies from different places like Africa, Europe, Asia, etc. were studied to increase the climate change adaptation strategies in urban centers. Thus, conferred that AI and ML techniques play a vital role in building climate-resilient urban areas and promoting sustainability. Policymakers outlined this study while formulating climate policies and frameworks.

AI tools can be used as a promising tool in addressing climate change-related issues in the energy sector to improve energy efficiency and reduce wastage, a transformative tool in transportation and health sectors, contributes to more accurate weather forecasting for providing EWS and predicting extreme weather events, predict renewable energy production, optimizing electrical consumption thereby enhancing better living conditions, optimizing traffic conditions, reducing transportation cost by implementing EVs, precision in agriculture makes smart-farmers, increasing the crop yield, reducing chemicals, reducing deforestation and promoting environmental protection and sustainable development. AI can mitigate climate change by increasing the efficiency of energy and providing accurate data for policymakers [16].

8 Challenges/Limitations of AI Models

The challenges/limitations associated with AI models may be grouped into the following categories:

The complexity of the climate system: The climate system is complex, and capturing all the processes and available data is difficult. In certain cases, historical data cannot completely determine the actual behavior of the impact happening. Also, the integration of multiple models is time-consuming. In the case of multiple events happening, to account for the combined effect of these events is difficult. Each event may be considered separately, which cannot capture the results completely [31].

Data Quality: High-quality data with good resolution and large datasets are required for model training for precise results, which is often unavailable. Inefficient data standards applied to the models cause uncertainties in the output [59].

Localization: Climate change impacts vary across the globe from small regions to larger areas. So, when the spatio-temporal data is used for smaller regions, downscaling it can cause errors [31].

Computational constraints: AI models require computers with good specifications, speed, and other tools and technologies to support modelers [31].

Model Transparency: Often, AI models are called black boxes due to the unknown calculations and processes happening during predictions. The lack of model interpretability makes it difficult to trust the outcomes of AI models [11].

Uncertainty: It can occur at many stages of modeling, like during the collection of data, observational errors, model validation, downscaling, etc. So, rather than focusing on these, we have to look forward to strategies for risk management [31, 35].

9 Conclusion

Artificial intelligence is a game-changer in climate science. Integrating AI into climate modeling is a productive way to move forward. It has great potential to advance knowledge in climate dynamics and to assist researchers and scientists in practical applications in climate modeling and forecasts with utmost precision. AI improves climate models, can downscale climatic projections, forecast extreme weather events, and support policy preparation, management of adaptation and mitigation strategies. The role of AI in real-time analysis, prediction, and optimization can tackle the issues, address the challenges, and provide innovative, sustainable solutions.

AI has proven itself as a powerful tool for application in various sectors, improving predictability and forecasting skills. Many studies and research are happening exponentially, and the evolution is very quick, so we expect a faster growth of AI applications in deriving solutions. The promise of AI is being trustworthy now, and the immense potential that resides in this tool can build resilient and sustainable societies in this fast-paced climate climate-affected environment.

References

1. Abbass, K., Qasim, M.Z., Song, H., et al.: A review of the global climate change impacts, adaptation, and sustainable mitigation measures. Environ. Sci. Pollut. Res. **29**, 42539–42559 (2022)
2. Al Shahrani, A.M., Alomar, M.A., Alqahtani, K.N., et al.: Machine learning-enabled smart industrial automation systems using internet of things. Sensors **23**, 324 (2022). https://doi.org/10.3390/s23010324
3. Asperti, A., Merizzi, F., Paparella, A., et al.: Precipitation nowcasting with generative diffusion models. Appl. Intell. **55**, 187 (2025). https://doi.org/10.1007/s10489-024-06048-y
4. Balaji, V., Couvreux, F., Deshayes, J., et al.: Are general circulation models obsolete? Proc. Natl. Acad. Sci. USA A **119** (2022). https://doi.org/10.1073/pnas.2202075119
5. Balakrishnan, T.S., Krishnan, P., Ebenezar, U.S., et al.: Machine Learning for Climate Change Impact Assessment and Adaptation Planning, pp. 1–6. Institute of Electrical and Electronics Engineers (IEEE) (2024)
6. Basher, S.A., Sadorsky, P.: How important are climate change risks for predicting clean energy stock prices? Evidence from machine learning predictive modeling and interpretation. J. Clim. Finance **10**, 100058 (2025). https://doi.org/10.1016/j.jclimf.2024.100058
7. Bhatia, S., Bhardwaj, D., Akash, A., Bhardwaj, L.: Prediction in stock market value using artificial intelligence. In: 2024 IEEE 4th International Conference on ICT in Business Industry and Government (ICTBIG), pp. 1–4. IEEE (2024)
8. Biswas, P., Rashid, A., Biswas, A., et al.: AI-driven approaches for optimizing power consumption: a comprehensive survey. Discov. Artif. Intell. **4**, 116 (2024). https://doi.org/10.1007/s44163-024-00211-7
9. Blizer, A., Glickman, O., Lensky, I.M.: Comparing ML methods for downscaling near-surface air temperature over the eastern mediterranean. Remote Sens. (Basel) **16** (2024). https://doi.org/10.3390/rs16081314
10. Bochenek, B., Ustrnul, Z.: Machine learning in weather prediction and climate analyses—applications and perspectives. Atmosphere (Basel) **13** (2022). https://doi.org/10.3390/atmos13020180
11. Bostrom, A., Demuth, J.L., Wirz, C.D., et al.: Trust, and trustworthy artificial intelligence: a research agenda for AI in the environmental sciences. Risk Anal. **44**, 1498–1513 (2024). https://doi.org/10.1111/risa.14245
12. Brindha Devi, V., Prabavathi, R., Subha, P., et al.: Climate prediction using AI. In: 4th International Conference on Power, Energy, Control and Transmission Systems: Harnessing Power and Energy for an Affordable Electrification of India, ICPECTS 2024. Institute of Electrical and Electronics Engineers Inc. (2024)
13. Calvin, K., Dasgupta, D., Krinner, G., et al.: IPCC, 2023: Climate Change 2023: Synthesis Report. Contribution of Working Groups I, II, and III to the Sixth Assessment Report of the Intergovernmental Panel on Climate Change [Core Writing Team, H. Lee and J. Romero (eds.)]. IPCC, Geneva, Switzerland (2023)
14. Camps-Valls, G., Fernández-Torres, M.Á., Cohrs, K.H., et al.: Artificial intelligence for modeling and understanding extreme weather and climate events. Nat. Commun. **16** (2025a)
15. Camps-Valls, G., Fernández-Torres, M.-Á., Cohrs, K.-H., et al.: Artificial intelligence for modeling and understanding extreme weather and climate events. Nat. Commun. **16**, 1919 (2025). https://doi.org/10.1038/s41467-025-56573-8
16. Chen, L., Chen, Z., Zhang, Y., et al.: Artificial intelligence-based solutions for climate change: a review. Environ. Chem. Lett. **21**, 2525–2557 (2023)
17. Crane, K., Li, L., Subramanian, P., et al.: Climate change and mental health: a review of empirical evidence, mechanisms, and implications. Atmosphere (Basel) **13**, 2096 (2022). https://doi.org/10.3390/atmos13122096
18. Cromp, S., Rabiei, B., Elling, M., et al.: Climate PAL: Climate Analysis through Conversational AI

19. Cuomo, S., Di Cola, V.S., Giampaolo, F., et al.: Scientific machine learning through physics–informed neural networks: where we are and what's next. J. Sci. Comput. **92** (2022). https://doi.org/10.1007/s10915-022-01939-z
20. Deitrick, R., Haqq-Misra, J., Kadoya, S., et al.: Functionality of ice line latitudinal EBM tenacity (FILLET). Protocol Version 1.0. A cuisines intercomparison project. Planet. Sci. J. **4** (2023). https://doi.org/10.3847/PSJ/acba05
21. Delfani, P., Thuraga, V., Banerjee, B., Chawade, A.: Integrative approaches in modern agriculture: IoT, ML and AI for disease forecasting amidst climate change. Precis. Agric. **25**, 2589–2613 (2024). https://doi.org/10.1007/s11119-024-10164-7
22. Dikshit, A., Pradhan, B., Matin, S.S., et al.: Artificial Intelligence: A new era for spatial modeling and interpreting climate-induced hazard assessment. Geosci. Front. **15** (2024)
23. Dutta, S., Innan, N., Yahia, S.B., Shafique, M.: AQ-PINNs: Attention-Enhanced Quantum Physics-Informed Neural Networks for Carbon-Efficient Climate Modeling (2024)
24. Edwards, P.N.: History of climate modeling. Wiley Interdisc. Rev. Clim. Change **2**, 128–139 (2011). https://doi.org/10.1002/wcc.95
25. Eyring, V., Gentine, P., Camps-Valls, G., et al.: AI-empowered next-generation multiscale climate modeling for mitigation and adaptation. Nat. Geosci. (2024). https://doi.org/10.1038/s41561-024-01527-w
26. Fares, O.H., Butt, I., Lee, S.H.M.: Utilization of artificial intelligence in the banking sector: a systematic literature review. J. Financial Serv. Mark. **28**, 835–852 (2023). https://doi.org/10.1057/s41264-022-00176-7
27. Hausfather, Z., McSweeney, R.: How do climate models work? In: Carbon Brief (2018). https://www.carbonbrief.org/qa-how-do-climate-models-work/#inout. Accessed 5 Apr 2025
28. Heavens, N.G., Ward, D.S., Natalie, M.M.: Studying and projecting climate change with earth system models. Nat. Educ. (2013)
29. Huntingford, C., Jeffers, E.S., Bonsall, M.B., et al.: Machine learning and artificial intelligence to aid climate change research and preparedness. Environ. Res. Lett. **14** (2019). https://doi.org/10.1088/1748-9326/ab4e55
30. Jevinger, Å., Zhao, C., Persson, J.A., Davidsson, P.: Artificial intelligence for improving public transport: a mapping study. Public Transport **16**, 99–158 (2024). https://doi.org/10.1007/s12469-023-00334-7
31. Jones, A., Kuehnert, J., Fraccaro, P., et al.: AI for climate impacts: applications in flood risk. NPJ Clim. Atmos. Sci. **6** (2023). https://doi.org/10.1038/s41612-023-00388-1
32. Kannari, L., Kantorovitch, J., Piira, K., Piippo, J.: Energy cost driven heating control with reinforcement learning. Buildings **13**, 427 (2023). https://doi.org/10.3390/buildings13020427
33. Kayusi, F., Kasulla, S., Malik, S.J., et al.: Advanced AI, machine learning and deep learning techniques for climate change studies: a review. South Asian Res. J. Agric. Fish. **6**, 101–108 (2024). https://doi.org/10.36346/sarjaf.2024.v06i06.001
34. Khan, S., Adnan, A., Iqbal, N.: Applications of artificial intelligence in transportation. In: 2022 International Conference on Electrical, Computer and Energy Technologies (ICECET), pp. 1–6. IEEE (2022)
35. Materia, S., García, L.P., van Straaten, C., et al.: Artificial intelligence for climate prediction of extremes: State of the art, challenges, and future perspectives. Wiley Interdisc. Rev. Clim. Change (2024)
36. Molina Catricheo, C.A., Lambert, F., Salomon, J., van 't Wout, E.: Modeling global surface dust deposition using physics-informed neural networks. Commun. Earth Environ. **5** (2024). https://doi.org/10.1038/s43247-024-01942-2
37. Ngarambe, J., Yun, G.Y., Santamouris, M.: The use of artificial intelligence (AI) methods in the prediction of thermal comfort in buildings: energy implications of AI-based thermal comfort controls. Energy Build **211**, 109807 (2020). https://doi.org/10.1016/j.enbuild.2020.109807
38. Nicholls, Z., Meinshausen, M., Lewis, J., et al.: Reduced complexity model intercomparison project phase 2: synthesizing earth system knowledge for probabilistic climate projections. Earth's Future **9** (2021). https://doi.org/10.1029/2020EF001900

39. Olawade, D.B., David-Olawade, A.C., Wada, O.Z., et al.: Artificial intelligence in healthcare delivery: prospects and pitfalls. J. Med. Surgery Public Health **3**, 100108 (2024). https://doi.org/10.1016/j.glmedi.2024.100108
40. Pan, X., Chen, D., Pan, B., et al.: Evolution and prospects of Earth system models: challenges and opportunities. Earth Sci. Rev. **260** (2025)
41. Peddagolla, N., Muniganti, P.: Implementing Artificial Intelligence for Hydroinformatics and Enhanced Climate Forecasting to Promote Sustainable Decision Making and Improve Ecological Understanding
42. Rauf, M., Kumar, L., Zulkifli, S.A., Jamil, A.: Aspects of artificial intelligence in future electric vehicle technology for sustainable environmental impact. Environ. Challenges **14**, 100854 (2024). https://doi.org/10.1016/j.envc.2024.100854
43. Reichstein, M., Benson, V., Blunk, J., et al.: Early warning of complex climate risk with integrated artificial intelligence. Nat. Commun. **16**, 2564 (2025). https://doi.org/10.1038/s41467-025-57640-w
44. Ridzuan, N.N., Masri, M., Ansari, M., et al.: AI in the financial sector: the line between innovation, regulation, and ethical responsibility. Information **15**, 432 (2024). https://doi.org/10.3390/info15080432
45. Jain, R.: Role of artificial intelligence in banking and finance. J. Manag. Sci. **13**, 1–4 (2023). https://doi.org/10.26524/jms.13.27
46. Saki, S., Soori, M.: Artificial intelligence, machine learning and deep learning in advanced transportation systems, a review. Multimodal Transp. 100242 (2025). https://doi.org/10.1016/j.multra.2025.100242
47. Salcedo-Sanz, S., Pérez-Aracil, J., Ascenso, G., et al.: Analysis, characterization, prediction, and attribution of extreme atmospheric events with machine learning and deep learning techniques: a review. Theor. Appl. Climatol. **155**, 1–44 (2024)
48. Saxena, A.K., Ness, S., Khinvasara, T.: The influence of AI: the revolutionary effects of artificial intelligence in the healthcare sector. J. Eng. Res. Rep. **26**, 49–62 (2024). https://doi.org/10.9734/jerr/2024/v26i31092
49. Sharma, N., De, P.K.: Climate Change and AI in the Financial, Energy, Domestic, and Transport Sectors, pp. 1–21(2023)
50. Shaw, T.A., Arias, P.A., Collins, M., et al.: Regional climate change: consensus, discrepancies, and ways forward. Front. Clim. **6** (2024)
51. Shobanke, M., Bhatt, M., Shittu, E.: Advancements and future outlook of artificial intelligence in energy and climate change modeling. Adv. Appl. Energy **17**(2025)
52. Singh, S., Babu, K.V.S., Singh, S.: Machine learning approach for climate change impact assessment in agricultural production. In: Visualization Techniques for Climate Change with Machine Learning and Artificial Intelligence, pp. 317–335. Elsevier
53. Srivastava, A., Maity, R.: Assessing the potential of AI–ML in urban climate change adaptation and sustainable development. Sustainability **15**, 16461 (2023). https://doi.org/10.3390/su152316461
54. Wang, G.G., Lu, D., Gao, T., et al.: Climate-smart forestry: an AI-enabled sustainable forest management solution for climate change adaptation and mitigation. J For Res (Harbin) **36**, 7 (2024). https://doi.org/10.1007/s11676-024-01802-x
55. Waqas M, Humphries UW, Chueasa B, Wangwongchai A (2024) Artificial intelligence and numerical weather prediction models: a technical survey. Natural Hazards Research
56. Zucker, S., Batenkov, D., Rozenhaimer, M.S.: Physics-informed neural networks for modeling atmospheric radiative transfer. J. Quant. Spectrosc. Radiat. Transf. **331** (2025). https://doi.org/10.1016/j.jqsrt.2024.109253
57. Artificial Intelligence for Climate Action in Developing Countries: Opportunities, Challenges and Risks Information Note Acknowledgements UNFCCC TEC Information Note: Artificial Intelligence for Climate Action in Developing Countries: Opportunities, Challenges and Risks
58. UNCC : The Paris Agreement
59. Climate Change and AI Recommendations for Government Action Global Partnership on AI Report In collaboration with Climate Change AI and the Centre for AI and Climate

Quantum Computing and AI: Advancing Climate-Resilient Urban Planning

Mamta

Abstract Urban planning is fundamentally changing as cities worldwide face increasing climate risks, rapid urbanization, and the demand for sustainable development. Incorporating advanced computation tools, like quantum computing or artificial intelligence, invites a glimpse of planning simulations against complex climate scenarios, efficient resource allocation, and resilient infrastructure development. These include mitigation through energy-efficient development, low-carbon transport, and integrating renewable energies and adaptation, such as resilient coastlines, green–blue infrastructure, and climate-resilient design. Effective climate action in urban areas is required, supported by cross-sectoral planning, participatory approaches, and strong governance to ensure that interventions are context-specific and socially inclusive. Case studies from around the world demonstrate that when climate sensitivity is embedded in policy, spatial planning, and community engagement, vulnerability can be significantly reduced and cities can enhance adaptive capacity. As urban systems become more interconnected, it will be essential for quantum and AI-driven solutions to be leveraged for achieving low-carbon, resilient, and inclusive urban futures.

Keywords Urban planning · Climate resilience · Quantum computing · Artificial intelligence · Adaptation · Sustainable cities

1 Urban Climate Challenges and the Quantum-AI Revolution

Urban areas represent the core locations that both generate and experience the worst consequences of the climate emergency [1]. They function as concentration points of economic operations and residents while producing more than 70% of global CO_2 emissions and dealing with deadly climate effects. Quantum computing combined

Mamta (✉)
Department of Computer Science and Engineering, Chandigarh University, Punjab, India
e-mail: mamta.airesearch@gmail.com

A. L. Srivastav et al. (eds.), *Climate Resilience: Impact of Quantum Computing and Artificial Intelligence on Urban Planning*, Studies in Systems, Decision and Control 637, https://doi.org/10.1007/978-3-032-06791-3_8

with artificial intelligence enables cities to make a fundamental change in their approach to controlling these challenges, so they transition from basic adaptation to active resilience [2].

1.1 The Multifaceted Climate Crisis in Urban Environments

Urbanization amplifies climate risks through interrelated environmental, infrastructural, and socioeconomic pressures:

(i) **Environmental Degradation and Extreme Weather**

- **Heat Islands**: Heat Islands affect urban environments since concrete-dominated landscapes generate and emit heat, which raises urban temperatures by 3–5 °C above rural areas [3]. Excessive heat in cities such as Phoenix and Delhi causes deadly heat waves, which are prevented from cooling at night by the high demands of energy-intensive cooling systems.
- **Flood Vulnerabilities**: Extreme flooding occurs because urban areas have too many non-absorbent structures and weak drainage systems during heavy rainfall. Coastal megacities, including Mumbai and Miami, experience twofold dangers from ocean surges and rising sea levels, which threaten $1 trillion worth of properties before 2050 [4].
- **Air and Water Strains**: Industrial emissions combined with vehicular pollution seriously damage both air quality and water resources through dehydration and water depletion of aquifers. An extreme water shortage during 2019 in Chennai shows how unpredictable climate impacts urban water resource operations [5].

(ii) **Infrastructural and Economic Pressures**

- Energy distribution networks have limited capacity to maintain power balance between growing electricity consumption demands and expanding population rates, and market-wide electricity adoption. Blackouts occurred after heat waves affected Los Angeles and Tokyo, which revealed the weaknesses in centralized energy distribution systems [6].
- Transportation networks have to manage two major challenges: traffic congestion leads to increased pollution, and extreme weather events damage supply chains. Several railway transportation routes were destroyed during the 2021 European floods, which resulted in a direct cost of 2.5 billion euros for the economy [7].

1.2 Quantum Computing and AI: A Synergistic Framework for Resilience

Quantum technologies linked to AI help cities solve climate problems by using all three features: improved simulations, smart adjustments, and smart learning methods [8].

(i) **Precision Climate Modeling**: This enables Normal city-level temperature tasks to run beyond the processing power of standard computers. By using quantum processes, superposition helps quantum technology operate through millions of changes in data sets faster than classical computers. This enables:

- **Microclimate Predictions**: Neighborhood-level heat risk is calculated by the system based on building shapes and structures, plus plant locations, together with the movement of air within the area [9].
- **Extreme Weather Forecasting**: Cyclone trajectory accuracy is improved by quantum-enhanced neural networks, buying critical hours for evacuation planning.

(ii) **Dynamic Resource Optimization**: Real-time energy, water, and transportation network balancing is required by urban systems. Quantum annealing and AI-driven analytics are enabled by:

- **Grid Resilience**: The distribution grid of renewable power gets adjusted to handle sudden power usage demands. The Barcelona hybrid quantum–classical model decreased solar power cuts by 18% during heat wave periods [10].
- **Water Management**: The implementation of quantum algorithms together with AI technology enables the prediction of leaks in pipes and reservoir water release management optimization for times of drought [11]. The SMART Water Grid of Singapore implements these systems to lower non-revenue water losses down to a rate of 5% at a rate significantly below the worldwide average of 35% [12].

(iii) **Adaptive Infrastructure Design**: Historical climate data is analyzed by machine learning to recommend adaptive materials and designs [13]. The discovery of this is accelerated by quantum simulations:

- **Carbon-Capture Concrete**: External organizations test carbon-absorbing concrete products in Amsterdam's canal-side building projects.
- **Phase-Change Building Skins**: Smart towers in Dubai use phase-change building skins that help save heating costs by absorbing and emitting heat, which lowers the cooling load by 40%.

2 Quantum Computing Applications for Climate Resilience

The pressing nature of innovative data-based solutions becomes more important because cities continue to expand and climate threats become stronger day by day. Quantum computing demonstrates potential for major urban environmental changes in climate change anticipation and adaptation, as well as impact mitigation through its rapid processing and analysis of complex data [14]. This new quantitative computational approach bases its operations on quantum mechanics principles to facilitate significant development in climate modeling and resource optimization, and energy management, which form fundamental blocks for developing sustainable urban climate defenses.

2.1 Climate Modeling and Simulation

Effective urban climate resilience planning is founded on accurate climate modeling. Supercomputers have great power but struggle when trying to reproduce detailed interactions between urban microclimates and the specific behaviors of people, as well as the movement of air [15]. Quantum computers can handle large datasets all at once, which helps build precise time-sensitive urban models that show the entire urban environment.

(i) **Transforming Weather and Climate Forecasts**

Atmospheric chemistry, ocean currents, and land-atmosphere feedback loops can be simulated by quantum algorithms with a level of detail previously unattainable. Planners can develop more specific area forecasts by using this system [16]. Quantum models now show building impacts such as wind changes, rainfall patterns, and enhanced urban heat effects in isolated areas. Designing buildings and parks with infrastructure depends on this information to make them withstand climate change now and in the future.

(ii) **Simulating Extreme Events**

Extreme weather conditions affect cities more than other country areas because they encounter frequent flooding, hurricanes, and dangerous heat waves [17]. By running many scenario simulations at once, quantum computers assist cities in finding all their weak points for creating strong emergency responses. These storm modeling exercises show how flooding and power loss affect many parts of society, so planners can pick which flood protection measures and emergency cooling spots should receive their funding first.

(iii) **Material Innovation for Climate Mitigation**

The quantum simulation of materials for climate adaptation and mitigation is another frontier [18]. The atomic structure and behavior of new materials can be modeled by quantum computers, with the discovery of advanced composites for energy-efficient buildings, carbon capture, and water purification being accelerated. The development of urban infrastructure that is both sustainable and robust in the face of evolving climate risks is supported by this capability.

2.2 *Optimization of Resource Allocation in Urban Planning*

The allocation of energy, together with water and green spaces, and transportation networks in urban areas represents a sophisticated task that requires meeting different population needs. Current optimization techniques prove insufficient when addressing the uncertain challenges that climate change introduces into urban resource management practices [19]. Quantum computing achieves this progress through its ability to execute numerous allocation scenario evaluations simultaneously in parallel computing to find optimal solutions that enhance efficiency and maximize both resilience and fairness.

(i) **Water Systems Management**

Water scarcity, together with flooding, creates significant problems that affect numerous cities. Quantum algorithms optimize operations by defining sensor positions and release intervals, and stormwater distribution patterns to minimize both water scarcity and excess conditions. Water spikes can be anticipated by seasonal cities benefiting from quantum prediction models, and this data can be used for early conservation efforts that require infrastructure adjustments [20]. Quantum optimization technologies assist during heavy weather by directing water infrastructure management to protect against flooding and conserve water resources for later use.

(ii) **Transportation and Mobility Planning**

During emergencies, quantum models help cities establish safety routes through their transportation systems, which act as both emission sources and life-saving pathways. Through quantum computing, operations managers can dynamically optimize transportation routes for public transit systems, as well as for emergency responders and individual cars at the moment. The approach decreases traffic jams and reduces air pollutant levels while keeping escape paths available during urgent situations. Quantum models serve communities through their capability of analyzing transportation equality patterns between neighborhoods, thus protecting marginalized groups from being neglected.

(iii) **Land Use and Green Infrastructure**

The correct placement of green areas and urban forests with permeable surfaces stands as a fundamental tool for heat control and superior air quality, and flood safety. Quantum computing systems evaluate geographic features together with population statistics and climate prediction models to establish perfect site locations for park development and green roof installation, and tree distribution. Human population groups benefit more as the strategic distribution secures environmental advantages and prioritizes disadvantaged communities across the nation [21]. Quantum models help urban planners develop strong decisions through future simulation, which enables them to withstand diverse climate projections.

(iv) **Waste and Resource Recovery**

Cities produce significant total waste volumes, which can be reused to generate power. Quantum optimization helps cities develop efficient pick-up routes and operate better material separation systems while searching for optimal placements of waste recycling operations and waste energy facilities. Urban environments experience reduced waste-based landfill usage as well as decreased pollution levels when building a circular economy; this contributes indirectly to climate resilience.

2.3 *Energy Grid Management and Optimization*

Making the change to renewable power is vital for climate resilience, but it brings extra problems to urban power networks. Because solar and wind power operate on a variable basis, experts must develop complex management systems to maintain reliable power distribution. Quantum computing provides optimal real-time solutions for grid operations management, combined with storage and distribution processes [22]. Figure 1.1 shows how quantum computing brings renewable energy resources into a single system, which optimizes their management alongside storage facilities and both supply-side and demand-side actions within a contemporary urban network.

(i) **Integrating Distributed Renewables**

When cities add more solar power systems, they need more complex strategies to match power output with usage needs at every moment. Quantum algorithms evaluate weather predictions together with consumer behavior alongside power grid restrictions before handling renewable power distribution. The implementation of distributed renewables reduces both energy system vulnerability to climate impacts alongside energy costs and decreases dependence on fossil fuels (Fig. 1).

(ii) **Grid Resilience and Fault Detection**

Stable power delivery in urban areas remains at risk because of weather-related events as well as temperature extremes, along with digital intrusions against grid systems. Quantum machine learning tools can spot system abnormalities through predictive failure detection, thus enabling operators to act ahead of failures as well as initiate quick response procedures. The system cuts operational interruptions while safeguarding important utilities so hospitals and water facilities, together with emergency response teams, receive uninterrupted power during emergencies.

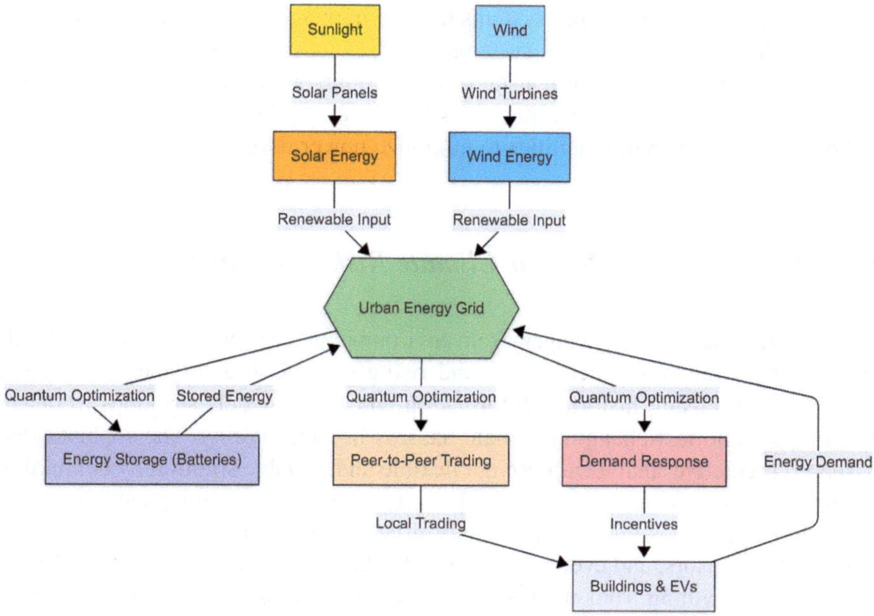

Fig. 1 Quantum-optimized urban energy grid

(iii) **Energy Storage and Demand Response**

Storing energy efficiently, together with dispatch capabilities, becomes vital for achieving renewable energy management. Quantum computing technology improves battery operations, together with pumped hydro storage, along with other power storage techniques, to deliver energy when it is required most. Advanced quantum models enable programs that encourage users through incentives to move their consumption towards peak periods, which leads to enhanced grid stability.

(iv) **Peer-To-Peer Energy Trading**

Through quantum algorithms, urban microgrids enable households and businesses to transact their renewable energy supply directly with each other. These changes make energy available to everyone and cut down distribution losses, and spawn economic possibilities that help achieve the city's climate targets.

3 AI Solutions for Climate-Resilient Urban Planning

Artificial Intelligence (AI) achieves a transformation of urban planning because it allows municipalities to forecast upcoming situations and develop suitable responses to climate change adaptations [23]. AI supplies data analytics with machine learning capability, automating systems that strengthen climate-related urban operations' resilience across various systems. This part describes four fundamental AI approaches that create climate-resistant cities through predictive modeling for threat assessments, as well as AI-run initiatives for dispersing heat and mastering water, and optimizing residential facilities to minimize power usage.

3.1 Predictive Analytics for Climate Risk Assessment

Modern cities face growing dangers from environmental disruptions, including floods along with wildfires and heatwaves, and rising seas. Risk assessment conducted traditionally by using historical information and fixed models struggles to detect the complex and fast-changing climate hazards in their entirety. The sophisticated AI-based predictive analytics present a flexible and scalable solution by combining disparate data sources and continuously learning from new data. Fundamental to AI's predictive potential is the ability to analyze large datasets from satellites, weather stations, IoT sensors, and even social media feeds. These streams serve as real-time sensors for everything from environmental factors to infrastructure status to human activity. Machine learning algorithms can uncover subtle relationships and novel trends that would not be possible to identify by traditional analysis.

AI-enhanced flood prediction models, for instance, can analyze rainfall intensity, soil moisture, river levels, urban drainage capacity, and so on, to predict floods even more accurately and earlier. In cities like Jakarta, such systems have allowed authorities to set up flood defenses and warn the community days before the water rises, preventing damage and death on a massive scale [24]. Likewise, machine learning models for vegetation dryness, wind direction, and temperature have helped predict wildfire risk in California, thus informing preventative strategies like controlled burns and resource allocation. In addition to near-term threats, AI enables long-term climate risk forecasting. AI-powered simulations predict erosion patterns and the impact of rising sea levels for the cities along the coast, helping cities make infrastructure investments and set zoning rules. Urban planners in Copenhagen, for example, are using AI to predict how different sea wall designs will fare under different storm surge scenarios, to maximize protection and minimize cost. However, the problems still existed. Representative, high-quality data are essential so AI models don't perpetuate biases that could stigmatize at-risk groups. Computational costs are high and require hardware and skill. Furthermore, the explainability of AI predictions is important for fostering trust and meaningful decision-making from policymakers and the public.

3.2 AI-Driven Urban Heat Island Mitigation

Urban Heat Islands (UHIs) are a common problem in cities around the globe in which compact urban landscapes and a lack of vegetation lead to high temperatures at the local scale and intensify heat stress, energy demand, and health issues. AI brings new solutions for the understanding, mapping, and mitigation of UHI through the combination of environmental measures and urban design principles [25]. By analyzing thermal satellite imagery, drone surveys, and ground-based sensors, AI produces highly detailed heat maps of the city that show temperature differences at the neighborhood or even street level. These maps allow urban planners to locate hotspots that are most in need of interventions. In Phoenix, for instance, AI-based heat vulnerability assessments helped determine where to strategically locate cooling centers and green space in low-income neighborhoods, leading to measurable reductions in heat-related illness. AI also contributes significantly to enhancing green infrastructure, including parks, green roofs, and urban forests. By modeling how various tree species cool, how dense their foliage is, and how they can be configured in parks, AI is used to most effectively deliver the benefits of shade and evapotranspiration, while factoring in spatial constraints and biodiversity objectives. AI is being used to increase urban tree coverage to make cities more livable, like in Melbourne, where an AI-driven urban forestry program compared control areas to see how much temperatures could be cut in hotspots, lowering the average summer temperature by more than 1 °C.

On top of foliage, AI can encourage the use of heat-reducing, permeable, or reflective materials. Computer vision methods evaluate how reflective pavements and rooftops are, reinforcement learning produces recommendations of where to put cool coatings or permeable surfaces for maximum cooling. Los Angeles' "Cool Streets" initiative, guided by an AI model, has reduced the temperature of road surfaces by as much as 10 degrees Celsius in treated areas. Dynamic shading technologies provide another opportunity for UHI mitigation. AI manages smart awnings, solar panels, and adjustable louvers that change position in real time according to the sun, weather, and moving pedestrians. In Dubai, the energy consumption of outdoor cooling has been drastically decreased due to these systems, increasing comfort and sustainability. Importantly, AI advances equitable UHI management by combining demographic and health indicators with environmental metrics. This means the most vulnerable communities—those who are most exposed to heat places, and often in the greatest proportion are affected by heat, get prioritized interventions, driving climate resiliency, alongside social justice.

3.3 Intelligent Water Management Systems

Urban water management is becoming complicated because of fluctuating climate, increasing population, and the ageing of infrastructure. Cities are confronting two difficult challenges: water scarcity in times of drought and flooding from severe storms. Computational intelligent water management systems give smart, proactive solutions, data related to drive efficiency, reduction of loss, and climate change resilience. These systems are built on top of a network of IOT sensors that are constantly monitoring the rainfall, reservoir levels, pipe pressure, water quality, and consumer usage. This real-time data integrates with AI analytics engines that track anomalies, predict demand, and maximize loads [26]. As shown in Fig. 1.2, an AI-enabled urban water management system is being implemented, and real-time sensor data converges into one data analytics engine for leak detection, storm water management, demand forecast, and water quality monitoring. With the system, repairs happen automatically, prices change in real time, and the government can respond quickly to floods. Through the use of predictive analytics and automation, cities can maximize water efficiency, decrease losses, and improve climate resiliency. One of the applications with a great influence is leakage detection. AI algorithms detect leaks with flow and pressure data, generating alerts in minutes rather than hours and days for utilities to send repair crews before losses rise. The Smart Water Grid of Singapore is one such example of such a capability, with one of the world's lowest levels of non-revenue water losses. When it rains hard, AI models adjust the storm water infrastructure—pumps, valves, retention basins—in real time to stop flooding. Using reinforcement learning, the system can learn the best response while flooding is mitigated, without external intervention.

Those storms have been wreaking less flood damage thanks to Amsterdam's AI-optimized water management system. Another important task is demand prediction. The AI forecasts the consumption of water in an hourly or day-ahead schedule using history, weather prediction, and social factors [27]. This allows utilities to better match supply and demand and to charge dynamic rates that encourage conservation during periods of high demand. Intelligent Demand Management in Cape Town was AI-driven and averted water deprivation in times of drought crisis.

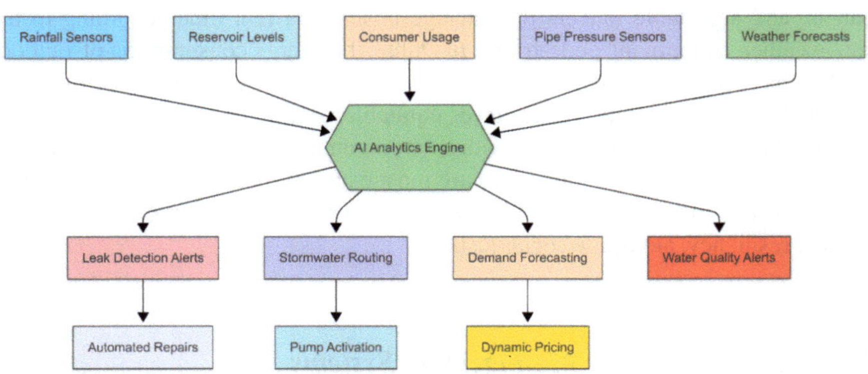

Fig. 1.2 Components of an AI-driven urban water management system

3.4 Smart Building Design and Energy Efficiency

City buildings are some of the biggest consumers and polluters of energy. Artificial intelligence (AI) is transforming building design and operation by delivering adaptive, efficient, and occupant-centric spaces that enhance occupant experience in changing environmental and human situations. Generative design tools powered by AI enable architects and engineers to trial thousands of building permutations in the pursuit of energy efficiency, natural light, ventilation, and comfort for users. These tools emulate the performance of buildings in different climates, enabling us to choose materials, orientations, and layouts that use a minimum of energy and are resilient. When constructed, smart building management systems (BMS) rely on AI for real-time monitoring and control over lighting, climate (HVAC), and more. By recognizing the occupant behavior and considering weather forecasts with these kinds of systems, they can control the settings to save energy without influencing the comfort. For instance, the Edge in Amsterdam uses artificial intelligence to keep its energy use 70% lower than conventional office buildings. Another AI application extending building life and saving costs is predictive maintenance [28]. Through sensors that are mounted in elevators, HVAC units, and structural components, data is fed into AI models that predict when elements will fail so they can be repaired with minimal disruption.

New construction for the AI age, Smart materials are here to fill in the gaps of our AI-driven building systems. Self-healing concretes and adaptive insulation materials controlled by AI can adaptively fix cracks or modify thermal response properties under environmental stimulation. Buildings with this technology installed in Tokyo have shown signs of a prolonged life and minimal maintenance [29]. AI makes incorporating renewable energies and energy storage in buildings easy. Intelligent systems will manage the interplay among rooftop solar panels, battery storage, and electric vehicle charging to balance supply and demand within local areas. Buildings can also engage in peer-to-peer energy trading, in which they sell any excess power back to the grid or neighboring buildings. Prominent examples include Shanghai Tower, where AI manages lighting and elevator scheduling and provides more efficient insulation, decreasing energy consumption by 21% and carbon emissions by 35%. Likewise, Hudson Yards (New York) utilizes AI-supported microgrids and recycling of resources for resilient urban districts.

4 Integration of Quantum Computing and AI

Embedded quantum computing and AI are transforming the paradigm of urban climate resilience and sustainability. As cities face increasing environmental challenges, the collective force of these technologies is facilitating a new wave of solutions that are more exact, adaptable, and scalable than ever [30]. This section examines the intersection of quantum computing and AI through three important lenses:

quantum machine learning for environmental use cases, real-time data processing for urban infrastructure, and advanced decision support systems for climate adaptation.

4.1 Quantum Machine Learning for Environmental Applications

QML, the frontier of data analysis and predictive modeling for challenging environmental problems, is a new approach to data analysis and prediction. Even the power and flexibility of conventional machines are limited by the statistical size and complexity of the climatic and urbanist datasets [31]. Quantum computers, using states of matter to represent data as bits, can process and analyze information in ways classical computers cannot, due to phenomena like superposition and entanglement. This becomes especially important when addressing climate science, since climate models that cover the atmosphere dynamics, ocean currents, and land-use change have an exponential scaling of the number of variables used when the resolution is increased.

Recent studies suggest the potential of QML to expedite decarbonization, optimize energy systems, and improve climate data forecasting. Quantum algorithms are, for example, already working to improve the accuracy and speed of weather forecasting, renewable energy forecasting, load prediction, and carbon price modelling. For energy systems, it allows for distributed frequency control in micro-grids, for energy-efficient optimization of HVAC systems, and for controlling the charging of the electric vehicle infrastructure. In agriculture, quantum neural networks are used to forecast crop disease and optimize planting strategies, and in transportation, hybrid quantum–classical approaches are employed to decide on the optimal placement of electric vehicle charging stations and energy trade between vehicles and microgrids. QML is making great advances in the domain of climate monitoring. Image classification methods to interpret high-resolution satellite imagery for earth observing purposes, specifically to be able to discriminate vegetation from other land types and to monitor land cover changes, have been performed by using quantum-assisted approaches. The advances are important for tracking deforestation, urban sprawl, and the health of ecosystems. For hazard prediction, quantum QML models, such as QSVM and VQC, have been shown to predict earthquakes, asteroids, and other extreme events better and earlier than their classical counterparts. However, QML is still a nascent field, and further studies are necessary to reach quantum supremacy and to develop for deployment at large. Challenges are the current restrictions of quantum resources, the search for efficient quantum algorithms to cope with noisy data, and the processing of quantum and classical processors. Yet the vector is undeniable: as QC matures as a technology, the synergy with AI is likely to open up new frontiers for applications in environmental modeling, material simulation, and climate data analysis, helping cities to foresee and mitigate climate risks in ways that would have been impossible up to now.

4.2 Real-Time Data Processing for Urban Infrastructure

The operation of contemporary urban infrastructure produces massive streams of real-time data collected by IoT sensors, smart meters, traffic cameras, environmental monitors, and others [32]. Reacting to and processing this information in real-time is important for energy optimization, traffic flow, water control, and emergency response. Quantum computing combines with AI to provide a breakthrough in the ability to process these huge data sets and obtain insights that can be acted on immediately, avoiding the lag time traditional systems produce.

In the smart city scenario, traffic management, decongestion pollution control—themes that impact the health and wellbeing of citizens—are only some of the fields where quantum-enhanced AI systems are currently being investigated. For instance, quantum algorithms could process live traffic data, simulate millions of routing possibilities, and change traffic signals on the fly to reduce bottlenecks and emissions. Also in public transportation, the use of quantum-inspired algorithms has had an impact by organizing bus routes and schedules more efficiently, saving time and fuel." Simulations have demonstrated that such traffic solutions can cut traffic by 40% across a city and, in turn, help deliver extensive benefits to both mobile societies and the environment.

Energy control is another important example. With the move to smart grids and increasing amounts of renewable energy in the mix, balancing supply becomes all the more difficult. Quantum AI can provide real-time analysis of data from DERs and forecast fluctuations in renewable generation down to the second, ensuring that the grid can operate as efficiently and effectively as possible, even balancing PG on-site. It is an effective elimination of intentional outages and environmentally friendly, thanks to its reliable electricity supply. For water management, quantum-enhanced AI processes sensor data to identify leaks, predict demand, and maintain stormwater infrastructure, avoiding water shortages or floods. Quantum AI could be used in public safety and emergency response. By analyzing live feeds from surveillance cameras, social media, and emergency sensors, they are being used to identify abnormal patterns that predict incidents and to coordinate rapid responses. For example, quantum AI may enable emergency services to dispatch more efficiently in disaster situations, resulting in faster response times and more lives saved. While more and more urban infrastructure is interconnected and data-rich, marrying quantum computing with AI will help us create cities that are not only smarter but also more resilient to climate shocks and stresses.

4.3 Decision Support Systems for Climate Adaptation

Successful adaptation to climate change depends on accurate, timely, and locally relevant decision-making. The fusion of QIP with AI is transforming the decision support systems of tomorrow, where the simulation of urban scenarios, the assessment of adaptation strategies, and the resource allocation planning are all carried out in urban scenarios with an amazing performance and precision. Digital twins—digital models of cities that replicate their physical, social, and economic systems-are leading this transformation. When fueled with the additional horsepower of quantum computing, these digital twins could run very detailed, long-term simulations that model the likely impacts of natural disasters on the climate, such as sea-level rise, extreme heat events, or stress on infrastructure under different scenarios [33]. Machine learning interprets the results, highlighting vulnerabilities and recommending interventions from the deployment of green infrastructure to the targeting of flood defences.

Participatory urban planning is also supported by quantum-enhanced decision support systems. Systems that draw from a wide array of data sources citizen science, demographics, and sensor input, act to ensure adaptation measures are just and address the concerns of the whole community. For instance, quantum AI can discover concealed patterns of vulnerability, for example, identifying where investments need to be directed in the most underserved neighborhoods and how resilience strategies should not inadvertently widen social inequality. "Policy and governance listen too. In the policy and governance space, quantum-analytic tools empower city leaders to test the trade-offs and long-term advantages of specific adaptation pathways. These models can forecast the consequences of policy decisions decades into the future, making it easier to develop short-term policies that point the way to long-term sustainability. In agriculture, for example, quantum machine learning can process data on soil health, weather patterns, and crop genetics to guide drought-resistant planting in areas sensitive to climate change, helping to secure food in vulnerable regions.

Bringing together quantum computing and AI in decision support is not devoid of obstacles. Algorithmic transparency, privacy, and open collaboration across disciplines are necessary for trust and maximum impact. But early test projects in cities like Singapore and Amsterdam have shown that, when used properly, a smart city system can have real, measurable benefits, from reducing waste of power to making cities more prepared for natural disasters to lowering their carbon emissions. The relationship between quantum computing and AI is a transformative shift for urban climate resilience. With the ability to model the environment with higher fidelity, optimize infrastructure in real-time, and provide advanced decision support, this combination allows cities to deliver beyond comments and see the possibility of a sea-change. As quantum technologies mature, the relationship between quantum and AI will form the nexus of smarter, safer, and more sustainable future cities.

5 Case Studies

5.1 Smart Cities Implementation Examples

Booming innovations in the field of technology, sustainability, resiliency, and urban planning are underway in cities around the world [34]. These cases regularly illustrate in what way smart city frameworks impact climate issues and enhance the quality of life.

(i) **Singapore: The Smart Nation Initiative**

Singapore's Smart Nation and Technology-Using Tech to Create A Sustainable Urban Environment Singapore's Smart Nation is a prime example of how a city can harness advanced technologies to ensure its rapid urbanization does not come at the environmental expense. With a shortage of land, temperatures on the rise, and flood risks to manage, the city-state deployed a network of IoT sensors to track air quality, energy use, and water levels in real time. It's a 3D dynamic model of the city, a digital twin, simulating climate impacts such as periods of extreme rainfall and heatwaves, and allowing planners to test infrastructure designs and evacuation routes. For example, flood-prone neighborhoods were identified from the digital twin, and smart drainage systems were implemented that decreased the number of flood events by 40%. The city also employs AI to optimize public transit, reducing commutes by 25% and decreasing emissions from private cars.

(ii) **Amsterdam: Climate-Adaptive Urban Design**

Using AI and quantum-inspired algorithms for a carbon-neutral Amsterdam Circular economy is getting help from AI and quantum-inspired algorithms to achieve carbon neutrality. A project called "Rainproof Amsterdam" employs quantum-optimized drainage systems that control stormwater and avert flooding during heavy rains. Artificial intelligence processes data from smart meters to stabilize energy grids so that 80% of homes use renewable energy. Amsterdam's "Cool Streets" program, in which reflective pavements and AI-controlled shading systems are used, cooled surface temperatures by up to 10 °C in heat-sensitive neighbourhoods. These initiatives have placed Amsterdam at the forefront of climate-resilient urban planning: heat-related hospital admissions in the city dropped 35% in 2023.

(iii) **Songdo, South Korea: A Tech-Driven Urban Laboratory**

(Songdo International Business District, developed on reclaimed land, epitomizes the melding of smart technology and sustainability.) The city's centralised AI system controls everything from tossing waste, pneumatic trash chutes that shoot rubbish to sorting centres, to traffic lights that respond to real-time congestion data. Buildings have solar panels and AI-controlled HVAC systems, reducing energy consumption by 45%. Songdo's green infrastructure with roof gardens and automatic irrigation helps cool urban heat islands and encourages biodiversity. Even as it is prospering, there are concerns that Songdo will struggle to generate community engagement, reminding us of the need to reconcile technological efficiency with social inclusivity.

(iv) **Copenhagen: Carbon–Neutral Urban Ecosystems**

Copenhagen wants to be the world's first carbon-neutral capital by 2025, and data has become an unlikely superhero in its quest. The Climate Adaptation Plan applies AI to model sea-level rise and storm surges, informing the construction of adaptation barriers such as the Enghave Park water reservoir, which holds 23 million liters of stormwater. Society is based entirely on renewable energy; artificial intelligence is used to optimize bus and train schedules to the point where 95% of journeys will be made on time. Fueled by waste-to-energy plants, Copenhagen's district heating system provides carbon-neutral heat to 98% of the city's households. That has led to a 65% reduction in the amount of carbon entering Chisinau's coffers since 2010, and it's a reminder that systemic change is possible with integrated planning.

5.2 Quantum-AI Solutions for Extreme Weather Events

The countries, on average, are getting warmer, and many extreme weather patterns like hurricanes, floods, and heat waves are becoming more severe due to climate change [35]. Cities are deploying quantum-AI systems to forecast, plan for, and mitigate these disasters with a precision and rapidity never before possible.

(i) **New York City: Flood Prediction and Mitigatio**n

The New York "Climate Ready NYC" project uses quantum machine learning to forecast coastal flooding. Simulating storm surges and tidal patterns with quantum algorithms, the city predicts flooding risks for each neighborhood three days in advance. In 2026, during Hurricane Ida, the system singled out vulnerable subway stations, leading to preemptive closures that avoided $500 million in damage. AI-powered drones now monitor the city's 520-mile coastline daily for erosion and prioritize areas needing repair. 50% of the accuracy of flood prediction has been enhanced through integrating quantum computing, enabling the use of mobile flood barriers in vulnerable areas such as Lower Manhattan.

(ii) **Tokyo: Heatwave Management**

Tokyo's AI-Quantum Heat Resilience Project grapples with deadly urban heat islands. Quantum algorithms take in satellite thermal data and building energy use to generate real-time, hyper-local heat maps every 15 minutes. In the 2025 heatwave, it detected areas of the city that went over 40 °C, and misting stations and cooling buses were sent to assist vulnerable people. AI also helped optimize power grids to prevent blackouts, which in turn reduced heat-related deaths by 30%. The city's "Green Curtain" project, which deploys artificial intelligence to choose fast-growing vines to cover building facades, has helped reduce temperatures by 8° inside public schools.

(iii) **Rotterdam: Storm Surge Defense**

Rotterdam's blok-kade Maeslantkering, a giant storm barrier that can be opened and closed, for example, is controlled by a quantum-AI system that calculates tide levels and wind speeds in real time. During a 2024 North Sea storm, the wall shut automatically 12 hours before human forecasts advised it, saving the city from disastrous flooding. The AI model, the product of an FIT PhD thesis, built on 100 years of tidal data, cut false alarms by 40%, saving the gate's mechanical systems for a real emergency. The city also deploys quantum simulations to try different future sea-level rise scenarios as it designs floating neighborhoods and amphibious homes.

(iv) **Mumbai: Monsoon Preparedness**

Mumbai's "Flood AI" uses quantum computing and IoT sensors to predict monsoon flooding. The technology uses rainfall intensity, drainage, and tide data to produce street-level flood warnings. In 2023, it re-routed emergency services across dry zones in a notorious monsoon year, cutting response times by 55%. The city's AI-supported disaster recovery drones dropped supplies to 10,000 marooned residents, demonstrating how technology can help build resilience in low-resource settings.

5.3 Sustainable Urban Transportation Systems

CO_2 released by transportation makes up 20% of all global emissions. Smart cities are reinventing mobility with AI-driven networks, EVs, and inclusive public transit [36].

(i) **Singapore: Congestion Pricing and EV Integration**

Singapore's AI-powered Electronic Road Pricing (ERP) system can automatically adjust toll charges based on actual traffic conditions, leading to a 45% reduction in peak-hour congestion. The quantum-AI platform of the city focuses on the optimization of EVCS locations, taking into account grid capacity and driver behaviors. Sixty percent of new cars sold in 2024 were electric, thanks to a 500 percent increase in charging infrastructure. AI-powered self-driving electric shuttles cover 30% of public transit routes, reducing emissions by 18,000 tons per year.

(ii) **Oslo: Zero-Emission Public Transit**

All of Oslo's public transportation, meanwhile, runs on renewable energy, with AI regulating the schedules of its trams and buses to ensure the least amount of waiting time. The city's "Green Wave" system already coordinates traffic lights for electric buses to keep them moving and cut energy use by 20%. Oslo is also a quantum master that uses quantum algorithms for bike-sharing networks, which increased ridership by more than 70% since 2022. It's left Oslo, the first major city where more than 90% of trips are not fossil-fuelled.

(iii) **Barcelona: Superblocks and Mobility Hubs**

Barcelona envisions restricting cars in 9-block zones through its "superblocks" initiative to reclaim streets for pedestrians and bikers. Through artificially intelligent analysis of traffic patterns, it creates these zones, cutting car usage by 30% and NO_2 levels by 45%. In the city's mobility hubs, buses, bikes, and e-scooters are all bundled into one app, and quantum algorithms predict demand, moving vehicles around so availability is even. During heatwaves, the system will recommend that cyclists take shaded routes, providing an example of how climate adaptation can be bundled into urban mobility.

(iv) **Curitiba, Brazil: Bus Rapid Transit (BRT) Innovation**

Curitiba's BRT system, an inspiration for growing cities, employs AI to coordinate its 340-strong fleet. At stops, real-time sensors measure passengers and dynamically schedule to alleviate crowding. Hybrid biofuel and electric buses have cut emissions by 60% compared with diesel models. The city's quantum-fueled routes have reduced the average commute to 28 minutes, and as a result 80% of its citizens opt for mass transportation over private vehicles.

6 Challenges and Future Directions

A new era of opportunity and complexity for quantum computing and artificial intelligence (AI), when combined with efforts to achieve climate resilience and sustainable urban development, is unfolding. The promise of these technologies to transform how cities forecast, respond to, and reduce climate risks is vast. However, this promise is met with several technical, practical, and ethical challenges. Identifying these constraints and navigating the way forward is critical to realizing the full potential of quantum and AI solutions for climate and society.

6.1 Current Limitations and Technical Barriers

(i) **Quantum Computing Challenges**

- Quantum computing is very much in its infancy.
- Current devices are "noisy intermediate-scale quantum" (NISQ) machines that are noisy and prone to environmental errors.
- Sustaining quantum coherence without decay is challenging since small independent perturbations disrupt the calculations.
- Scalability: Constructing fault-tolerant, scalable quantum computer systems with 1000 + qubits remains the major challenge/community homework we need to address.

- Quantum computing's utility for complex climate modeling is limited by current capacity.

(ii) **AI Challenges**

- AI models need large-scale, high-quality datasets, which are often missing or not consistent in environmental and urban domains.
- Scarcity of data is particularly prevalent in the Global South, where sensor coverage and digital infrastructure fall short.
- A bias toward data-rich areas may miss vulnerable, less-documented communities.
- Urban systems incorporate volatile human, political, and cultural elements that AI has trouble modelling.
- In advanced AI, opacity ("black box") impedes trust from policymakers and the public.

(iii) **Quantum-AI Integration Issues**

- There are still many quantum algorithms that are experimental with unknown real-world behavior.
- Quantum and classical systems have to be coupled, creating "bottlenecks for the transfer of data" that must be bridged by "specialized interfaces".
- Environmental sustainability becomes a concern when systems scale with the accompanying increased resource and energy requirements.

(iv) **Institutional and Financial Barriers**

- High costs of quantum hardware and limited availability of skilled professionals.
- Expensive and has few experts in quantum hardware.
- The lack of mature, commercial applications will make deployment on a large scale difficult.
- Potential to compound global inequalities, as benefits may flow to more affluent areas.

6.2 *Emerging Technologies and Research Opportunities*

(i) **Hybrid Quantum–Classical Algorithms**

- Bridging the gap between quantum and classical computing.
- Effectively used for solving large, complicated process optimization and simulation problems.
- Excellent for climate modeling and urban planning.

(ii) **Quantum Machine Learning (QML)**

- Applies quantum parallelism for pattern recognition, clustering, and high-dimensional data analysis.
- Surpasses state-of-the-art methods in various problems.
- Allows better forecasts of weather, extreme events, and climate variations.

(iii) **Optimization of Smart Urban Systems**

- Dynamic management in real-time of energy grids, transport, and water networks.
- Quantum-AI systems sift through data from millions of sensors.
- Resilience of the city is increased through adaptive decision-making to cope with daily variations and extreme events.

(iv) **Explainable AI (XAI) developments**

- Boosts transparency by clarifying the decision-making process behind models.
- Highlights important features, adds confidence intervals, and returns a human-readable explanation.
- Establishes confidence and promotes stakeholder interaction in AI-based systems.

(v) **Research on the Sustainability of PPPs**

- Design of energy-efficient quantum devices with limited use of rare materials.
- Lifecycle assessments are used to find ways to reduce their impact.
- Encourages the development of environmentally sensitive high-performance computing.

(vi) **Global Collaboration**

- Cross-border collaborations take open source tools, shared data, and common regulations further.
- Promotes even distribution of technological benefits.
- Supports research and deployment by sharing knowledge and resources.

(vii) **Education and training of the workforce**

- Increasing demand for experts in quantum computing and AI.
- Interdisciplinary academic areas between theory and practice.
- Nurtures the next generation of professionals equipped to meet the global challenges of climate and technology.

6.3 Policy Recommendations for Implementation

(i) **R&D Investment**: Dedicate funding to basic and applied research in quantum and AI in climate resilience-specific technologies, including pilot projects and testbeds [37].
(ii) **Public–Private Partnership**: Promote partnerships to speed innovation and scale models that work.
(iii) **Standardization and Benchmarking**: Determine quantum performance, energy use, and environmental effects, and provide comparable metrics for benchmarking.
(iv) **Data Infrastructure**: Develop comprehensive sensor networks, integrated data platforms, and open data initiatives; fill data coverage gaps and data deserts in underserved areas.
(v) **Algorithm Transparency**:
(vi) Require AIs and quantum-AIs used in crucial decision-making to be explainable, interpretable, and have an independent review process.
(vii) **In-depth long-term integration of sustainability**: Lifecycles, energy efficiency, and green procurement standards into all stages of tech development.
(viii) **Engagement of the Public**: Advocate for stakeholder participation, citizen science, and clear risk–benefit communication [38].
(ix) **International Collaboration**: Achieve regulatory convergence and share ethical compass through global partnerships.
(x) **Education and Workforce Development**: Form interdisciplinary training, certification programs, and lifelong learning to generate relevant expertise.

7 Conclusion

The fusion of quantum computing with artificial intelligence (AI) represents a quantum shift in climate-resilient urban planning. This convergence is not about just a tech update, but a paradigm shift in the way cities decide, plan, and respond to climate change. Quantum computing's clear advantage in dealing with complex multi-variable nonlinear datasets has already shown applications in climate modelling, urban microclimate simulation, and resource optimisation that will soon be capable of competing within the near-edge and edge computing markets. Contrary to classical systems, quantum algorithms can simulate complex environmental systems, providing policymakers with the means to foresee risks, test adaptive strategies, and make decisions that are sustainable in the long term. At the same time, AI is injecting intelligence into the nuts and bolts of urban systems—helping cities cope with challenges as enormous as climate change and as fiddly as making sure that street sweepers pick up detritus efficiently. Its edge in adapting to different sources of data and ever-changing patterns is what keeps urban areas dynamic and functioning. Combining AI with quantum can lead to real-time scenario modeling and powerful DSS (Decision Support) that can foster a proactive rather than reactive approach to dealing with climate. Furthermore, developments such as digital twins and quantum-aided machine learning enable open governance and inclusive planning, a means of opening access to climate solutions for everyone. This awesome synergy is not only transforming city resilience, but it is also creating equitable adaptation for all its citizens and providing the foundations for great, sustainable, and habitable cities in the future.

References

1. Munawar, G., Surendro, K.: Utilizing quantum algorithms to achieve carbon neutrality in urban areas: a systematic review. Alex. Eng. J. (2024). https://doi.org/10.1016/j.aej.2024.09.043
2. Srivastava, A., Maity, R.: Assessing the potential of AI–ML in urban climate change adaptation and sustainable development. Sustainability (2023). https://doi.org/10.3390/su152316461
3. Ho, K., Chen, K., Lee, L., Burt, F., Yu, S., Lee, P.: Quantum computing for climate resilience and sustainability challenges. In: 2024 IEEE International Conference on Quantum Computing and Engineering (QCE), vol. 02, pp. 262–267 (2024). https://doi.org/10.1109/QCE60285.2024.10289
4. Ajagekar, A., You, F.: Quantum computing and quantum artificial intelligence for renewable and sustainable energy: a emerging prospect towards climate neutrality. Renew. Sustain. Energy Rev. (2022). https://doi.org/10.1016/j.rser.2022.112493
5. Son, T., Weedon, Z., Yigitcanlar, T., Sanchez, T., Corchado, J., Mehmood, R.: Algorithmic urban planning for smart and sustainable development: systematic review of the literature. Sustain. Cities Soc. (2023). https://doi.org/10.1016/j.scs.2023.104562
6. Bibri, S., Huang, J., Jagatheesaperumal, S., Krogstie, J.: The synergistic interplay of artificial intelligence and digital twin in environmentally planning sustainable smart cities: a comprehensive systematic review. Environ. Sci. Ecotechnol. **20** (2024). https://doi.org/10.1016/j.ese.2024.100433
7. Almulhim, A.: Building urban resilience through smart city planning: a systematic literature review. Smart Cities (2025). https://doi.org/10.3390/smartcities8010022
8. He, W., Chen, M.: Advancing urban life: a systematic review of emerging technologies and artificial intelligence in urban design and planning. Buildings (2024). https://doi.org/10.3390/buildings14030835
9. Rahman, S., Alkhalaf, O., Alam, M., Tiwari, S., Shafiullah, M., Al-Judaibi, S., Al-Ismail, F.: Climate change through quantum lens: computing and machine learning. Earth Syst. Environ. (2024). https://doi.org/10.1007/s41748-024-00411-2
10. Gao, P., Adnan, M.: Overview of emerging electronics technologies for artificial intelligence: a review. Mater. Today Electron. (2025). https://doi.org/10.1016/j.mtelec.2025.100136
11. Hao, H., Wang, Y., Chen, J.: Empowering scenario planning with artificial intelligence: a perspective on building smart and resilient cities. Engineering (2024). https://doi.org/10.1016/j.eng.2024.06.012
12. Bonab, A., Fedele, M., Formisano, V., Rudko, I.: Urban quantum leap: a comprehensive review and analysis of quantum technologies for smart cities. Cities (2023). https://doi.org/10.1016/j.cities.2023.104459
13. Rajkumar, N., Viji, C., Latha, P., Vennila, V., Shanmugam, S., Pillai, N.: The power of AI, IoT, and advanced quantum based optical systems in smart cities. Opt. Quant. Electron. (2024). https://doi.org/10.1007/s11082-023-06065-0
14. Lanbaran, N., Naujokaitis, D., Kairaitis, G., Jenciūtė, G., Radziukynienė, N.: Overview of startups developing artificial intelligence for the energy sector. Appl. Sci. (2024). https://doi.org/10.3390/app14188294

15. Li, S.: Analysis of the impact of quantum social science on future urban governance. Quantum Soc. Sci. (2025). https://doi.org/10.6914/qss.01015
16. Hadiyana, T., Ji-Hoon, S.: AI-driven urban planning: enhancing efficiency and sustainability in smart cities. ITEJ (Information Technology Engineering Journals) (2024). https://doi.org/10.24235/itej.v9i2.124
17. Khatoon, A., Riaz, R.: Quantum computing impacts on smart city cybersecurity through resilient defense framework. Ubiquitous Technol. J. (2025). https://doi.org/10.71346/utj.v1i1.8
18. Bonab, A., Fedele, M., Formisano, V., Rudko, I.: In complexity we trust: A systematic literature review of urban quantum technologies. Technol. Forecast. Soc. Chang. (2023). https://doi.org/10.1016/j.techfore.2023.122642
19. Ashwani, S., Tripathy, A., Karna, S., Jahanve, P., Rajagopal, S.: Quantum computing for climate change: a comprehensive review of current applications, challenges, and future directions. In: 2024 15th International Conference on Computing Communication and Networking Technologies (ICCCNT), pp. 1–7 (2024). https://doi.org/10.1109/ICCCNT61001.2024.10724347
20. Bibri, S.: Data-driven smart sustainable cities of the future: urban computing and intelligence for strategic, short-term, and joined-up planning. Comput. Urban Sci. **1** (2021). https://doi.org/10.1007/s43762-021-00008-9
21. Gupta, R.: Quantum computing and AI. Int. J. Multidisc. Res. (2024). https://doi.org/10.36948/ijfmr.2024.v06i03.17789.
22. Al-Raeei, M.: The smart future for sustainable development: Artificial intelligence solutions for sustainable urbanization. Sustain. Dev. (2024). https://doi.org/10.1002/sd.3131
23. Sanchez, T., Shumway, H., Gordner, T., Lim, T.: The prospects of artificial intelligence in urban planning. Int. J. Urban Sci. **27**, 179–194 (2022). https://doi.org/10.1080/12265934.2022.2102538
24. Anwar, M., Sakti, L.: Integrating artificial intelligence and environmental science for sustainable urban planning. In: IAIC Transactions on Sustainable Digital Innovation (ITSDI) (2024). https://doi.org/10.34306/itsdi.v5i2.666
25. Varzeshi, S., Fien, J., Irajifar, L.: Integrating technology and urban resilience: a comprehensive analysis of smart city initiatives in Sydney. Sustainability (2024). https://doi.org/10.3390/su162410967
26. Prodanović, V., Bach, P., Stojković, M.: Urban nature-based solutions planning for biodiversity outcomes: human, ecological, and artificial intelligence perspectives. Urban Ecosyst. (2024). https://doi.org/10.1007/s11252-024-01558-6
27. Nammouchi, A., Kassler, A., Theocharis, A.: Quantum machine learning in climate change and sustainability: a short review. Proceedings of the AAAI Symposium Series (2024). https://doi.org/10.1609/aaaiss.v2i1.27657
28. Kesavan, H., Choudhary, S., Khan, S., R, S., Sharma, A.: AI based urban resilience planning: opportunities and challenges. J. Environ. Earth Sci. (2024). https://doi.org/10.30564/jees.v6i2.6681
29. Sun, L., Li, H., Nagel, J., Yang, S.: Convergence of AI and urban emergency responses: emerging pathway toward resilient and equitable communities. Appl. Sci. (2024). https://doi.org/10.3390/app14177949
30. Schwabe, M., Pastori, L., De Vega, I., Gentine, P., Iapichino, L., Lahtinen, V., Leib, M., Lorenz, J., Eyring, V.: Opportunities and challenges of quantum computing for climate modelling (2025)
31. Almulhim, A., Cobbinah, P.: Framing resilience in Saudi Arabian cities: on climate change and urban policy. Sustain. Cities Soc. (2024). https://doi.org/10.1016/j.scs.2024.105172
32. Reddy, C.: Exploring advanced techniques in artificial intelligence for environmental monitoring and climate change management. Int. J. Multi. Res. (2025). https://doi.org/10.36948/ijfmr.2025.v07i01.34529
33. Venigandla, K., Vemuri, N., Aneke, E.: Empowering smart cities with AI and RPA: strategies for intelligent urban management and sustainable development. Int. J. Sci. Res. Manag. (IJSRM) (2024). https://doi.org/10.18535/ijsrm/v12i04.ec02

34. Camponeschi, C.: The resilience of urban entrepreneurialism: Challenging the 'neoliberal turn' of municipal climate planning. Environ. Plann. C: Polit. Space **41**, 130–147 (2022). https://doi.org/10.1177/23996544221117942
35. Il'ina, I., Zhdanchikov, P.: Special aspects of urban planning in the context of quantum technologies development. Regional Econ. Theory Practice (2024). https://doi.org/10.24891/re.22.8.1404
36. Atadoga, A., Ike, C., Asuzu, O., Ayinla, B., Ndubuisi, N., Adeleye, R.: The intersection of AI and quantum computing in financial markets: a critical review. Comput. Sci. IT Res. J. (2024). https://doi.org/10.51594/csitrj.v5i2.816
37. Ye, X., Niyogi, D.: Resilience of human settlements to climate change needs the convergence of urban planning and urban climate science. Comput. Urban Sci. **2** (2022). https://doi.org/10.1007/s43762-022-00035-0
38. Kumar, D., Bassill, N.: Analysing trends of computational urban science and data science approaches for sustainable development. Comput. Urban Sci. (2024). https://doi.org/10.1007/s43762-024-00142-0

Role of Bioenergy and Green Fuels to Achieve Climate Resilience

Nguyen Minh Viet, Tran Thi Viet Ha, Nguyen Thi Lan Huong, Nguyen Thi An Hang, Pham Tien Thanh, Nguyen Tien Dung, Nguyen Ngoc Vinh, Pham Huynh Thanh Trang, and Nguyen Minh Phuong

Abstract The increasing impacts of climate change have recently intensified the global pursuit of sustainable and resilient energy solutions. Bioenergy and green fuels have emerged as prospective replacements to fossil fuels, offering significant potential to reduce greenhouse gas emissions, enhance energy security, and promote environmental sustainability. As renewable alternatives derived from biological sources, bioenergy and green fuels are vital in bridging the transition toward a low-carbon and circular economy. This chapter examines the diverse role of bioenergy and green fuels in fostering climate resilience. Initially, it delineates the basic principles and typologies of bioenergy resources and green fuel technologies. After that, the chapter analyzes the climate-related advantages of bioenergy and green fuels, focusing on carbon neutrality, waste valorization, and opportunities for rural economic development. Furthermore, the technical, financial, and policy-related impediments constraining widespread implementation are also addressed, and the key advancements propelling the sector are highlighted in this chapter. By integrating scientific, technological, and policy perspectives, this work aims to provide a holistic understanding of bioenergy and green fuels' current and future role, then move forward to support informed decision-making and strategic planning for sustainable energy transitions worldwide.

N. Minh Viet · P. H. Thanh Trang
VNU Key Laboratory of Advanced Material for Green Growth, VNU University of Science, Hanoi, Vietnam

T. Thi Viet Ha (✉) · N. T. An Hang · P. Tien Thanh · N. Tien Dung · N. Ngoc Vinh
Faculty of Advanced Technology and Engineering, VNU Vietnam Japan University, Hanoi, Vietnam
e-mail: ttv.ha@vju.ac.vn

N. Thi Lan Huong
Thuyloi University, Hanoi, Vietnam

N. Minh Phuong
Faculty of Chemistry, VNU University of Science, Hanoi, Vietnam

A. L. Srivastav et al. (eds.), *Climate Resilience: Impact of Quantum Computing and Artificial Intelligence on Urban Planning*, Studies in Systems, Decision and Control 637, https://doi.org/10.1007/978-3-032-06791-3_9

1 Introduction

Climate change threatens ecosystems, economies, and human well-being worldwide. Building climate resilience—enhancing the capacity of natural and human systems to absorb, adapt to, and recover from climate-related shocks—is essential for sustainable development. Transitioning to low-carbon energy sources is a critical component of this endeavor. Bioenergy and green fuels, derived from renewable biological resources, offer promising pathways to reduce greenhouse gas emissions and bolster climate resilience [45].

Bioenergy encompasses a range of technologies that convert biomass into usable energy forms, such as electricity, heat, and transportation fuels. Studies indicate that sustainable bioenergy deployment can contribute significantly to climate change mitigation by displacing fossil fuels and reducing net greenhouse gas emissions [16]. For instance, bioenergy with carbon capture and storage has been identified as a potential negative emissions technology, capable of removing carbon dioxide from the atmosphere while providing energy [43].

Green fuels, including advanced biofuels from non-food biomass, offer additional benefits. They can reduce reliance on fossil fuels in sectors that are challenging to decarbonize, such as aviation and heavy transport [35]. Moreover, developing bioenergy systems can stimulate rural economies, enhance energy security, and promote sustainable land management practices [36].

However, the expansion of bioenergy must be managed carefully to avoid adverse environmental and social impacts. Concerns have been raised regarding land-use changes, biodiversity loss, and competition with food production [53]. Therefore, rigorous sustainability assessments and governance frameworks are necessary to ensure that bioenergy contributes positively to climate resilience.

This paper explores the multifaceted role of bioenergy and green fuels in achieving climate resilience. It examines their potential contributions to greenhouse gas mitigation, energy security, and sustainable development while addressing the challenges and trade-offs associated with their implementation.

2 Understanding Bioenergy and Green Fuels

2.1 Biofuel

Unlike other forms of renewable energy, biomass can be directly transformed into liquid fuels—known as biofuels—which can be used to meet the demands of the transportation sector. Biofuels are renewable liquid or gaseous fuels produced from various biological sources, including crops, municipal waste, and byproducts from agriculture and forestry. They are designed to serve as partial or complete alternatives to conventional petroleum-based fuels, particularly in transportation engines. Because biofuels are biodegradable and derived from renewable resources, they offer environmental advantages, such as a closed carbon cycle and reduced emissions compared to fossil fuels. In the Americas, ethanol is the most well-known biofuel, while in Europe, Asia, and increasingly in Brazil, other biofuels, mainly biodiesel, are more widely used. Both ethanol and biodiesel are considered first-generation biofuels. New biofuels, such as cellulosic ethanol and bio-derived dimethyl ether (bioDME), are emerging and classified as second and third-generation biofuels.

2.1.1 First-Generation Biofuels

First-generation biofuels are derived from biomass sources commonly used as food, such as corn, soybeans, and sugarcane. The production of these biofuels involves processes like fermentation or chemical conversion, which transform the sugars, starches, or oils found in the raw materials into liquid fuels. In terms of technology and market presence, first-generation biofuels are already well-developed. For example, corn-based ethanol is the most prevalent type in the United States and is frequently blended with gasoline distributed nationwide.

Regarding fuel properties, bioethanol has a high octane number, which enhances engine performance and efficiency when blended with gasoline. However, it also has a lower energy content per liter than traditional gasoline. Biodiesel, on the other hand, is compatible with conventional diesel engines and has favorable lubricity, which can reduce engine wear. Biodiesel also has a higher flash point than petroleum diesel, making it safer to handle and transport [29]. Nevertheless, both bioethanol and biodiesel can present challenges, such as their tendency to absorb water (bioethanol) or the formation of deposits in engines at low temperatures (biodiesel) [11].

The main applications of first-generation biofuels are in the transportation sector, where they are used either as pure fuels or as blending components with conventional fossil fuels. For instance, bioethanol is commonly mixed with gasoline at various concentrations (E10, E15, or E85, representing 10%, 15%, or 85% ethanol, respectively) [4]. Similarly, biodiesel is blended with petroleum diesel in ratios such as B5 or B20. These blends help reduce tailpipe emissions of carbon monoxide, particulates, and certain hydrocarbons, improving urban air quality [50]. Beyond transportation, first-generation biofuels have also found applications in small-scale power generation and heating in some regions [24].

However, the production of first-generation biofuels is closely linked to several sustainability challenges. One primary concern is their competition with food crops for arable land, water, and other agricultural inputs, which can increase food prices and threaten food security, especially in vulnerable regions [53]. Additionally, expanding the cultivation of energy crops for biofuel production may lead to indirect land-use changes, such as deforestation or conversion of grasslands, which can result in significant releases of carbon stored in soils and vegetation, undermining some of the climate benefits attributed to these fuels [28]. Another environmental concern is the high demand for fertilizers and pesticides associated with intensive monoculture cropping systems, which can contribute to water pollution and biodiversity loss [11].

Despite these limitations, first-generation biofuels are essential in the global transition to renewable energy. Still, their future expansion is expected to be limited as attention shifts to more sustainable alternatives, such as second- and third-generation biofuels.

2.1.2 Second-Generation Biofuels

Second-generation biofuels, often advanced biofuels, are produced from non-food-based materials like agricultural byproducts, perennial grasses, dedicated energy crops, and forestry residues. By relying on these more sustainable resources, they seek to overcome the drawbacks associated with first-generation biofuels. The raw materials used for producing second-generation biofuels are generally grouped into three categories: homogeneous feedstocks (such as pure wood chips), quasi-homogeneous sources (like agricultural or forest byproducts), and non-homogeneous materials (for example, municipal solid waste).

The production of second-generation biofuels involves more complex technologies compared to first-generation processes. Lignocellulosic biomass is composed of cellulose, hemicellulose, and lignin, making it more breakdown-resistant. As a result, advanced pretreatment methods, such as enzymatic hydrolysis or thermochemical conversion (including gasification and pyrolysis), are required to release fermentable sugars or to produce synthesis gas, which can then be converted into liquid fuels like cellulosic ethanol, Fischer–Tropsch diesel, or bio-oil [42]. These biofuels generally have a smaller carbon footprint than their first-generation counterparts. They offer significant potential for greenhouse gas mitigation, mainly when produced sustainably and integrated with practices such as crop residue management and soil carbon sequestration [16].

Second-generation biofuels minimize competition with food resources by using waste products as raw materials and improve sustainability. They are particularly attractive for decarbonizing hard-to-electrify sectors, including aviation and long-haul transport. They have been supported by policy incentives and demonstration projects in the United States, European Union, and several other countries [3, 16]. However, commercial-scale deployment still faces high production costs, technological barriers, supply chain logistics, and further conversion efficiency improvements [10]. Further advancements in research and technology are necessary to optimize these crops' production, transportation, and conversion into usable energy. Additionally, cultivating even second-generation biofuel crops may still result in compromises regarding existing land use, affecting habitats, wildlife, and ecosystem services. It will require resources like labor, equipment, fuel, water for irrigation, and fertilizers, thereby competing with other agricultural needs.

2.1.3 Third-Generation Biofuels

Third-generation biofuels are mainly sourced from algae and cyanobacteria, organisms capable of generating alcohols and lipids suitable for fuel production. Various types of marine biomass—such as seaweed, water hyacinth, diatoms, duckweed, kelp, and salvinia—have shown significant potential for biofuel applications, particularly biodiesel creation. Algae are considered an ideal raw material for biofuels because they possess a high oil yield, grow rapidly, and do not significantly compete with conventional agriculture for land, freshwater, space, sunlight, or nutrients.

Algae can be cultivated on non-arable land and utilize saline or wastewater, thus avoiding competition with food crops and freshwater resources. These microorganisms can convert sunlight and carbon dioxide directly into lipids, carbohydrates, and other energy-rich compounds, which can then be processed into various types of biofuels such as biodiesel, bioethanol, biogas, and even jet fuels [7]. The productivity of algae is remarkably high, some species can yield up to 20 times more oil per unit area compared to traditional oilseed crops, making them an attractive feedstock for large-scale fuel production [52].

A unique advantage of third-generation biofuels is their potential for carbon dioxide mitigation, as algal cultivation systems can be integrated with industrial facilities to capture and utilize carbon dioxide emissions as a nutrient source, thus contributing to greenhouse gas reduction (Andres 2010). Furthermore, algal biofuel production is often associated with the concept of biorefineries, where not only fuels but also valuable co-products, such as animal feed, fertilizers, pharmaceuticals, and bioplastics, can be generated, improving overall process economics and sustainability [19]. Algae utilize CO_2 during photosynthesis, which could help achieve a carbon-negative impact. They can produce several biofuels such as bioethanol, biodiesel, syngas, and biohydrogen. Regarding yield, algae can make as much as 61,000 L of biodiesel per hectare. Particular species, like Chlorella, are especially valued for their high lipid concentration and impressive productivity.

Despite these promising attributes, the commercial viability of third-generation biofuels remains constrained by several factors, including high costs of cultivation, harvesting, and downstream processing, as well as challenges in maintaining optimal growth conditions for algae on a large scale. Collecting microalgae poses challenges because of their tiny size and delicate nature. The subsequent processing steps require significant energy and incur high costs, which hinder their large-scale commercial application. Technological improvements are necessary to scale production and lower expenses.

2.2 *Feedstocks Used and Conversion Technologies*

2.2.1 Sugar and Starch Crops

Sugar and starch crops are the primary first-generation feedstocks for biofuel production, especially bioethanol. Sugar crops include sugarcane (Saccharum officinarum), sugar beet (Beta vulgaris), and sweet sorghum (Sorghum bicolor), while starch crops include maize (Zea mays), cassava (Manihot esculenta), wheat (Triticum aestivum), and potato (Solanum tuberosum) [38].

Sugar crops store energy primarily as sucrose, which can be directly extracted and fermented to ethanol with high conversion efficiency and relatively simple technology. In Brazil, sugarcane is the leading feedstock for ethanol production, not only due to its high biological yield but also because its byproduct, bagasse, can be used for power generation or as a feedstock for other processes, enhancing the overall energy efficiency of the production chain [27].

Starch crops accumulate carbohydrates in the form of starch, which enzymes must first hydrolyze into simple sugars before fermentation to ethanol. Corn is the principal feedstock for ethanol production in the United States due to large-scale agricultural infrastructure and stable yields. Although ethanol yield per hectare from corn or cassava can be substantial, the production process is more complex than sugar crops because it requires an additional hydrolysis step [12].

In general, producing biofuels from crops rich in sugar and starch provides notable benefits, especially regarding established technologies and high conversion efficiency. However, significant challenges include the competition these crops pose to the food supply and the necessity to improve their genetics, cultivation practices, and processing methods to ensure that bioenergy production remains sustainable.

2.2.2 Fiber and Grass Cellulosic Crops

Fiber and grass cellulosic crops such as switchgrass (Panicum virgatum), miscanthus (Miscanthus × giganteus), and agricultural residues (e.g., corn stover, wheat straw) are significant sources of lignocellulosic biomass for second-generation biofuels. Typically, these crops contain more fiber components, such as cellulose, hemicellulose, and lignin, while having lower amounts of carbohydrates, proteins, and oils. They can be utilized for energy through several approaches, including burning them directly for heat or electricity, converting cellulose into ethanol, applying thermochemical methods to create fuel additives, or using anaerobic digestion to generate methane. However, challenges include the need for efficient pretreatment technologies and logistics for large-scale biomass collection.

2.2.3 Oil Crops

Oil crops such as soybean (Glycine max), rapeseed (Brassica napus), sunflower (Helianthus annuus), and oil palm (Elaeis guineensis) play a central role as feedstocks for biodiesel production due to their high oil content and well-established cultivation systems. Oil is obtained by pressing the seeds to release their contents, and then the extracted oil undergoes transesterification to produce biodiesel. The triacylglycerol (TAG) oils stored in these plants mainly consist of C16 and C18 hydrocarbon chains, which closely resemble the hydrocarbons found in diesel fuel. While vegetable oils can sometimes be blended with diesel and used directly in specific engines, doing so can shorten engine lifespan. Therefore, most biodiesel produced commercially today is made by transesterifying TAGs into their methyl ester forms, known as fatty acid methyl esters (FAME). The freezing point of each type of crop oil varies based on the length and degree of unsaturation of its fatty acid chains, enabling different oils to be mixed and tailored to create biodiesel with specific properties [51].

While oil crops like soybean dominate in the United States and rapeseed in Europe, oil palm stands out in Southeast Asia for its remarkably high oil yield per hectare; however, this expansion has raised significant concerns about deforestation, peatland drainage, greenhouse gas emissions, and loss of biodiversity, especially in tropical regions [1, 32]. Additionally, the increasing allocation of arable land to biofuel oil crops intensifies the food-versus-fuel debate and can impact local food security [28].

2.2.4 Crop Residues, Manures, and Organic Wastes

Crop residues (such as rice straw, corn stover, and wheat straw), animal manures, and various organic wastes (including municipal solid waste, food waste, and industrial by-products) are increasingly recognized as valuable feedstocks for second-generation biofuel production due to their abundance, low cost, and potential to mitigate waste management challenges [40]. These materials primarily consist of lignocellulosic biomass or organic matter that can be converted to biofuels like bioethanol, biogas, and biodiesel through a variety of biochemical (e.g., anaerobic digestion, fermentation) and thermochemical (e.g., pyrolysis, gasification) processes.

Agricultural residues such as corn stalks and cobs, wheat straw, and other small-grain straws are abundant resources that can be converted into renewable fuels. However, a significant challenge in collecting these crop residues is ensuring the long-term sustainability of cropping systems. Typically, these materials are returned to the fields to enhance soil organic matter and to support soil and water conservation.

Manure results from the digestive processes of livestock consuming plants. While it is commonly spread on fields to enrich soil nutrients, it can also be processed through anaerobic digestion. This technique has long been used to convert manure and other organic waste into methane and similar gases. The resulting methane can be used as a direct heat source to power gas turbines or can be further purified as a substitute for natural gas [14]. Large quantities of food processing byproducts and municipal wastes, such as restaurant fats, leaves, grass cuttings, and other yard debris, are common in urban areas. These materials can be gathered and transformed into energy through various conversion technologies [14].

Utilizing these residues and wastes for bioenergy reduces environmental pollution and greenhouse gas emissions associated with their disposal and helps minimize competition with food crops for land use. Nevertheless, the efficient collection, transport, and processing of these heterogeneous materials remain significant technical and economic challenges, and their nutrient content (especially in manures) must be managed to avoid negative impacts on soil fertility and local ecosystems [44].

2.2.5 Algae

Algae, recognized as a third-generation biomass source, stand out from other types of biomass thanks to their strong environmental and economic sustainability, reliable supply, lack of lignin, high photosynthetic capacity, rapid growth, and their positive effect on reducing greenhouse gas emissions. They can uniquely convert atmospheric CO_2 into valuable biomass, grow in wastewater environments, and produce biofuels without significantly impacting food resources, biodiversity, or agriculture. As a result, algae play an essential role in utilizing and treating wastewater, helping to mitigate environmental concerns related to waste disposal. These organisms can also thrive in saline or coastal waters, artificial seawater, and non-arable lands such as deserts or semi-arid regions, reducing pressure on agricultural land for food production. Compared to other advanced feedstocks like cellulosic sources, algal genomics and fundamental biology research are more developed and rapidly progressing. Algae are known for their high photon conversion rates and ability to accumulate substantial amounts of carbohydrates suitable for bioethanol production using inexpensive raw materials. Aquatic algae are buoyant, unlike land plants, so they do not require structural polymers like hemicellulose or lignin. This simplifies the conversion process to bioethanol by removing the need for complex chemical and enzymatic pretreatments. Furthermore, algae can be harvested much faster than other feedstocks, making them a promising solution to the growing demand for raw materials in ethanol production [31]. However, large-scale commercialization faces challenges including high costs of cultivation, harvesting, dewatering, lipid extraction, and energy-intensive processing steps, which currently limit economic viability [33].

2.3 Role in the Circular Economy

The circular economy operates on three core principles, one of which is the elimination of waste by designing products to be both renewable and biodegradable. Biofuels are integral to the circular economy by transforming renewable biological resources and organic waste streams into valuable energy carriers, closing material loops, and minimizing environmental impact. Unlike fossil fuels, biofuels can be produced from agricultural residues, food waste, manure, and non-food crops, supporting the reuse and valorization of by-products that might otherwise contribute to pollution or landfilling. Their production processes can integrate nutrient recycling and carbon sequestration, especially when residues and digestates are returned to soils, promoting soil fertility and reducing the need for synthetic fertilizers [17]. Additionally, biofuel production from waste streams such as municipal solid waste or wastewater further enhances circularity by capturing embedded energy and reducing emissions [25]. By coupling energy generation with resource recovery and waste management, biofuels help build resilient, low-carbon energy systems, advance rural economies, and contribute to the sustainable transition of industrial sectors, making them a key enabler of circular bioeconomy principles. However, maximizing these benefits depends on careful feedstock selection, sustainable land management, and advances in biofuel conversion technologies.

3 Climate Benefits of Bioenergy and Green Fuels

Bioenergy and green fuels offer climate-related advantages, making them increasingly relevant in the global effort to combat climate change and transition toward more sustainable energy systems. Among these advantages, three core dimensions stand out: carbon neutrality, waste valorization, and creating opportunities for rural economic development.

3.1 Carbon Neutrality

The bioenergy sources are derived from recently living biomass, such as crops, agricultural residues, forest materials, and organic waste. When these types of biomass are grown, they can absorb CO_2 from the atmosphere through the photosynthesis process under the sunlight. Besides, the carbon sinks can be enhanced through biomass cultivation. After that, it is later used for energy production via combustion, fermentation, or gasification, which means carbon is released back into the atmosphere. This creates a process in which the CO_2 emissions from bioenergy use are somehow offset or "neutralized" by the carbon uptake during photosynthesis. This neutral carbon exchange contributes to lower net greenhouse gas emissions than fossil fuels, which is well known as a type of fuel that can release carbon stored underground for millions of years. There is no doubt that greenhouse gas emission reductions occur when using bioenergy and green fuels compared to fossil fuels.

Furthermore, advances in second- and third-generation biofuels can enhance the carbon-saving potential of these technologies. Additionally, if combined with carbon capture and storage (CCS) technologies—resulting in what is known as Bioenergy with Carbon Capture and Storage (BECCS)—bioenergy can even deliver negative emissions, helping to actively remove CO_2 from the atmosphere, a critical strategy highlighted in many climate mitigation scenarios outlined by the Intergovernmental

Panel on Climate Change (IPCC) [5]. As the report of IPCC, renewable energy deployment, energy efficiency improvements, and the transition to low-carbon technologies are key components of mitigation efforts [5]. Several scientific studies indicate that deploying BECCS can significantly increase its climate change mitigation potential, and it is an excellent option for GHG emission reduction and carbon dioxide removal [9, 15, 37, 48]. BECCS can be defined as a synergistic approach that integrates the use of bioenergy sources with carbon capture and storage technologies. BECCS promises to be a versatile and sustainable solution that targets energy demand and carbon reduction goals. It can act as a negative emission technology and is crucial in many climate change mitigation pathways, limiting global warming to 1.5–2 °C [48].

3.2 *Waste Valorization*

Biofuels have attracted significant attention and have been widely researched and applied. This growing interest is entirely justified, especially as the "waste-to-energy" approach becomes increasingly relevant and necessary when humans are facing the global energy crisis. One of the significant benefits of biofuels is waste valorization. Waste valorization is the process of reusing, recycling, or composting waste materials and converting them into more valuable products, such as materials, chemicals, fuels, or other energy sources [13]. In recent years, thermochemical conversion approaches, including torrefaction, pyrolysis, liquefaction, gasification, etc., have been widely developed and applied to valorize biowastes into high-value fuels, materials, and chemicals [21, 46, 54, 55]. The biowaste mentioned here includes agricultural residues, food waste, and other organic by-products generated from farming, food processing, household activities, etc. However, traditional biowaste alone is unsuitable for replacing fossil energy sources because it depends on geographic, climatic, and social factors, competes with food, and has a low energy density [18]. Then, the solution here is to transfer it into the biofuels. Waste valorization through biofuel production can help reduce raw biomass, thus minimizing environmental pressures and providing significant climate benefits. Bioenergy and green fuels significantly contribute to waste valorization, simultaneously becoming the solution to energy generation and waste management challenges. Countries in the world can increase energy security while reducing the climate and health impacts of mismanaged waste by exploiting biowaste, then directly achieve Sustainable Development Goal 12 (Responsible Consumption and Production) and Sustainable Development Goal 13 (Climate Action) of the Sustainable Development Goals established by the United Nations in 2015.

Valorization of waste into biofuels is an eco-friendly and economically feasible option. However, these wastes must be pretreated to enhance the conversion rate, reduce energy usage, and increase the quantity and quality of the products generated [41]. The contents of the biomass, including cellulose, hemicellulose, and lignin, can be degraded via the pretreatment step, which enhances the digestibility and biodegradability of the biomass. Thus, it is necessary to find new and improved pretreatment methods to break down crop residues and other agricultural waste more effectively, then convert them into useful products.

3.3 Opportunities for Rural Economic Development

Beyond environmental and climate mitigation benefits, bioenergy and green fuel development also benefit rural economic growth, particularly in regions where agriculture and forestry are primary livelihood sources. A wide array of employment and higher income for rural communities can be achieved with the bioenergy value chain, from feedstock cultivation, collection, and processing to transportation and distribution. For example, biofuel feedstock production has been considered a possible avenue to boost rural development in Sub-Saharan Africa [47]. It was evidenced that the households in Sub-Saharan Africa who engaged in biofuel feedstock production have higher incomes than households not engaged in such activities, particularly in sugarcane areas. The outcomes may vary between crops, types of involvement, modes of production, and institutional arrangements [47]. Another example that can be mentioned is in Colombia, where biodigesters are introduced to produce biogas and biofertilizer, and it is evident that they offer significant socio-economic benefits [34]. It is also recommended that a comprehensive support program and policy incentives are needed to ensure the successful and sustainable adoption of biodigesters in rural areas.

The economic opportunities from selling the biomass, such as agricultural residues, forestry by-products, or animal manure, can contribute to the rural economies, reduce urban migration, and enhance energy access in remote areas such as high mountains and islands. This contributes to poverty reduction (SDG 1), energy equity, and community resilience. Women and marginalized groups especially benefit from rural bioenergy initiatives. For instance, the widespread deployment of biogas digesters in rural households has been shown to reduce indoor air pollution and the burden of collecting firewood, tasks often shouldered by women, thereby improving health outcomes and freeing time for education and economic activities. Moreover, local ownership and participation in bioenergy projects can empower communities, foster overall development, build small-scale entrepreneurship, and create job opportunities in biomass production chains.

4 Challenges and Concerns

4.1 Technical Challenges and Environmental Concerns

Although biofuels have been and are being researched a lot today, however, this type of fuel still encounters some technical challenges that need to be improved and modified. These challenges encompass:

(i) Feedstock-related challenges

It is necessary to determine the advanced pretreatment strategies in biofuel production to release fermentable sugars and valuable products necessary for microbial fermentation in the biomass. It is crucial, especially in the case of the agro-waste biomass, with the main components of lignin, cellulose, and hemicellulose. Pretreatment of lignocellulosic biomass is essential to break down complex structures like lignin and hemicellulose, but often requires harsh chemical or physical processes that can generate inhibitors and increase costs. The preprocessing and logistics are also challenging because biomass is bulky and degrades over time, and drying, grinding, and transporting biomass increase costs and energy use.

(ii) Conversion technology challenges

Thermochemical conversion methods like pyrolysis and gasification operate at high temperatures and need tight process control, yet syngas purification and catalytic fuel upgrading remain inefficient and economically challenging. Besides, these processes often generate a complex mixture of gases and by-products that require extensive downstream treatment. Additionally, the catalysts used in upgrading can degrade quickly due to coking or poisoning, leading to high replacement costs and reduced process stability.

(iii) Scale-up challenges

Scaling up biofuel production from laboratory to commercial scale presents significant technical challenges, particularly heat and mass transfer limitations affecting reaction efficiency and consistency. It can be easy to see that even when the processes that work well in small-scale settings, they still can become inefficient or unstable when scaled up due to uneven temperature distribution or inadequate mixing. Additionally, integrating biofuels into existing fuel infrastructure requires the fuels to meet strict standards for properties like energy density, viscosity, and emissions. Without compatibility, biofuels may cause engine damage or require costly modifications to current systems.

In recent years, one of the main drivers for promoting bioenergy systems has been the strong interest in reducing anthropogenic GHG emissions. The positive impacts of bioenergy and green fuels on the environment through reducing waste and producing green energy are evident. However, environmental implications related to bioenergy and green fuels still need to be pointed out. As discussed, bioenergy and green fuels are mainly produced via thermochemical conversion. These processes can emit carbon dioxide, methane, and nitrous oxide, all of which are GHGs. Besides, incomplete combustion and poor gasification in real-world conditions may lead to unhealthy air pollutants such as volatile organic compounds and carbon monoxide emissions [22]. In addition, the energy consumption required for the combustion process compared to the benefits gained from biofuels needs to be carefully calculated and considered.

4.2 Economic Challenges

As a new and still under-researched fuel, bioenergy and green fuels are currently facing significant economic barriers that hinder their widespread adoption and commercialization. One of the foremost challenges lies in the high capital investment required to develop the necessary infrastructure, including the construction of specialized conversion facilities and the deployment of advanced technologies for efficient fuel production. These initial costs can be particularly prohibitive for emerging economies or small-scale producers. Beyond infrastructure, there are substantial ongoing operational expenses that further elevate the overall cost of production. These include the energy demands of the conversion processes, the need for skilled technicians to operate and monitor complex systems, and regular maintenance to ensure the longevity and efficiency of the facilities. Additionally, supply chain limitations—such as the availability and transportation of biomass feedstock—can introduce further financial strain and logistical complexity. These economic challenges create a competitive disadvantage for bioenergy and green fuels compared to conventional fossil fuels, which benefit from established infrastructure, mature technologies, and often, government subsidies.

4.3 Policies and Regulatory Barriers

The transition of the energy system towards renewable resources is a key target of national and international agreements, and converting waste to energy is a highly promoted way now. Some of the global policies related to this approach are the Paris Agreement, the Global Methane Pledge, the United Nations' SDGs, and some programs of the Asia–Pacific Economic Cooperation [23, 26, 39, 49]. Countries and regions have laws and regulations regarding the development and application of biofuels. For instance, the United States promotes biofuel production through the Renewable Fuel Standard Program to reduce greenhouse gas emissions and expand the nation's renewable fuels sector while reducing reliance on imported oil. This program was authorized under the Energy Policy Act of 2005 and expanded under the Energy Independence and Security Act of 2007 [20]. In another way, the European Union enforces strict sustainability criteria under the Renewable Energy Directive, targets for increased renewable energy use, and supports cooperation between EU countries towards this goal [8]. Strict emission standards coupled with waste disposal regulations in developed countries can lead to high costs of undertaking waste-to-biofuel projects and hence discourage investors.

However, developing countries face challenges implementing such policies due to limited infrastructure, financial resources, or technical expertise. This can lead to several consequences. For example, the lack of policies and guidelines from the government can make the biomass collection and sorting of input materials for biofuel production inefficient; it can also make stakeholders unsure of the legal or official procedures for biofuel production. Other forms of renewable energy, such as solar and wind, benefit from substantial subsidies and incentives from the government and private economic sector, which places biofuel/green fuel at a competitive disadvantage. The lack of financial support policies encouraging the willingness to undertake biofuel projects is also challenging. Moreover, the absence of long-term, stable policy frameworks creates uncertainty, further discouraging innovation and commitment in the biofuel sector.

5 Conclusions and Future Recommendations

Despite the potential of bioenergy and green fuels, realizing these climate-related advantages depends heavily on how these systems are designed, implemented, and governed. There is a truth that socio-enviro-economic objectives are challenging to balance. With the rapid development of research and science, while bioenergy and green fuels represent a promising path toward a more sustainable and circular energy future, their growth is currently hindered by various technical, economic, environmental, and policy-related challenges. These barriers must be addressed holistically, from the complexity of biomass feedstock pretreatment and conversion inefficiencies to the high capital costs and inconsistent regulatory frameworks. Moving forward, a multi-pronged approach is essential. Continued investment in research and development is needed to improve feedstock processing technologies, enhance catalyst stability, and optimize conversion pathways. Governments should implement stable, long-term policy frameworks and provide financial incentives to attract private sector investment and mitigate risk. Moreover, international collaboration and knowledge exchange can support capacity-building, particularly in developing countries, ensuring that bioenergy solutions are inclusive and globally scalable. By aligning technological innovation with supportive policy mechanisms and environmental safeguards, the potential of bioenergy and green fuels to contribute significantly to a low-carbon future can be fully realized.

Acknowledgements This research is funded by the Japan International Cooperation Agency (JICA) under grant number VJU.JICA.RSG.25.01.

References

1. Atabani, A.E., Silitonga, A.S., Badruddin, I.A., Mahlia, T.M.I., Masjuki, H., Mekhilef, S.: A comprehensive review on biodiesel as an alternative energy resource and its characteristics. Renew. Sustain. Energy Rev. **16**, 2070–2093 (2012)
2. Clarens, A.F., Resurreccion, E.P., White, M.A., Colosi, L.M.: Environmental life cycle comparison of algae to other bioenergy feedstocks. Environ. Sci. Technol. **44**, 1813–1819 (2010)
3. Andrew Carroll, C.S.: Cellulosic biofuels. Annu. Rev. Plant Biol. **60**, 165–182 (2009)
4. Solomon, B.D., Barnes, J.R., Halvorsen, K.E.: Grain and cellulosic ethanol: History, economics, and energy policy. Biomass Bioenergy **31**, 416–425 (2007)
5. Change, T.I.P.O.C.: Climate Change 2014: Mitigation of Climate Change. Contribution of Working Group III to the Fifth Assessment (2014)
6. Edenhofer, O., Pichs-Madruga, R., Sokona, Y., Farahani, E., Kadner, S., Seyboth, K., Adler, A., Baum, I., Brunner, S., Eickemeier, P., Kriemann, B., Savolainen, J., Schlömer, S., Von Stechow, C., Zwickel, T., Minx, J.C. (eds).: Report of the intergovernmental panel on climate change. In: Climate Change 2014—Mitigation of Climate Change. Cambridge University Press
7. Chisti, Y.: Biodiesel from microalgae. Biotechnol. Adv. **25**, 294–306 (2007)
8. Commission, E.: Renewable Energy Directive [Online]. Available: https://energy.ec.europa.eu/topics/renewable-energy/renewable-energy-directive-targets-and-rules/renewable-energy-directive_en. Accessed.
9. Salas, D.A., Boero, A.J., Ramirez, A.D.: Life cycle assessment of bioenergy with carbon capture and storage: a review. Renew. Sustain. Energy Rev. **199**, 114458 (2024)
10. Daniel Klein-Marcuschamer, P.O.-P., Simmons, B.A., Blanch, H.W.: The challenge of enzyme cost in the production of lignocellulosic biofuels. Biotechnol. Bioeng. **109**, 1083–1087 (2012)
11. David Pimentel, T.W.P.: Ethanol production using corn, switchgrass, and wood; biodiesel production using soybean and sunflower. Nat. Resour. Res. **14**, 65–76 (2005)
12. Edgard Gnansounou, A.D.: Ethanol fuel from biomass: a review. J. Sci. Ind. Res. **64**, 809–821 (2005)
13. Farah Kanani, M.D.H., Gilroyed, B.H., Pelletier, N.: Waste valorization technology options for the egg and broiler industries: a review and recommendations. J. Clean. Prod. **262**, 121129 (2020)
14. Farm-Energy.: Feedstocks for Biofuel Production [Online] (2019). Available: https://farm-energy.extension.org/feedstocks-for-biofuel-production/. Accessed
15. Farooq Sher, S.H., Omerbegović, N.S., Wang, B., Hai, I.U., Rashid, T., Teoh, Y.H., Yildiz, M.J.: Bioenergy with carbon capture and storage technology to achieve net zero emissions–a review. Renew. Sustain. Energy Rev. **210**, 115229 (2025)
16. Felix Creutzig, N.H.R., Berndes, G., Bolwig, S., Bright, R., Cherubini, F., Chum, H., Corbera, E., Delucchi, M., Faaij, A., Fargione, J., Haberl, H., Heath, G., Lucon, O., Plevin, R., Popp, A., Robledo-Abad, C., Rose, S., Smith, P., Stromman, A., Suh, S., Masera, O.: Bioenergy and climate change mitigation: an assessment. Global Change Biol. Bioenergy **7**, 916–944 (2015)
17. Francesco Cherubini, A.H.S.: Life cycle assessment of bioenergy systems: state of the art and future challenges. Biores. Technol. **102**, 437–451 (2011)
18. Gil, A.: Challenges on waste-to-energy for the valorization of industrial wastes: electricity, heat and cold, bioliquids and biofuels. Environ. Nanotechnol. Monit. Manag. **17**, 100615 (2022)
19. Giorgos Markou, I.A., Georgakakis, D.: Microalgal carbohydrates: an overview of the factors influencing carbohydrates production, and of main bioconversion technologies for production of biofuels. Appl. Microbiol. Biotechnol. **96**, 631–645 (2012)
20. Goverment, T.U.S.: Renewable Fuel Standard Program [Online] (2005). Available: https://www.epa.gov/renewable-fuel-standard-program. Accessed
21. Hehuan Peng, W.C., Huang, M., Xia, S., Zhu, L., Fang, X., Ma, Z.: Advancing biomass gasification by the dry and wet torrefaction pretreatment. Energy **324**, 136118 (2025)
22. Ibukun Oluwoye, M.A., Gore, J., Dlugogorski, B.Z.: Products of incomplete combustion from biomass reburning. Fuel **274**, 117805 (2020)

23. IEA.: Global Methane Tracker 2022 [Online] (2022). Available: https://www.iea.org/reports/global-methane-tracker-2022/the-global-methane-pledge. Accessed
24. Jason Hill, E.N., Tilman, D., Tiffany, D.: Environmental, economic, and energetic costs and benefits of biodiesel and ethanol biofuels. Proc. Natl. Acad. Sci. **103**, 11206–11210 (2006)
25. Milledge, J.J., Heaven, S.: A review of the harvesting of micro-algae for biofuel production. Rev. Environ. Sci. Bio/Technol. **12**, 165–178 (2013)
26. Delbeke, J., Runge-Metzger, A., Slingenberg, Y., Werksman, J.: The Paris agreement. In: Towards a climate-neutral Europe (2019)
27. José Goldemberg, S.T.C., Guardabassi, P.: The sustainability of ethanol production from sugarcane. Energy Policy **36**, 2086–2097 (2008)
28. Joseph Fargione, J.H., Tilman, D., Polasky, S., Hawthorne, P.: Land clearing and the biofuel carbon debt. Science **319**, 1235–1238 (2008)
29. Knothe, G.: Biodiesel and renewable diesel: a comparison. Prog. Energy Combust. Sci. **36**, 364–373 (2010)
30. Abioye, K.J., Rajamanickam, R., Ogunjinmi, T., Paul, S., Selvasembian, R., Ighalo, J.O.: Advancements in biomass waste conversion to sustainable biofuels via gasification. Chem. Eng. J. **505**, 159151 (2025)
31. Chaudhary, L., Pradhan, P., Soni, N., Singh, P., Tiwari, A.: Algae as a feedstock for bioethanol production: new entrance in biofuel world. Int. J. Chem. Tech. Res. **6**, 1381–1389 (2014)
32. Koh, L.P., Wilcove, D.S.: Is oil palm agriculture really destroying tropical biodiversity. Conserv. Lett. **1**, 60–64 (2008)
33. Lam, M.K., Lee, K.T.: Microalgae biofuels: A critical review of issues, problems, and the way forward. Biotechnol. Adv. **30**, 673–690 (2012)
34. Mario Luna-Delrisco, C.A.G., Mendoza-Hernández, S., Vanegas-Trujillo, E., Da, L., Meneses, R., Rio, J.-D., Castillo-Meza, L.E., Santos-Ballardo, D.U., Montoya, J.P.G.: Evaluating the socio-economic drivers of household adoption of biodigester systems for domestic energy in rural Colombia. Sustain. Energy Technol. Assess. **73**, 104146 (2025)
35. Staples, M.D., Malina, R., Barrett, S.R.: The limits of bioenergy for mitigating global life-cycle greenhouse gas emissions from fossil fuels. Nat. Energy **2**, 16202 (2017)
36. Martin Banse, H.V.M., Tabeau, A., Woltjer, G., Hellmann, F., Verburg, P.H.: Impact of EU biofuel policies on world agricultural production and land use. Biomass Bioenerg. **35**, 2385–2390 (2011)
37. Hayat, M.A., Alhadhrami, K., Elshurafa, A.M.: Which bioenergy with carbon capture and storage (BECCS) pathways can provide net-negative emissions? Int. J. Greenhouse Gas Control **135**, 104164 (2024)
38. Mustafa Balat, H.B., Cahide, Ö.Z.: Progress in bioethanol processing. Prog. Energy Combust. Sci. **34**, 551–573 (2008)
39. Nations, U.: 2030 Agenda for Sustainable Development [Online] (2017). Available: https://sdgs.un.org/goals. Accessed
40. Nicolae Scarlat, M.M., Dallemand, J.-F.: Assessment of the availability of agricultural crop residues in the European Union: potential and limitations for bioenergy use. Waste Manag. **30**, 1889–1897 (2010)
41. Omojola Awogbemi, D.V.V.K.: Valorization of agricultural wastes for biofuel applications. Heliyon **8**, e11117 (2022)
42. Alvira, P., Tomas-Pejo, E., Ballesteros, M., Negro, M.J.: Pretreatment technologies for an efficient bioethanol production process based on enzymatic hydrolysis: a review. Bioresour. Technol. **101**, 4851–4861 (2010)
43. Smith, P., Davis, S.J., Creutzig, F., Fuss, S., Minx, J., Gabrielle, B., Kato, E., Jackson, R.B., Cowie, A., Kriegler, E., Van Vuuren, D.P., Rogelj, J., Ciais, P., Milne, J., Canadell, J.G., Mccollum, D., Peters, G., Andrew, R., Krey, V., Shrestha, G., Friedlingstein, P., Gasser, T., Grübler, A., Heidug, W.K., Jonas, M., Jones, C.D., Kraxner, F., Littleton, E., Lowe, J., Moreira, J.R., Nakicenovic, N., Obersteiner, M., Patwardhan, A., Rogner, M., Rubin, E., Sharifi, A., Torvanger, A., Yamagata, Y., Edmonds, J., Yongsung, C.: Biophysical and economic limits to negative CO_2 emissions. Nat. Clim. Change **6**, 42–50 (2016)

44. Sims, R.E., Mabee, W., Saddler, J.N., Taylor, M.: An overview of second generation biofuel technologies. Bioresour. Technol. **101**, 1570–1580 (2010)
45. Raphael Slade, A.B., Gross, R.: The global bioenergy resource. Nat. Clim. Change **4**, 99–105 (2014)
46. Kavitha, S., Gondi, R., Kannah, R.Y., Kumar, G., Banu, J.R.: A review on current advances in the energy and cost effective pretreatments of algal biomass: enhancement in liquefaction and biofuel recovery. Bioresour. Technol. **369**, 128383 (2023)
47. Mudombi, S., Ochieng, C., Johnson, F.X., Von Maltitz, G., Luhanga, D., Dompreh, E.B., Romeu-Dalmau, C., Gasparatos, A.: Fuelling rural development? the impact of biofuel feedstock production in southern Africa on household income and expenditures. Energy Res. Soc. Sci. **76**, 102053 (2021)
48. Hanssen, S.V., Daioglou, V., Steinmann, Z.J., Doelman, J.C., Van Vuuren, D.P., Huijbregts, M.A.: The climate change mitigation potential of bioenergy with carbon capture and storage. Nat. Clim. Change 1023–1029 (2020)
49. Shaohua Jiang, G.M., Shahzadi, I., Jebli, M.B., Iqbal, N.: Renewable energy, trade diversification and environmental footprints: evidence for Asia-Pacific Economic Cooperation (APEC). Renew. Energy **187**, 874–886 (2022)
50. Sheehan, J., Camobreco, V., Duffield, J., Graboski, M., Shapouri, H.: Life Cycle Inventory of Biodiesel and Petroleum Diesel for Use in an Urban Bus (1998)
51. Long, S.P., Karp, A., Buckeridge, M.S., Davis, S.C., Jaiswal, D., Moore, P.H., Moose, S.P., Murphy, D.J., Onwona-Agyeman, S., Vonshak, A.: Crop feedstocks for biofuels and bioenergy. Scope (2015)
52. Mata, T.M., Martins, A.A., Caetano, N.S.: Microalgae for biodiesel production and other applications: a Review. Renew. Sustain. Energy Rev. **14**, 217–232 (2010)
53. Timothy Searchinger, R.H., Houghton, R.A., Dong, F., Elobeid, A., Fabiosa, J., Tokgoz, S., Hayes, D., Tun-Hsiang, Y.U.: Use of United States croplands for biofuels increases greenhouse gases through emissions from land-use change. Science **319**, 1238–1240 (2008)
54. Yi Wang, H.Q., Menghao Zuo, L., Zuo, X.W., Song, P., Guo, S., Shen, B.: Pyrolysis mechanism of isolated microalgal composition and their potential as liquid biofuels: Neutral lipids, phospholipids, glycolipids, and sterols. Energy Convers. Manag. X **26**, 100960 (2025)
55. Gan, Y.Y., Ong, H.C., Chen, W.-H., Sheen, H.-K., Chang, J.-S., Chong, C.T., Ling, T.C.: Microwave-assisted wet torrefaction of microalgae under various acids for coproduction of biochar and sugar. J. Clean. Prod. **253**, 119944 (2020)

Role of AI in Developing Smart Urban Cities and Resilience to Climate Change

Lata Rani, Hurmat, Balpreet Kaur, and Rashami Arora

ABSTRACT As we landed in the twenty-first century there is an urge to develop a resilient as well as sustainable urban environments owing to global warming. Artificial Intelligence (AI) can be a significant tool to deal with different factors that arises due to climate change. This chapter will explore the fundamental role of AI in the development of smart urban cities and enhancing their resilience to climate-related stresses. AI plays a significant role in urban area to efficient predictive modelling, data analysis as well provide a support to decision making in various sectors i.e. waste management, urban planning, energy and transportation. AI system act as a heart of different smart cities by playing its significant role in different areas i.e. by controlling traffic system, air quality monitoring, minimize energy consumption, climate change adaptation etc. Furthermore, AI is helpful to identifies vulnerable urban zones and provides help to intimate public about extreme weather events such as floods, heatwaves, and storms. Urban resilience is further reinforced through AI-assisted infrastructure planning that integrates climate risk assessments and sustainable development goals. Besides this AI pays an indispensable role to optimize resource allocation and enhances circular economy practices by improving recycling systems (Birkmann et al. in Sustain. Sci. 5:185–206, 2010). Even though AI processes a transformative power still it faces certain challenges i.e. algorithmic bias, data privacy, the digital divide etc. Therefore, to ensure equitable and inclusive smart city development with sustainability; management of these problems is mandatory. Further research and interdisciplinary collaboration can speed up AI's full potential in shaping the cities as per climate conditions to service the public.

Keywords Artificial Intelligence (AI) · Energy consumption · Sustainability · Climate change · Data privacy

---

L. Rani · Hurmat (✉) · B. Kaur
Chitkara School of Pharmacy, Chitkara University, Solan, Himachal Pradesh, India
e-mail: hurmat@chitkarauniversity.edu.in

R. Arora
Chitkara College of Pharmacy, Chitkara University, Punjab, India

A. L. Srivastav et al. (eds.), *Climate Resilience: Impact of Quantum Computing and Artificial Intelligence on Urban Planning*, Studies in Systems, Decision and Control 637, https://doi.org/10.1007/978-3-032-06791-3_10

1 Introduction

The twenty-first century is witnessing two of the most transformative global trends: rapid urbanization and accelerating climate change. Urban areas, now home to over 55% of the world's population and projected to reach 68% by 2050, are the epicentres of economic growth, technological advancement, and social development [8]. However, this urban expansion often occurs in ways that increase environmental degradation, resource consumption, and vulnerability to climate hazards. Urbanization contributes significantly to climate change. Cities account for nearly 70% of global carbon dioxide emissions, largely due to energy consumption in buildings, transportation, and industrial processes. Ecological balance is disturbed due to certain factors i.e. deforestation, unethical use of land, invulnerable surfaces making etc. At the same time, a number of unexpected urban growth is responsible for insufficient infrastructure that may not handle certain circumstances like floods, fire, heatwaves etc. These conditions are really dangerous for public health by different ways i.e. water, food, infrastructure, economic sources etc. Underprivilege population is affected dangerously because they may have less sources to cope up with climate changes [24]. Therefore, there is requirement of policies to handle building resilience as well as climate variation simultaneously. At this point government, different technologies and efficient policies play their significant role. Furthermore, Artificial Intelligence work as a transformative implementation to deal with these complex issues by optimization of resources, model prediction as well it provides a support in decision making related to different climatic conditions i.e. type and design of infrastructure.

1.1 The Requirement of Smart Urban Cities

To deal with detrimental effects of boom of urbanisation and climate variations there is an urge to add concept of smart cities for the modernization of urban areas. Smart cities mean the utilization different digital tools i.e. of IoT (Internet of Things), AI, cloud computing etc. to enhance the sustainability and efficiency of urban areas. These all techniques may enhance the experience of the citizens by adding innovations in different sectors i.e. water disposal methods, transportation, water and energy management etc. Furthermore, if we want to create smart cities then there is need of efficient planning to change traditional infrastructure into modern one with combination of digital system. Because these smart cities will be prepared by keeping in mind not only to provide shelter to public but also to cope up with various environmental effects and build resilience to climate variation outcomes. As an illustration, advance transportation system is able to manage traffic issues and shared mobility. Similarly, by managing accurate balance among supply of energy as per demand by using smart grids that enables inhibition of carbon footprints [5].

Smart cities are required to showcase a significant transformation in urban life by the utilization of digital technologies. These techniques are able to open data platforms to provide different personalized services to its public, enhance transparency among government and public, sensor networking etc. Apart from these different climate modelling tools, mitigation strategies, green building technologies as well as climate focused projects should be introduced to deal with global warming and various alarming environmental changes. These services are already used in different countries i.e. Amsterdam, Barcelona, Singapore, and Songdo to demonstrated the potential of smart urbanization with efficient life style for the public.

1.2 Role of Artificial Intelligence in Urban Innovation

Artificial Intelligence plays its significant role in the planning and management to enhance the experience of the public. AI is helpful to manage previous data of the social platforms that enables to generate sensors or different smart tools. This data is helpful to predict different environmental conditions or ecosystem to support self-regulation. Moreover, this driven data enables decision makers to make efficient decision to boost sustainability and efficiency of urban cities. If we focus on transportation sector specifically AI is helpful for route optimization, predictive maintenance, digital vehicle coordination and smart traffic management techniques etc. Algorithms provides support to estimate traffic flows that facilitate to adjust traffic signals to handle traffic congestion. This is also helpful to manage behaviour of the public regarding traffic rules [15].

On the other hand, AI facilitate to introduce renewable energy resources to increase the efficiency of the energy systems. AI is utilized to smart grid management to balance the public demand and service of energy. As an illustration, deep learning models can estimate the peak load to enhance system reliability via minimizing energy waste and automatic power distribution. Moreover, algorithms & AI-powered surveillance systems are able to earlier detection of natural hazards that ultimately helpful to enhance the public safety and boost planning to manage these kind of emergency responses [3]. AI tools are also used to estimate environmental impacts and help to augment decisions as per these conditions. AI also works via natural language processing (NLP)-based Chabot's and platforms to collect information from public that will enable to provide better services to the public in the future. There is a plethora of machine learning models that works on the basis of AI to prior estimate about chances of floods, heatwaves, thunderstorm, heavy rain, extreme poor air quality etc. These predictions may work as a supporting system support to design strategies to deal with these unforeseen situations. Moreover, island mapping as well as air forecasting information generated by using AI is facilitate to create healthier urban sites and green infrastructure efficiently. Even though, utilization of AI to improve urban life also faces different challenges i.e. ethical concerns, data privacy, poor government policies, algorithmic bias etc. Therefore, there is an urge to generate different regulations to enhance the cybersecurity or liability in AI system. AI plays a significant role at both structural as well as technological level in smart cities formation [26].

1.3 Different AI Technologies

There are different fields of AI techniques (Fig. 1) that are helpful in the process of modernization of urban cities. These techniques are able to enhance the efficiency and sustainability of smart cities.

AI techniques Foremost, is machine learning (ML), that enable predictive analytics of traffic forecasting, energy requirement, disease outbreaks and crime mapping. ML system works on the basis of previous data and history to estimate unexpected problems before they arise as well provides solutions to handle them.

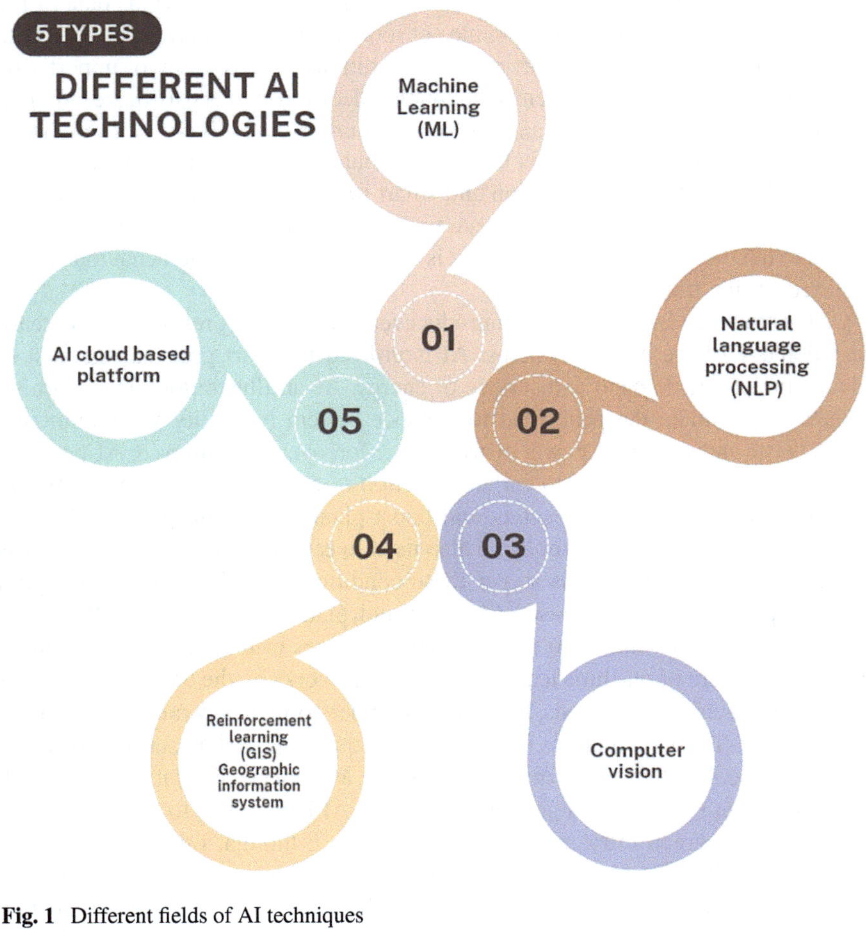

Fig. 1 Different fields of AI techniques

Moreover, ML is a deep learning tool and works on the basis of AI networking to create image patterns and their identification. ML is also utilized for the identification of autonomous vehicles and diagnosis of infrastructure i.e. recognition of bridge cracks [25].

Apart from ML, NLP (Natural Language Processing) plays a significant role to evaluate data collected as feedback from public through different methods i.e. social media, emails, chatbots etc. This information is really beneficial for public service conveyance, e-governance and civic engagement.

Another important technique is computer vision, that is used to visualize data of sensors and city cameras. It plays a significant role to investigate traffic violations, pollution causes as well as to manage public safety. Reinforcement learning is also evaluated for estimating its potential to control traffic signals as well as logistics to facilitate optimal actions as per feedback or trials. Moreover, AI- based models mostly connected with GIS (geographic information systems) to deal with environmental changes, infrastructure variations or disasters in smart cities. Other AI cloud-based platforms i.e. edge computing are also essential for data processing like for traffic cameras IoT devices are utilized [19]. These techniques boost the efficacy of the digital platforms. If these techniques are used in combination may made an unmatchable ecosystem of smart cities by optimization and adaptation of different advance applications.

1.4 Integration with IoT, Big Data, and Cloud Computing

AI plays a significant role in the system of urban cities by employing unique digital techniques includes Cloud Computing, IoT, Big Data. These three technologies are mainstay of urban smart cities and able to enhance the climate resilience. Various IoT tools includes, environmental sensors, smart meters, traffic controllers act as a sensory shield of the urban areas. The data provided by these tools is utilized to detect infrastructure faults, traffic congestion and to evaluate public health indicators. Big data is a digital platform that is used to collect data from imagery satellite to social platforms. AI algorithm is utilized efficiently to convert complex data sets into simple and fruitful insights i.e. modelling flood risks, estimating energy demand, heat detection etc. Moreover, cloud computing is able to provide information about infrastructure for processing and storage. It facilitates prompt actions for emergency conditions. It ensures efficient analytics and responsible for unmatchable applications includes adaptive traffic systems, accurate coordination to ensure prompt response in case of emergency. Furthermore, owing to the model of cloud platforms, cities remain attentive to changing environmental conditions & and prepared to response accordingly.

In combination, these digital techniques are rewarding routine requirements of modern cities as well as teaching them to deal with upcoming challenges. By combining AI with IoT, Big Data, and Cloud platforms, urban cities may become more sustainable, smarter with robust ecosystems [29].

2 Urban Data Sources and Infrastructure

The availability, accessibility, and quality of urban data lays down the foundation of AI-driven smart urban systems. AI applications draw insights from a wide variety of data sources that reflect how cities and people function. These include information from sensor networks that monitor air quality, noise levels, and temperature; satellite images and remote sensing tools; logs from public transport and traffic systems; patterns of electricity and water usage; energy consumption in buildings; and even trends and conversations on social media. All this data helps AI understand and improve urban life more effectively. Data streams—often varied and unstructured—are gathered from both fixed infrastructure like weather stations and traffic cameras, as well as moving sources such as GPS-enabled vehicles and mobile apps. This mix of static and dynamic inputs helps to create a completer and more real-time picture of what is happening in the environment. Cities must invest in digital infrastructure such as high-speed broadband, edge computing nodes, secure cloud storage, and interoperable data platforms to make this data effective. The demand for transparency and collaboration is increasing day by day due to urban data hubs and open data portals. The transparency and collaboration are a basis for government agencies, private firms, and researchers to co-develop AI solutions. To make the decision for urban planning, disaster management and climate adaptation, data quality, privacy, and ethical use play an important role especially as AI models. The need for a well-structured data infrastructure may not only fuels AI innovation but also strengthens a city's capacity that how to respond to complex challenges such as heatwaves, floods, and carbon emissions. This powerful combination of rich data and advanced digital infrastructure allows cities to shift from simply reacting to problems, to predicting them before they happen—and eventually, to adapting in real-time for smarter, more responsive urban management.

3 Applications of AI in Smart City Development

3.1 Intelligent Transportation Systems (ITS)

AI is valued for it's a plethora of applications in smart cities development as explained in Fig. 2.

For smart city development Intelligent Transportation Systems (ITS) is one of the most impactful applications of Artificial Intelligence. By combining AI technologies such as machine learning, computer vision, and edge computing, urban mobility networks become more adaptive, data-driven, and efficient by the development of ITS. To manage the traffic various AI-driven platforms process real-time data from road sensors, GPS devices, traffic cameras, and mobile applications to optimize and send signal timings to any road congestion or to reroute traffic dynamically. A real-life example using this technology is seen in the cities like Singapore and Stockholm

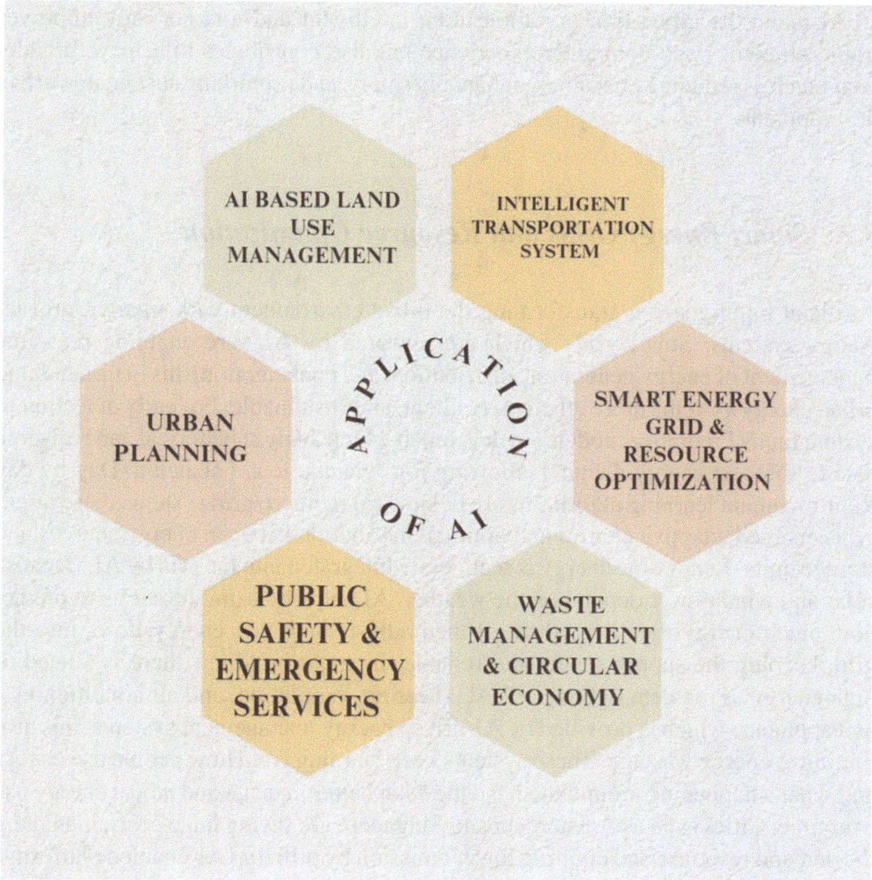

Fig. 2 Applications of AI in smart city development

which employ AI algorithms to manage urban traffic flows this subsequently reduce carbon emissions significantly.

AI also enhances predictive analytics which help transport departments to anticipate peak travel times, potential traffic accidents, and sudden spikes in demand these features enable smoother and more efficient transportation planning.

AI systems also play a great role in public transportation by analysing ridership patterns and urban events. This helps to schedule bus and train effectively so that it provides satisfaction to the user and enhance service efficiency. To monitor traffic violations, to identify road hazards, and to support pedestrian safety, computer vision is used in smart surveillance. One of the examples which shows the role of AI in transforming urban mobility is the rise of autonomous and connected vehicles. These vehicles are made to understand deep learning and real-time object detection which enables them to navigate complex environments also to avoid collisions, and interact with smart infrastructure like adaptive traffic lights [17].

AI made the urban transportation more intelligent and this not only improves traffic efficiency and commuter experience but also contributes to achieve broader goals such as reducing emissions, enhancing safety, and promoting sustainable urban development.

3.2 Smart Energy Grids and Resource Optimization

Artificial Intelligence is transforming the urban environment with smarter, greener energy systems. Smart grids which are powered by AI—are enabling real-time management of energy generation, distribution, and consumption, this in turn making urban energy systems more efficient, resilient, and sustainable. For early detection of system faults Predictive models are developed which analyse historical and real-time data to forecast energy demand, allowing for dynamic load balancing. Day by day Reinforcement learning algorithms are being used in micro grid systems these micro grids enabled adaptive control and optimal distribution between energy sources and storage units. Renewable energy is seamlessly integrated into the grid by AI. Because solar and wind power depend on the weather, AI helps by using forecasts to predict how much energy they'll produce. It then adjusts how that energy flows into the grid, keeping the supply steady and reliable. In smart buildings there is a need of high energy to regulate lighting, HVAC (heating, ventilation, and air conditioning), and appliance which is provided by AI-driven energy management systems, this also minimize energy wastage. These systems keep learning from how people use energy and what's happening around them, so they can better manage and adjust energy use over time. Cities such as Amsterdam and Singapore are saving huge costs, managing the demand response and emitting lower emission by utilizing AI-enabled platforms.

AI doesn't only contribute to electricity but also contributes to broader resource optimization by managing urban water supply and wastewater systems. Machine learning models are well advanced to detect leaks, predict usage patterns, and to ensure efficient allocation, especially during periods of stress or scarcity. To ensure either sustainability targets are fulfilled or not, these intelligent systems work together with IoT sensors to monitor resource flows [23].

Overall, AI help to make energy and resource systems more efficient and dependable moreover it is also responsible for cutting down carbon emissions in cities and supporting the creation of climate-resilient infrastructure—both of which are essential for building smart and sustainable urban spaces.

3.3 Waste Management and the Circular Economy

More development also leaves more waste, for this Artificial Intelligence is significantly enhancing urban waste management systems AI system make them more efficient, data-driven, and aligned with the principles of the circular economy. Traditional waste collection systems are not totally removed but they are often reactive and inefficient for transforming the traditional method AI introduces proactive, optimized strategies that reduce environmental impact and operational costs. The various AI developed waste management systems involves the use of smart bins, GPS-equipped trucks, and environmental sensors. By placing GPS AI algorithms can optimize collection routes, predict fill levels, and dynamically schedule pickups. All this pre-planned preparation will lead to reduced fuel usage, lower emissions, and improved service delivery.

AI-powered image recognition systems are revolutionizing waste sorting in recycling plants. The different types of materials such as plastics, metals, glass, and organics are identified and separated with high accuracy and speed by using computer vision and robotic automation. This system not only reduces contamination in recyclable streams but also increases the volume of waste that can be reused or repurposed. Various Start-up's and cities around the world such as San Francisco and Seoul, are trying to use and navigate AI-based smart recycling stations that reward citizens for correct disposal. This initiative will make the public aware and reinforce public engagement with waste reduction goals [10].

In a circular economy, AI helps by tracking how products are used over time, finding better ways to reuse resources, and making it easier to recover materials for future use. AI supports the generate urban composting loops, repair hubs, or industrial symbiosis zones to the city planners so that waste generation pattern and consumption behaviour will be mapped easily. Predictive models can even estimate how much waste might be generated during festivals, weather events, or as the population grows. This helps cities plan ahead and manage resources more effectively. Furthermore, environmental compliance and detection of illegal dumping is also monitored by AI through drone-based surveillance and anomaly detection models. These innovations not only contribute to cleaner, healthier urban environments but also reduces the burden on landfills and ecosystems.

By using AI in waste management and circular economy efforts, cities can shift away from the traditional 'produce-use-dispose away' approach and move toward systems that reuse materials. The habit of reusing the waste will cut down on pollution, and add lasting environmental and economic value.

3.4 Public Safety and Emergency Services

The rate of crimes are increasing day by day so the safety and security of people is a concerning issue. To solve this Artificial Intelligence is playing a transformative role in enhancing public safety and emergency response systems in smart cities. With the help of AI, along with surveillance systems, sensors, and emergency plans, cities are getting better at spotting potential issues early, stopping problems before they happen, and reacting quickly when they do. Many unusual activities such as crowd formation, abandoned objects, or suspicious behavior is analyzed via CCTV networks this Computer vision algorithms deployed through CCTV will triggers immediate alerts to law enforcement or emergency responders. This initiative drastically reduces response time and helps prevent threats. In crime prevention and law enforcement, AI uses past crime data, local social and economic patterns, and location trends to spot high-risk areas and help police plan patrols more effectively. Cities like Chicago and Los Angeles has transparent governance framework as they have implemented such systems, though their use raises important ethical considerations regarding bias and accountability.

During any natural disasters there is a need of instant response to handle the situation for this AI is enabling faster decision-making during natural disasters, fires, or accident by revolutionizing emergency response. The real time data is collected from IoT sensors, drones, and social media which is all together analysed by AI systems to assess the severity of incidents, GPS enables to locate victims, and recommend optimal deployment of ambulances, fire brigades, or evacuation routes. AI-powered chatbots and virtual assistants play a key role during emergencies by sharing alerts quickly and guiding people on how to stay safe. AI is also playing an emerging role in healthcare emergencies it aids in ambulance dispatch, triage prioritization, and hospital bed allocation. By Integrating AI with GIS mapping, these systems will help to create dynamic, city-wide dashboards for emergency managers to coordinate responses across multiple agencies.

Moreover, for resolving any kind of barrier such as language, natural language processing (NLP) tools are used to analyse calls and social media posts for distress signals or early warning signs of public safety threats, including riots or cyberattacks. AI enhances the precision and agility of urban emergency systems by fusing real-time analytics with historical intelligence.

In short, AI helps cities become and safer by moving from simply reacting to problems to predicting and preventing them. This makes urban areas more responsive and better equipped to handle emergencies [28].

3.5 Urban Planning and Land Use Management

Artificial Intelligence is redefining the future of urban planning and land use management by enabling data-driven, dynamic, and sustainable city design stronger. Traditional method includes the dependence on manual data collection and static zoning models but now with the effect of AI urban planning is being enhanced through various AI technologies that includes satellite imagery, traffic flows, population density, and socio-economic indicators. These AI technologies are used to make more intelligent decisions. For the growth of urban cities Machine learning models are used which would assess the impact of infrastructure investments, and forecast future housing, transport, and green space needs. For the identification of underutilized land or to optimize zoning regulations and for modelling the effects of urban expansion on ecosystems and climate resilience, AI-powered geospatial analysis is planned for these identifications. Tools like remote sensing and computer vision allow cities to keep an eye on how land is being used, spot illegal buildings, and track encroachments as they happen. These technologies are especially important for rapidly growing cities dealing with unplanned development and environmental damage. Furthermore, AI provide scenario for urban development and also provide what if situation to test the urban development models under future climate or population conditions. This helps policymakers balance different priorities and make smarter choices about how to allocate resources, reduce disaster risks, and promote fairness across neighbourhoods. For example, AI can show how various zoning plans affect urban heat spots or flood risks, supporting better climate-ready city planning. For any new development Citizen engagement is also a priority, which is also enhanced through AI-enabled platforms. These platforms analyse public sentiment, preferences, and complaints to align urban planning initiatives with community needs. Natural Language Processing (NLP) tools mine data from surveys, social media, and digital forums, enabling more participatory, make helps to make better planning processes.

By bringing AI into land use and city planning, cities can grow in a way that's smart, fair, and sustainable—making sure the needs of people, infrastructure, and the environment are all balanced as urban areas expand quickly.

4 AI as a Tool for Climate Change Adaptation and Resilience

4.1 Early Warning Systems for Extreme Weather Events

Weather conditions are deteriorating due to various climatic changes and for providing early detection Artificial Intelligence has become an indispensable tool in enhancing early warning systems (EWS) [22]. Traditionally meteorological models are used but these models often lack the spatial precision and real-time responsiveness which is needed for localized urban risk management. However, to meet the growing demands, AI-driven systems address this gap by processing huge volumes of data from satellites, weather stations, IoT sensors, and historical climate records. From these sources AI driven system produces high-resolution and real-time forecasts. For earlier and more accurate predictions of events such as floods, heatwaves, cyclones, and wildfires it is important to identify complex, non-linear patterns in climatic data which has been done by Machine learning algorithms. By learning from past storm patterns, neural networks can help forecast where a storm might hit and how strong it could be. This gives authorities a better chance to warn people and organize evacuations in time. In urban regions also AI integrates local topography, drainage systems, and land-use patterns to simulate flood scenarios and assess neighbourhood-level vulnerabilities.

In the situation of emergency public trust and cooperation is utmost important and these AI-enhanced EWS not only improve the lead time of alerts but also reduce false positives, hence increasing public trust and institutional responsiveness. In case of any language barrier which hinder the community response Natural language processing (NLP) is used to give or understand alert messages to different communities, languages, and platforms so that effective communication and behavioural response take place. In addition to NLP AI chat box and mobile apps can spread information in real-time instructions to residents, helping coordinate evacuation, medical aid, and emergency supplies.

Cities such as Jakarta, New York, and Chennai are ensuring the future situations and risks so they are already deploying AI-integrated climate monitoring systems to anticipate disasters and mobilize resources. If AI-driven early warning systems is used effectively and accurately, it can help save lives, reduce damage to homes and infrastructure, and speed up recovery after disasters. They've become a key part of how cities prepare for and respond to climate challenges.

4.2 Risk Assessment and Vulnerability Mapping

Due to various climatic changes there is a need risk assessment and vulnerability mapping which is fulfilled by artificial intelligence. AI enables cities to identify, quantify, and proactively respond to the localized impacts of climate change. To change the traditional static assessments, AI systems include topography, population density, infrastructure condition, socio-economic factors, and historical disaster records to produce dynamic, high-resolution risk maps by Integrating AI with real-time and geospatial data. These maps play a major role in highlighting where the risk is, that is which communities, buildings, or assets are most susceptible to hazards like flooding, urban heat islands, droughts, or landslides. The correlations between environmental variables and social vulnerabilities are unveiled by Machine learning algorithms which enables planners to understand how climate risks intersect with poverty, age demographics, or housing types. There are some factor, like materials of building, green cover, AI models, which would predict which neighbourhoods will face the highest health risks during heatwaves. Likewise AI model will help to find an appropriate way to spread water during extreme rainfall in flood-prone zones on the basis of factors like runoff patterns, drainage efficiency, and elevation data [14].

The risk assessment done by AI can be used by Cities to prioritize interventions, like retrofitting infrastructure, escalating green buffers, relocating critical services, or improving drainage systems. These tools also play a crucial role in mapping climate risks, shaping insurance plans, and helping communities prepare for potential disasters. Various AI-enhanced platforms like Google's AI for Disaster Response and the World Bank's Think Hazard! Platform are being increasingly used by local governments to guide climate-smart urban planning.

AI helps us see not just where communities are vulnerable, but also tell how well they can adapt. This big-picture view makes sure that climate solutions aren't just effective they're also fair and inclusive. Due to this AI application, it enables cities to allocate resources more efficiently and to make prior safety measures to vulnerable populations, and build long-term urban resilience.

4.3 Climate-Resilient Infrastructure Design

As the climate is changing drastically, there arises a doubt that if urban development can withstand the growing threats posed by climate changes or not. So, to provide a solution Artificial Intelligence is reshaping how cities design and maintain climate-resilient infrastructure to ensure that urban area can withstand the growing threats. Traditional infrastructure planning was quite good but it often fails to account for the unpredictability and severity of climate extremes. Unlike the traditional model AI addresses this gap by using predictive analytics, simulation models, and optimization algorithms. AI model aims to design infrastructure systems that are adaptive, durable, and future-ready. climate projection data is integrated with geographic, hydrological, and engineering datasets so that AI models can simulate how infrastructure such as roads, bridges, drainage systems, and buildings will perform under different climate scenarios. For example, machine learning algorithms can identify materials and structural designs which are most resistant to heat stress, flooding, or seismic shifts. These algorithms guide engineers to make informed design choices. AI also helps decision-makers figure out where to invest by weighing the costs and benefits of different resilience strategies—like raising roads, adding more green spaces, or setting up flexible flood barriers.

Moreover, in urban planning, AI-powered tools like digital twins are being used to create virtual replicas of entire city blocks or infrastructure networks. These digital environments allow testing for the planners to test the performance of infrastructure under simulated extreme weather conditions. The observation from these testing enables the risk-informed decision-making. Such models are being implemented by some Cities like Copenhagen and Singapore to optimize storm water drainage, minimize surface runoff, and manage coastal inundation.

Additionally, through real-time data embedded sensor, AI enhances infrastructure monitoring and maintenance. These sensors are used in critical infrastructure to detect stress, vibration, or moisture levels and use AI to predict potential failures before they occur. Thus, these sensors enable preventive maintenance and reduces the long-term costs. By enabling an intelligent, adaptive, and climate-responsive infrastructure systems with AI supports cities in building urban environments will not only survive but thrive in the face of climate uncertainty.

4.4 AI for Water Resource Management and Flood Control

The climatic change also causes floods, so to manage urban water systems and mitigating flood risks Artificial Intelligence is playing an increasingly vital role. From the traditional method of reactive water management AI enables cities to move toward predictive, data-driven, and adaptive systems so that would reduce vulnerabilities to extreme hydrological events the optimising the use, storage, and distribution of water. In flood control, AI models can also analyse historical rainfall data, river levels, land elevation, drainage infrastructure, and real-time sensor inputs in the case of flood of control to forecast urban flooding with high spatial and temporal precision. By using deep learning along with hydrological models, experts can now better predict how stormwater will flow and pinpoint areas most likely to flood. This helps authorities take action early—whether it's pumping out excess water, releasing it gradually from reservoirs, or sending timely alerts to help people evacuate safely. By using such AI-enhanced flood prediction systems, Cities like Tokyo and Rotterdam tried to manage coastal and pluvial flood risks effectively [16].

To manage these water resources, AI plays a vital role by supporting smart metering, leakage detection, and consumption forecasting. For the prediction of any anomalies such as leaks or illegal connections, the consumption patterns are analysed by Machine learning algorithms, across vast municipal water networks. Apart from floods, during stress conditions, such as droughts also AI helps in optimizing water distribution by dynamically reallocating supply based on usage patterns and priority zones.

In modern urban planning, AI tools are being used to understand how water moves through a city—like how catchment areas behave, how groundwater gets replenished, and where rainwater flows on the surface. These insights help city planners make smarter decisions about where to invest in eco-friendly infrastructure, such as bioswales, rain gardens, and surfaces that let water soak through instead of running off. When these nature-based solutions are combined with AI-driven monitoring systems, the urban resilience is enhanced by following the process of absorbing excess water and reducing the strain on built drainage systems [30].

Overall, AI improves the waste resource management ability of cities. With the help of AI systems cities can manage water sustainably, mitigate flood risks, and adapt infrastructure. Many policies are made to increasingly volatile climate patterns which would make it an important asset for climate-resilient urban development.

5 Different Challenges AI Facing While Implementation

5.1 Data Privacy and Surveillance Concerns

The explosive development in AI is accountable to build smart, climate-resilient urban cities. The privacy and monitoring of data is a major concern. In order to maintain data privacy and monitoring effectively, AI-driven urban systems often rely on vast, continuous streams of data like traffic cameras, smart sensors, mobile applications, and utility meters. The challenge arises when the line between data collection for environmental management and surveillance begins to blur. For example, AI models which designed to optimize energy use or monitor heatwave exposure may mistakenly collect sensitive behavioural data, such as household occupancy patterns or individual movement trajectories. If strong legal protections aren't in place, these systems could be misused leading to invasive profiling, the mishandling of personal data, and unfair consequences for already vulnerable communities [7].

To ensure ethical implementation, a proactive approach to data governance is required. Transparent privacy policies, integrated anonymization and encryption protocols should be adopted by Cities. Also, the citizens should be engaged to check how data is collected and applied. Earning public trust isn't just about following the law—it's essential for creating cities that are both sustainable and democratic. As we work toward smarter and greener urban spaces, protecting people's privacy shouldn't come later it needs to be built into the design from the very beginning [2].

5.2 *Algorithmic Bias and Fairness*

As AI is emerging in each field such as transportation, energy management, public health, and climate adaptation, to drive the decision the cities increasingly dependent on AI. The issue of algorithmic bias and fairness emerges as a critical concern. AI systems work on pre-installed data and are only as fair and inclusive for the data on which its trained for. When historical data includes patterns of social inequality like income gaps, racial segregation, or uneven access to public services—AI systems can end up repeating or even worsening those same issues in how cities are planned and managed. For example, in case of flood prone areas AI-driven risk assessments excludes vulnerable populations from early warning systems or infrastructure investments due to lack of reliable geospatial data. Similarly, certain communities are neglected by predictive policing algorithms or resource allocation tools due to disproportionately target or careless design and audit for fairness.

Addressing algorithmic bias is not only a technical challenge but also a socio-political important issue. It requires inclusive data collection practices, diverse stakeholder engagement, and transparent model auditing. Making fairness a priority at every stage of the AI process that is from choosing the right data to putting the system into action, it helps ensure that smart city technologies aren't just efficient, but also fair and inclusive. In the context of climate resilience, to save the urban residents especially those most at risk equitable, AI systems are developed which benefit from innovation rather than being marginalized by it [11].

5.3 *The Digital Divide and Inclusivity*

Persistent digital divide is a fundamental barrier to the equitable deployment of AI in smart, climate-resilient cities. Despite technological advancements, not all communities have equal access to digital infrastructure, high-speed internet, or smart devices which are essential services provided by AI. Many low-income populations, elderly residents, migrants, and those living in informal or underserved settlements or often the same groups most vulnerable to climate change are disproportionately affected by this divide. The deepening social inequities are at risk due to AI-driven urban systems are designed without inclusivity in mind. For example, in optimization of transportation routes or emergency alert system, smart mobility apps which are reliant on smartphones may exclude residents without internet access or digital literacy. Similarly, the voice of marginalized communities is often ignored in AI citizen engagement platforms for participating in urban planning, leading to planning decisions that do not reflect their needs or risks.

To conclude construction and utilization of the digital divide is very important for both technological effectiveness as well as social justice. Cities must increase broadband utilization and availability, these must be equipped with digital infrastructure, and different awareness programs about digital literacy to utilize digital policies effectively can be installed. Cities can become truly strong and equitable to deal with accelerating climate challenges only by ensuring the strong and unbreakable connection among urban residents that may contribute on the basis of AI system [21].

5.4 Governance and Policy Implications

The demand in cities to make accurate regulatory frameworks to facilitate accountability, transparency, and public oversight is enhanced the effect of AI in decision making in different sectors i.e. climate adaptation, mobility and energy etc. But government is not able to control systems and tools of various cities interlinked with AI and utilization of its various benefits for society and environment. This may happen due to lack of quality education and awareness about AI and its applications as well as ethical standards. Due to lack of a strong framework there is different difficulties in implementation of AI technologies to get different benefits of these services. Most importantly owing to various global differences many cities rely on private companies to install AI based systems effectively that sometime leads to certain issues i.e. poor decision making or loss of privacy and data.

Knowledge of AI plays a significant role in policy making, collaboration among different public institutions, civil society etc. that ultimately helps to deal with various issues of society related to infrastructure. The alignment of AI deployment with public values and climate resilience goals can be ensured by Regulatory innovation such as AI ethics councils, public data trusts, and urban technology sandboxes. Ultimately, it is not only essential to manage AI's risks but also to unlock its transformative potential in shaping inclusive, sustainable urban futures which is fulfilled by effective governance [1].

6 Case Studies

A Digital Twin for Urban Planning

"Virtual Singapore" At the heart of Singapore's AI-driven urban is a dynamic 3D digital twin of the entire city-state. The real-time simulations of urban scenarios, aiding in disaster preparedness, infrastructure planning, and environmental monitoring is enabled by this platform. For example, in case of flood risk and urban heat island it allows authorities to make models and facilitate proactive measures against climate-related challenges.

AI in Transportation and Energy Management

The idea of creating virtual Singapore also involves the optimization of traffic flow and reduce emissions which was done by AI. There are different ERP systems employed to deal with traffic conditions very effectively to manage heavy traffic congestions. Furthermore, AI algorithms is also useful to monitor energy consumption in city buildings to ensure sustainability among urban sectors.

Enhancing Public Services by Using AI

To increase accessibility and public response Singapore government use AI services effectively for efficient public service. The name was sing pass digital identity system, a online system for more than 17,000 people equipped with different online services for efficient interaction among different government agencies. As an illustration AI-empowered Chabot's are employed for different enquiries. Moreover, bureaucracy is maintained by decreasing paper-based transactions that results in better outcomes.

Building Climate Resilience

AI is also utilized very efficiently in Singapore to create different techniques to handle climate variations. By using various AI tools weather pattern is estimated and resources are managed accordingly. Apart from these AI is helpful to plan sustainable and green infrastructure in Singapore as per the collected data. These tasks are supporting to deal with climatic variations and prior preparations for sudden environmental problems. To summarize, Singapore's Smart Nation Initiative is the suitable illustration that inform us about the smart AI based facilities to cope up with environmental or weather changes specifically in urban cities.

Singapore provides a blueprint for cities universally that want to find the different issues of urbanization and climate variations through comprehensive inclusion of AI in urban cities maintenance, to employ other public services as well as climate policies [18].

Urban AI Applications in Copenhagen and Amsterdam

- Copenhagen—AI based Sustainability and Mobility

By utilizing AI across various sectors Copenhagen aims to become carbon–neutral by 2025.

AI-empowered Smart Grid: For accurate distribution of energy among various cities different AI based systems are employed for proper distribution of energy and wind. Furthermore, Predictive algorithm is able to decreased dependence non-renewable energy owing to efficient balance between demand and supply of these natural resources.

Smart Traffic Management: To minimize traffic, AI based traffic signal system is used in Copenhagen. This system is very supportive to manage heavy traffic seamlessly as well as facilitate sustainable transportation modes.

Digital Street Lighting: For maintaining energy efficiency as well as to ensure public safety, more than 19,000 streetlights have been installed with LEDs equipped with automatic controls. On the basis of this smart system AI systems manage lighting system as per situation and circumstances.

- Amsterdam—AI for Urban Innovation and Climate Adaptation

Amsterdam's Smart City steps concentrate on combination of more than one innovation by employing AI tools to deal with urban area problems and climate flexibility [12].

Drukte Radar Project: The safety of public as well as resource distribution is evaluated via collecting data from different sources or finding as well as balancing crowd densities at ground level by using AI oriented services.

Smart Mobility Solutions: To enhance efficiency and minimize emissions, AI tools are implemented to balance traffic flow and facilitate public transportation schedules flawlessly.

Schoonschip Floating Community: To start a model for climate-resilient in urban cities this system is started via which innovative neighbourhood use AI-dependent smart grids to manage energy usage and its distribution among floating homes.

AI-Driven Climate Resilience in New York City

New York City (NYC) is known as a leader in installing AI based smart infrastructure to deal with climate changes specially in urban sectors. NYC is smartly dealing with the plethora of problems arises due to climate change via combination of predictive analysis, smart technologies as well as collaborative initiatives.

Prior Flood Management: To deal with increasing flood risks, NYC is using AI oriented techniques to manage flood problems and hazardous impacts. Latest data predicts that AI-powered models are able to minimize the cost of flood management by 6 to 36% under moderate carbon emission scenarios. By employing these tools, urban cities can manage smartly at more reasonable cost and infrastructure for efficient protection against flood problems.

Urban Heat Island Mitigation: To manage heat or temperature rise in urban island, combination of the AI and Internet of Things (IoT) devices can be employed to estimate and forecast local surface temperatures by NYC. Long Short-Term Memory (LSTM) deep learning models are installed, to predict about the details on temperature of the cities in advance and handled smatly. This type of information is significant for ensuring focused cooling steps and techniques and keeping vulnerable communities happy and safe during extreme heatwaves.

Energy Efficiency in Buildings: Furthermore, NYC also focused on enhancing energy efficiency within its building sector to boost energy efficiency. As an illustration, at 45 Broadway in Manhattan, HVAC energy consumption was reduced to 15.8% with the help of implementation of AI-powered systems. This small initiate responsible for annual savings about $42,000 as well as reduction of carbon dioxide emissions by 37 metric tons [20].

Collaborative Climate Initiatives: The NYC is promised to ensure efficient dealing with climate flexibility and this was demonstrated by significant steps like Climate Solutions Summit. Climate adaptation and significance of various AI tools, provides a base to various data-driven approaches and actionable strategies.

The New York City's has been set an example in front of other urban cities by using various strategies to implement AI into its climate management procedures. By implementation of digital technologies and boosting collaboration among different sectors, New York City is enhancing its ability to find, respond to, and minimize the outcomes of climate change.

7 Future Recommendations

Interdisciplinary Collaboration and Innovation

To speed up the process of smart building by using various AI driven tools it is required to collaborate among people of different fields to work together to fully utilization of policy and innovation simultaneously. The prominent challenges i.e. flooding, heatwaves, social imbalance are major barriers to install various policies efficiently. Therefore, to manage these problems there is urge to jump on the conventional disciplinary restrictions. AI systems are made smarter by employing data scientists, public health experts, Engineers, climatologists, sociologists and urban planners together to get better outcomes. This activity may facilitate unmatchable solutions to deal with current climatic issues and may able to match goal of sustainable and AI equipped smart environment. As an illustration, adaptive building planning, different smart infrastructure models and efficient sustainable planning is responsible for high tech smart city environment. Likewise, ethical issues such as data bias, surveillance, and digital exclusion. Are addressed by collaboration between technologists and social scientists [27].

In order to support sustained interdisciplinary engagement, various institutional frameworks are created such as urban innovation labs, public–private research consortia, and cross-sector advisory boards. These platforms ensures that they are inclusive, transparent, and accountable as well as they accelerate the development of AI solutions. In the end, creating resilient smart cities isn't just about technology—it's a shared effort that needs the knowledge and skills of people from many different fields.

Capacity Building and Skill Development

As AI becomes a bigger part of how smart, climate-ready cities work, it's just as important to invest in people so that they can make the best use of the new technology. The rapid evolution of AI technologies demands a parallel investment in upskilling urban stakeholders. The skills should be provided to everyone that is from municipal staff and urban planners to data analysts and community leaders. If the workforce is not well prepared the planned objective could never be accomplished It would result in risk of underperformance, misapplication, or public resistance in even the most advanced AI systems (Wolniak et al. 2024).

Capacity building must have multiple layers and is inclusive. To increase the literacy among people to use AI technology, data governance, and ethical oversight city governments need to start various training programs at the institutional levels. For the wider workforce, learning about AI, data science, and environmental modelling should go hand-in-hand with the goal of making cities more resilient. Community-level engagement is equally important; citizens should be equipped with digital literacy which would empower them to interact with AI-driven public services.

For designing adaptive learning ecosystems, Partnerships with universities, technical institutes, and industry can play a central role. Initiatives like city-sponsored AI fellowships, interdisciplinary urban tech curricula, and open-access training platforms can ensure that skill development keeps pace with technological innovation. At the end of the day, creating smart and resilient cities isn't just about technology—it's about people who are informed, skilled, and engaged in shaping the future of their communities [4].

Policy Frameworks for Sustainable AI Deployment

To use AI in cities in a truly sustainable way, we need strong, future-focused policies that encourage innovation while also making sure things stay fair, responsible, and environmentally sound. While AI offers unknown capabilities to enhance urban efficiency and climate resilience, its unchecked implementation can lead to ethical, social, and ecological risks. Therefore, need of a clear and adaptive policy architecture is essential to guide AI development and ensure its alignment with long-term urban sustainability goals.

For data governance, algorithmic transparency, public participation, and environmental impact assessment, Effective policy frameworks should be established. These rules also need to stay flexible so they can keep up with how fast AI is changing—while still holding on to core values like human oversight, inclusivity, and climate justice. Cities should be supported in using flexible, modular policies that can grow with their needs and be changed according to different settings—whether urban, peri-urban, or coastal [9].

Many platforms such as the UN-Habitat's Smart Cities Program or OECD's AI Principles can accelerate the development of best practices through international cooperation and knowledge exchange. Under supervised conditions the real-time testing of AI applications is experimented by Cities through regulatory sandboxes and urban technology labs. In the end, sustainable AI policy isn't just about avoiding risks it's about using AI to bring fair, green, and inclusive change in our cities [13].

8 Conclusion

Due to rapid urbanization and climate change cities are facing intensifying challenges. To overcome these challenges Artificial Intelligence (AI) emerges as a transformative force in reimagining urban resilience, sustainability, and inclusivity. Urban systems can become more intelligent, adaptive, and anticipatory by integrating AI with enabling technologies like IoT, Big Data, and Cloud Computing, urban systems. This integration makes the urban system capable of mitigating environmental risks, optimizing infrastructure, and enhancing quality of life.

However, the effective deployment of AI is far from a purely technical try. To implement these development goals deliberate attention to data ethics, algorithmic fairness, equitable access, and inclusive governance is required. Case studies from cities like Singapore, Amsterdam, Copenhagen, and New York City illustrate that if AI is applied accurately, it can drive not only operational efficiency but also systemic change toward climate resilience and social justice.

Looking forward, creating smart and sustainable cities will depend on how well we can bring different fields together that includes building people's skills, and designing flexible policies that keep technology working for the public good. AI is not a silver bullet, but when guided by ethical design and empowered governance, it can be a powerful catalyst in shaping urban futures that are not only smarter but fairer, greener, and more resilient.

References

1. Alsubayhay, A., Abdalla, M.: Enhancing citizen engagement in E-government services through AI-driven chatbots. Sebha Univ. Conf. Proc. **3**(3), 188–194 (2024)
2. Amnuaylojaroen, T.: Advancements and challenges of artificial intelligence in climate modeling for sustainable urban planning. Front. Artif. Intell. **8**, 1517986 (2025)
3. Bajwa, A.: AI-based emergency response systems: a systematic literature review on smart infrastructure safety (2025). Available at SSRN 5171521
4. Ben Dhaou, S., Isagah, T., Distor, C., Ruas, I.: Global Assessment of Responsible Artificial Intelligence in Cities: research and recommendations to leverage AI for people-centred smart cities (2024)
5. Bharadiya, J.P.: Artificial intelligence in transportation systems a critical review. Am. J. Comput. Eng. **6**(1), 35–45 (2023)
6. Birkmann, J., Garschagen, M., Kraas, F., Quang, N.: Adaptive urban governance: new challenges for the second generation of urban adaptation strategies to climate change. Sustain. Sci. **5**, 185–206 (2010)
7. Braun, T., Fung, B.C., Iqbal, F., Shah, B.: Security and privacy challenges in smart cities. Sustain. Cities Soc. **39**, 499–507 (2018)

8. Chatterjee, A., Chattopadhyay, R.N., Chatterjee, A., Chattopadhyay, R.N.: Growth of metropolises and megacities with focus on global south. Satellite Towns in Neo-metropolitan Development in India: Lessons from Selected Cities, pp.1–28 (2020)
9. Colding, J., Nilsson, C., Sjöberg, S.: Smart cities for all? Bridging digital divides for socially sustainable and inclusive cities. Smart Cities **7**(3), 1044–1059 (2024)
10. Cugurullo, F.: Urban artificial intelligence: from automation to autonomy in the smart city. Front. Sustain. Cities **2**, 38 (2020)
11. González-Sendino, R., Serrano, E., Bajo, J., Novais, P.: A review of bias and fairness in artificial intelligence (2024)
12. Guernieri, L.: An economical, sustainable and technological analysis of impact of the smart cities-the case study of Barcelona and Singapore (Master's thesis, Universidade NOVA de Lisboa (Portugal)) (2023)
13. Hacker, P.: Sustainable AI regulation. Common Market Law Rev **61**(2) (2024)
14. Jain, H., Dhupper, R., Shrivastava, A., Kumar, D., Kumari, M.: AI-enabled strategies for climate change adaptation: protecting communities, infrastructure, and businesses from the impacts of climate change. Comput. Urban Sci. **3**(1), 25 (2023)
15. Komninos, N., Kakderi, C., Panori, A., Tsarchopoulos, P.: Smart city planning from an evolutionary perspective. J. Urban Technol. **26**(2), 3–20 (2019)
16. Krishnan, S.R., Nallakaruppan, M.K., Chengoden, R., Koppu, S., Iyapparaja, M., Sadhasivam, J., Sethuraman, S.: Smart water resource management using artificial intelligence—a review. Sustainability **14**(20), 13384 (2022)
17. Kunzmann, K.R.: Smart cities: a new paradigm of urban development. Crios **4**(1), 9–20 (2014)
18. Lehmer, N., Anguelov, N.: AI for climate change adaptation: analyzing machine learning's role in combating California's wildfires. AI Soc. 1–11 (2025)
19. Li, W., Batty, M., Goodchild, M.F.: Real-time GIS for smart cities. Int. J. Geogr. Inf. Sci. **34**(2), 311–324 (2020)
20. Min, J., Yan, G., Abed, A.M., Elattar, S., Khadimallah, M.A., Jan, A., Ali, H.E.: The effect of carbon dioxide emissions on the building energy efficiency. Fuel **326**, 124842 (2022)
21. Mirzaei, R., Lakizadeh, E.: Developing a climate resilience framework for cities with an emphasis on innovative technology applications. In: Cities of Tomorrow: Urban Resilience and Climate Change Preparedness, pp. 269–289. Springer, Singapore (2024)
22. Reichstein, M., Benson, V., Blunk, J., Camps-Valls, G., Creutzig, F., Fearnley, C.J., Han, B., Kornhuber, K., Rahaman, N., Schölkopf, B., Tárraga, J.M.: Early warning of complex climate risk with integrated artificial intelligence. Nat. Commun. **16**(1), 2564 (2025)
23. Seelam, D.R., Kidiyur, M.D., Whig, P., Gupta, S.K., Balantrapu, S.S.: Integrating artificial intelligence in blue-green infrastructure: enhancing sustainability and resilience. In: Integrating Blue-Green Infrastructure into Urban Development, pp. 347–372) IGI Global Scientific Publishing (2025)
24. Siegel, F.R.: Adaptations of Coastal Cities to Global Warming, Sea Level Rise, Climate Change and Endemic Hazards. Springer (2019)
25. Suleimany, M., Gonbad, M.R.S., Naghibizadeh, S., Niri, S.D.: Artificial intelligence as a tool for building more resilient cities in the climate change era: a systematic literature review. In: Artificial Intelligence and Machine Learning Applications for Sustainable Development, pp. 60–81 (2025)
26. Turner, S.W., Uludag, S.: Toward smart cities via the smart grid and intelligent transportation systems. In: Smart Grid, pp.117–166 (2017)
27. Wolniak, R., Stecuła, K.: Artificial intelligence in smart cities—applications, barriers, and future directions: a review. Smart Cities **7**(3), 1346–1389 (2024)
28. Xi, R.: On emerging technologies: the old regime and the proactivity. Cardozo Int Comp. L. Rev. **8**, 75 (2025)
29. Yao, Y.: A review of the comprehensive application of big data, artificial intelligence, and internet of things technologies in smart cities. J. Comput. Methods Eng. Appl. 1–10 (2022)
30. Yereseme, A.K., Surendra, H.J., Kuntoji, G.: Sustainable integrated urban flood management strategies for planning of smart cities: a review. Sustain. Water Resour. Manag. **8**(3), 85 (2022)

Role of Advanced Techniques to Adapt Climate Resilience

Avnish Chauhan, Meena Kumari, Muneesh Sethi, Gaurav Pandey, Sushila Devi, Layza Sá Rocha, and Man Vir Singh

Abstract Climate change has been considered as one of the most important global concerns; it disrupts ecosystems, affects the livelihood of mankind and it causes high frequency and intensity of extreme weather phenomena. Climate resilience, the ability to anticipate, prepare for and respond to climate-induced stress, has become a key element of sustainable development. This chapter examines how advanced technologies are transforming climate resilience in the realm of agriculture, water management, and disaster preparedness, as well as in urban planning. It addresses how state-of-the-art tools including new and emerging remote sensing and geographic information systems technology, artificial intelligence, machine learning, the Internet of Things, and big data analytics can be applied to climate impact management and adaptation. The chapter also features biotechnological breakthroughs toward the development of climate-resilient crops and nature-based solutions that are synergistic with smart infrastructure. Illustrated with case studies and applied examples this an ideal reference text for graduates and researchers and an invaluable guide for practitioners who need to understand and apply these technologies. The chapter also

A. Chauhan (✉)
Department of Environmental Science, Graphic Era Hill University, Dehradun, UK, India
e-mail: avnishchauhan_in@yahoo.com

M. Kumari
Department of Environment Science, Rajkiya Kanya Mahavidyalaya, Shimla, HP, India

M. Sethi
Dean (Research and Development), COER University, Roorkee, Haridwar, Uttarakhand, India

G. Pandey
Forest Research Institute (FRI), Dehradun, Uttarakhand, India

S. Devi
Department of Physics, Rajkiya Kanya Mahavidyalaya, Shimla, HP, India

L. S. Rocha
Pharmaceutical Science Department, Federal University of Mato Grosso do Sul, Campo Grande, Brazil

M. V. Singh
School of Applied and Life Sciences, UIT, Uttaranchal University, Dehradun, UK, India

A. L. Srivastav et al. (eds.), *Climate Resilience: Impact of Quantum Computing and Artificial Intelligence on Urban Planning*, Studies in Systems, Decision and Control 637, https://doi.org/10.1007/978-3-032-06791-3_11

presents barriers of technology adaptation, data assimilation, policy coordination, and socioeconomic constrains. It highlights the necessity for inter-disciplinary work and closer cooperation between scientists, policy makers, and communities to effectively use advanced methods to contribute to adaptive and resilient climate strategies.

Keywords Climate resilience · Artificial intelligence (AI) · Machine learning (ML) · Remote sensing · Geographic information systems (GIS) · Internet of things (IOT) · Climate change mitigation

1 Introduction

The increasingly pervasive effects of climate change are transforming ecosystems, eroding livelihoods, and increasing the incidence of extreme weather events worldwide [1–4]. Indeed, standard mitigation and adaptation strategies have adequate difficulty in handling the fast-paced and complicated nature of environmental problems. Consequently there is increasing demand for their application in order to enhance climate resilience and promote sustainable development. Advances in technologies including Artificial Intelligence (AI), Machine Learning (ML), Remote Sensing, Geographic Information Systems (GIS), Internet of Things (IOT) and biotechnology have brought a paradigm shift in the manner climate risks are monitored, modelled and managed in key sectors like agriculture, water resources, urban infrastructure, and disaster management [5]. State-of-the-art technologies that offer informed assessments, augmented decision making and adaptive capacities through real-time surveillance, predictive analytics, and precision interventions are also being developed [6]. In this chapter, we examine how these advanced technologies can apply for constructing climate-resilient infrastructure, as well as discuss some challenges and prospects of their applications in developed and developing areas.

Resilience to climate is the capacity of ecosystems, communities and economies to anticipate, plan for, respond to, and recover from the impacts of climate [72]. It's about adapting to change, making appropriate responses to risk, and stimulating sustainable development and managing the impacts of climate change. The potential of resilience-building is critical in the process of sustainability in the long term, and reduces the vulnerability of communities to climate change effects.

A. **Relevance of Climate Change Adaptation**

Living with climate change to sustain ecological systems, economy and human settlement continuity [7, 8] are imperative. The rapid warming of the planet and the oceans, greatly accelerated in recent decades by greenhouse gas emissions, and coupled with sweeping changes following decades of increases in greenhouse gas emissions, rising sea levels, disappearing coastlines, accelerated species loss, ill-planned development, and an epidemic of extreme weather events are threatening food security, access to fresh water, infrastructure and our ability to keep the public safe and healthy, in other words, facing everything that humankind does. Failing to put effective adaptation strategies into practice could by 2050 create negative, irreversible impacts of climate change on the natural world and long-term socioeconomic instability [74].

People at risk—particularly in the least developed countries—are the main reason why climate adaptation is an urgent necessity. These groups are most likely to suffer the consequences of climate impacts, such as those in low-lying ocean areas, desert regions and the small island nations, which are dealing with popular landscapes that are suffering from serious flooding, drought and increasing temperatures. Adaptable infrastructure in addition to resilient water systems, and climate-oriented sustainable agricultural practices, are necessary investments to offset these vulnerabilities, and solidify long-term sustainability and business continuity [9, 10].

In addition, adaptation contributes to economic sustainability by mitigating the economic costs incurred from the impacts of climate, for instance, those caused by natural disasters. Investing strategically in climate-proof infrastructure, renewable energy systems, early warning systems can yield high returns in the form of reduced property damage and lower emergency post-disaster costs [11]. Concrete cases here include the building of dikes, the cultivation of drought-resistant crops, and the reforestation of vast areas—all measures that not only help to protect the environment but also provide important resources to essential sectors like agriculture, fisheries and tourism.

Adaptation is also key to protecting the natural environment and to maintaining a healthy ecosystem. The broad reconfiguration of natural habitats, due to global change, has resulted in the loss of species and the impairment of forests, wetlands, and coral reefs [12]. Ecosystem resilience and consequently the conservation of biodiversity can be significantly strengthened by nature-based solutions, for example through afforestation, restoring mangroves and establishing Marine Protected Areas.

Successful adaptation will depend on coordinated actions at all levels, from governments to business organisations to communities. Action among policy innovation, technological progress and climate finance must be action in concert. International institutions such as the Paris Agreement also stress the role in adaptation funding and capacity development helping vulnerable countries cope with the effects of climate change [13].

B. **The Advanced Users in the Service of Resilience**

The use of leading-edge technologies is critical to building resilience to the effects of climate change. Techniques like remote sensing, AI, climate modeling, and sustainable infrastructure practices are transforming the way climate risk, resource management, and policy formulation are being tackled [14] [73]. These state-of-the-art instruments enable accurate monitoring, long-term forecasting and informed decision-making and enhance the preparedness and adaptive capacity against natural disasters. These technologies, if utilized in various areas, can create an integrated and proficient approach for tackling the issues pertaining to the climate. This multi-faceted perspective does not only counteract the negative impact of climate change, but also paves the way to strong, pro-active adaptation (adaptation here defined as any improvement in the overall effectiveness, social equity and environmental sustainability of a flood management system).

2 Bibliography Study

This chapter will be provided with full review from Review articles, Book chapters and Research articles from "Scopus Database" searching between 2010 and 2025 in English language with search term Techniques to adapt climate resilience with Boolean operator "AND" combined search terms. Recent publications are taken into consideration in the review of the techniques and recent advances in this sensitive domain of climate resilience. A total of 126 unique publications available in Scopus-indexed journals and books were included in the present work.

Data were extracted from Scopus database under relevant keywords (for a bibliometric analysis). The data sets were analysed using VOS viewer 1.6.18 software and co-occurrence analysis with respect to all keywords unit full counting method and for threshold minimum number of occurrences of a keyword is 2 and out of 1376 keywords 222 met the threshold in 9 clusters i.e., 44 keywords in first cluster, 32 keywords in second cluster 30 keywords in third cluster, 27 keywords in fourth cluster, 26 keywords in fifth cluster 26 keywords in sixth cluster, 21 keywords in seventh cluster, 9 keywords in eighth cluster and 7 keywords in ninth cluster. In Fig. 1 we show a network and its overlay visualization with bigger frame sizes indicating a more prominent exposure and the more number of connecting lines the more networking potential as contributions authors can have to impact the scientific community. Those keywords occurring more often were represented in larger frames, indicating how they co-occurred in the analyzed publications so that the 12 terms are the top citing terms of the 126 papers in 2010–2025.

3 Understanding Climate Change and Its Impacts

Long term changes in temperature, precipitation and weather patterns are what is referred to when the term climate change is used difference sources. This is largely driven by human activities, for example, such as, greenhouse gas emission and deforestation [14]. One of the most significant challenges of our time is the climate change [15].

Specific Climate Change Threats: Perception and Analysis of Climate Change and Its Impacts are reflected in a number of serious environmental challenges such as:

- **Extreme weather events**: The latter have already been on the rise due to global temperatures soaring, hurricanes intensifying and occurring in higher frequency, bushfires ravaging communities, and storms becoming more violent [16]. These

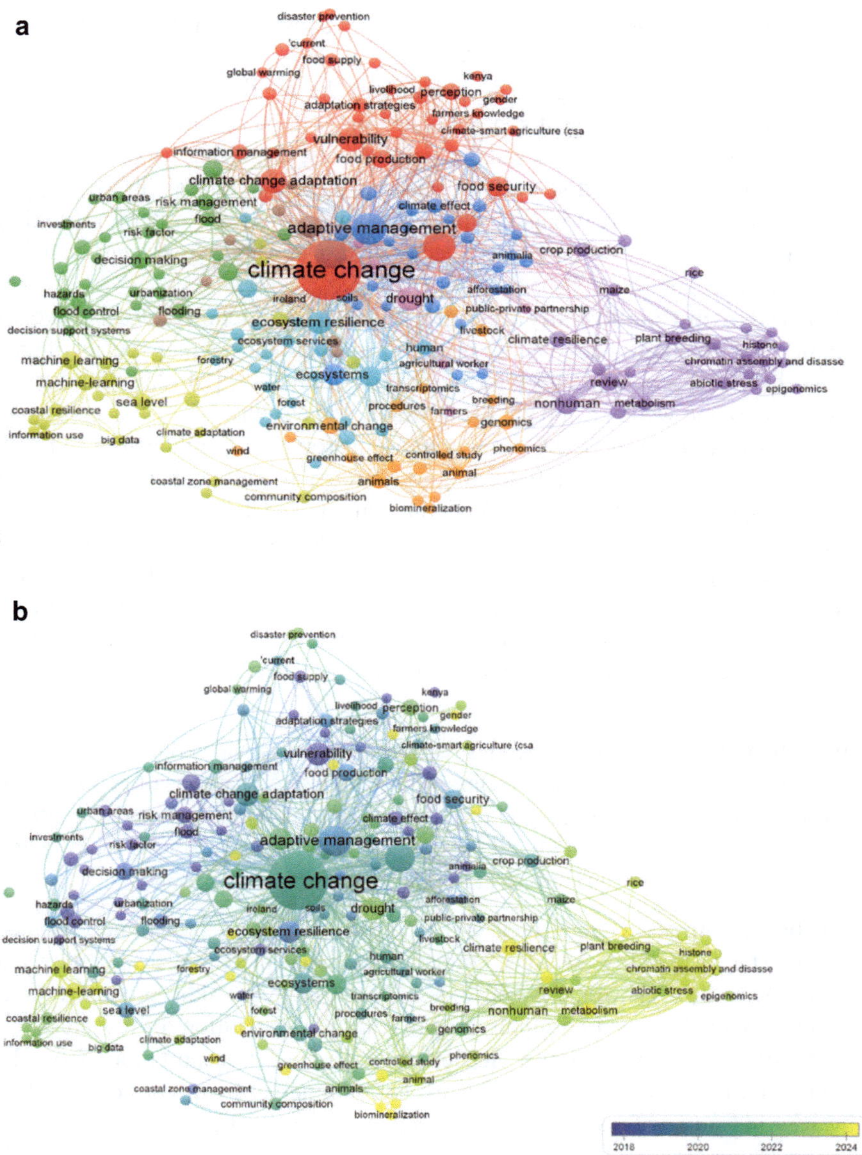

Fig. 1 Bibliometric analysis cantered on network clustering of 126 published articles from 2010 to 2025 which have high matches of selected keywords and their citations (1) network visualization (2) overlay visualization

are catastrophic events which devastate societies and damage infrastructure, often resulting in huge economic loss [17–21].

- **Water Level Rise**: Rising sea levels, due to the melting of ice caps and the expansion of warmer seawater, are threatening to flood coastal cities and island communities—flooding homes, damaging habitats, displacing people [22].
- **Extended droughts and scarcity of water**: Changes in rainfall are resulting in longer and more severe droughts, diminishing access to fresh water for drinking, farming and industry. These consequences, in turn, endanger food security, public health, and economic well-being [23].
- **Ecosystem and Biodiversity Changes**: Differences in temperature and precipitation are changing ecosystems, resulting in plant and animal species migration, habitat destruction and high biodiversity reduction [24, 25]. Coral reefs, woodlands and wetlands top the list of fragile ecosystems.
- **Agricultural impacts**: There are alterations in crop yields, soil fertility and pests and disease patterns, possibly leading to shortage as well as increase in market prices especially for those areas depending on rain-fed agriculture [17–19, 21].
- **Ecosystems and Biodiversity**: Species are threatened and ecosystems disrupted by increasing temperatures and habitat loss. Coral reefs—which protect shorelines from pounding breakers and waves by serving as a barrier—have been completely wiped out by ocean warming, putting at risk marine creatures and fish stocks. Water cycles and forest ecosystems are also affected because of deforestation and changing patterns of rainfall [2].

Agriculture and Food Security: Destruction of harvests and reduced agricultural output are the result of extreme weather events and altered patterns of precipitation. Shifting climates also let pests and diseases move to new regions, and that affects food supplies. Small pockets of soil erosion and water paucity are the main factors that impede sustainable agriculture [17–19, 19, 21].

- **Water Resources**: Glacial melt and changed patterns of precipitation impact freshwater availability. Warmer temperatures mean increased evaporation and along with it, even worse water stress. On the demand side, predicted water demand growth exceeding supply growth leads to competition for resources.
- **Infrastructure and Human Settlements**: Homes, transportation networks, public utilities and other areas of cities are pummelled by urban flooding, heat waves and storms. Food-producing coastal communities are being pushed out by encroaching seas and land erosion. Health risks are increased by poor quality air, water contamination, and vector-borne diseases [17–19, 21, 27].

4 Climate Resilience best Practices

Climate change is no longer a far-off threat—it's having serious impacts here and now on our most vital resources: our agriculture, water supply, infrastructure, and especially the health of our people and environment. With more frequent and severe extreme weather events, combined with ongoing sea-level rise, there is a pressing need for forward-thinking, innovative approaches to improve climate resilience. These challenges will not be addressed by conventional means only [22].

Comprehensive improvement, while traditional climate adaptation measures have made valuable contributions, they are generally not focused on the ambitious scale, complexity and dynamic nature of current climate threats [28]. Advanced approaches, however, comprise a transformative leap, coordinating new technologies, strategic approaches, and integrated methods to improve the ability of communities, infrastructure, and ecosystems to adapt to an increasing magnitude of climate-related risks.

What is Advanced Climate Resilience?

From being reactive to proactive. It is a proactive and transformational process that aims to:

- **Future-Proof**: Many organisations can provide future-proofing in their activities if they anticipate the impacts under future climate scenarios on their activities by using advanced climate modelling, analytics and risk assessment tools.
- **Reducing Exposure**: Address the drivers of community and ecosystem exposure to climate change, such as poverty, inequality and environmental degradation.
- **Enhance Adaptive Capacity**: Enhance the ability of people, communities and organizations to adapt to changes in climatic conditions, to access the knowledge and tools necessary to learn from experience, to understand how to innovate in the face of adversity, and to adjust to new conditions, situations and challenges in a flexible and proactive manner.
- **Advocate for system change**: Promote the systemic changes in policies, practices and behaviours that enable long term sustainability and resilience.

4.1 Data-Driven Technologies

How important is technology as you say that in a world that is becoming data driven, technology is becoming more and more important to build resilience to climate. These advanced technologies, such as AI and advanced analytics are leveraging real-time data analysis is playing a role in helping governments, industries, and communities to predict climate-related threats, adapt to ecological swings, and mitigate potential damages [29].

- **Satellite Imagery and GIS to Analyze Soil-Vegetation Relationships**

This tendency to create Data-forward processes has been consistently supported over the years by remote sensing and GIS-based tools. Remote sensing—the collection of information from an object or a phenomenon without making physical contact—produces huge amounts of data that come from satellites, aircraft and other sources. This information is generally present in form of spectral readings and pictures and gives spatially explicit information at different scales.

Remote sensing is only one example of spatial data that enters GIS, and GIS is an important framework for working with and viewing remotely sensed data. Geographical Information System (GIS) has strong analytical capabilities and can provide an analytic environment to support decision making by visualizing and analyzing the integration of these data with other spatial resources such as demographics, infrastructure and environmental variables. It allows hidden patterns, trends and relationships to be identified.

Remote sensing and GIS are point and counter point and has great influence for various applications [30]. With precision agriculture's optimal resource management coming from satellite-derived crop health analytics, and urban planning monitoring and modelling land use changes through aerial photography, it brings deep insights, and perspective. In addition, there has been growing reliance of all sectors of environmental monitoring, resource investigation and disaster management on the synergy between remote sensing and GIS to tackle some of the world's critical issues (Bello and Aina 2014). As data acquisition and processing capabilities further progress, so will the transformational possibilities for the way that we understand and manage the world around us.

- **AI and Big Data Analytics**

The data driven technologies making the fast industrial changes as well as the transformation of decision-making. The radical advancements in current industries are predominantly achieved through AI and Big Data Analytics, working in conjunction with each other and as complementary fields [31]. These methods are no competitive to each other, because still they can be combined and applied to obtain useful knowledge from large amounts of data. Big Data Analytics is a key enabler that allows Artificial Intelligence to operate at all. It is a field of study that offers all the necessary techniques and tools for analysing large scale daily data production [32]. The examination of patterns, trends as well as detection of anomalies in this information system, results in a significant insight in the knowledge of ongoing market dynamics and predictive forecasting capabilities. The AI tools essentially functions as a supplement, however, by consuming output from Big Data Analytics in order to create automated processes to grow forecast capabilities and customize service delivery. The fundamental AI part machine learning algorithms applies the given knowledge to achieve greater precision and to work more efficiently. These data stores make the learning potential of an AI system much more possible, which allow for the creation of high level complex operations such as fraud prevention systems and recommendation engines as well as autonomous vehicles.

Yet…, as their effectiveness is limited by the quality of data and ethical concerns [74] Prejudices in the dataset information induce further unfairness that results in discrimination. Responsible development and use of these technologies falls onto the shoulders of organizations, and that demands that they emphasize data governance, alongside transparency and accountability. Data is the power of AI Technology, when it is combined with Big Data Analytics. The symbiotic relationship between these resources increases possibilities for new business prospectuses and developments in various fields. When companies employ data responsibly, they can gain competitive advantage while increasing efficiency, which can ultimately be used to drive their operations forward.

- **Climate Modeling and Predictive Modeling/Soothills and Procedures**

Sophisticated tools are good enough to continue studying the adapting and tuning of climate changes, which pose an increasing threat to the Earth. These tools are important for controlling this global scale problem. Bahnhof processes large data sets that include atmospheric data as well as oceanic and land data for the production of comprehensive climate system simulations. Physical rules and statistical relationships push climate models towards producing future climatic conditions when pumped with different emission scenarios. The continual testing with enhancement also aids scientists in identifying the potential effects on temperature increase and height rise of oceans and extreme climate caused by human-induced activities. Statistical algorithms and machine learning techniques are used for predictive analysis to find patterns and trends in the historical climate data by researchers [33].

It can currently predict specific regional effects such as drought severity along with flood risk evaluations. The simulation of climate when coupled with predictive analytics gives policy-makers vital information for decision making. Using such results, analysts and policy-makers can create effective strategies (among others, carbon reduction plans, and investment in renewable energy development). Different segments can implement predictive analytics to develop targeted climate adaptation programs like dramatic infrastructure upgrades and disaster preparedness plans. Science-based technology is indeed a very important instrument in the fight against climate change which creates both current and future knowledge necessary to safeguard our planet [34].

4.2 Sustainable Agriculture Innovations

With climate change leading to the breakdown of the subsistent farming system, the sustainable agricultural innovations/interventions are becoming anchor as means of hope for better future. Either of the following methods can be used.

- **Climate-Smart Agriculture (CSA)**

Global food security is being challenged specifically by climate change, and demand for the optimized and sustainable use of inputs in agriculture. The concept CSA (Climate-Smart Agriculture) offers a comprehensive approach to increase agricultural productivity, climate adaptation and to minimize greenhouse gas emission [35]. By combining precision agronomy with sensors and data, CSA makes possible precise use of resources to save water and fertilizer, optimize crop yields. Conservation methods such as no till and cover cropping improve soil health, encourage carbon sequestration, and make for greater resilience in the face of extreme weather [36]. Advanced breeding and even genetic modification have resulted in drought-resistant crops that help farmers in dry stretches. CSA focuses on local context, local knowledge, enabling farmers and participatory approaches to adaptation. It contributes for the stability of the community by promoting diversified cropping systems [37]. CSA will need to be prioritized in agricultural institutions to create climate-resilient systems. With creative sustainable practices and people centred approaches, CSA ensures food security and environmental sustainability and stands to be a critical strategy for the future of agriculture amidst climate change [38].

- **Precision Agriculture and IoT Based Monitoring**

Smart agriculture and IoT-based monitoring have changed the face of agribusiness into intelligent management, which optimises resource use and even mitigates the impacts of climate [39]. In precision agriculture, GPS, remote sensing and variable rate technologies are employed to apply water, fertilizer and pesticides as well as tailor them to suit specific field requirements instead of applying uniformly and hence preventing losses and increasing yield. This approach is also improved thanks to IoT-enabled systems because horticultural data can now be monitored in real time, including, plant state, soil moisture and environmental data. Predictive intelligence and intervention are made possible by wireless sensors and by learning algorithms. Altogether these technologies enable farmers to enhance productivity, reduce environmental footprints, and establish more sustainable and climate smart agricultural systems [40].

- **Drought-Resistant and GM Crops**

Genetic modification (GM) approaches are valuable tools to increase the crop resilience to harsh environments, and to pest and diseases that is obligatory to guarantee the food security under the climate change pressure [41]. These methods, when coupled to traditional breeding practices, have resulted in the production of drought-resistant strains of crops that can maximise water-use efficiency such that crops require no watering during prolonged dry spells. This is particularly crucial given that global water shortages, exacerbated by climate change, jeopardize food security world-wide [42].

GM crops also enhance agricultural productivity by achieving more yield. Trait improvements, especially root growth, osmotic homeostasis, or stomatal closure, permit the plant to better tolerate stress, such as heat or water deficit. While concerns persist about the environmental and bio-safety of GM crops, a rigorous and transparent regulatory system resting on sound science is required to address this issue [43].

Genetically modified crops can be combined with drought tolerant plants to provide a long-term solution to a problem that can only worsen with a changing climate. Further studies and development in this field is needed to fortify the resilience of agriculture systems and ensure food availability for generations to come.

4.3 Green Infrastructure and Intelligent City

- **Urban Planning and Smart Water Management in a Sustainable Perspective**

The mounting risks associated with the changes in the climate call for comprehensive measures of dealing with this with the aim to safeguard human communities and ecosystems. Creating climate resilience requires more advanced agriculture systems, urban planning, and allocation of resources [44]. This article studies some of the leading innovations in sustainability in agriculture, cities and water that can address climate challenges.

Sustainable agricultural practices, including precision irrigation, drought-tolerant crops, and conservation tillage, support food security in the face of harsh weather. At the same time, green infrastructure—such as, urban forests, permeable pavements, and green roofs—provides nature-based solutions to mitigate the UHI effect, manage storm water, and enhance air quality.

Smart city projects also create resilience by maximizing the use of resources while building efficient infrastructure. Real-time data analysis assists with adapting to changing weather patterns, improving energy, transportation, and waste systems. Technological options for water management, such as monitoring for leaks and smart irrigation, can result in greater water use efficiency and alleviate scarcity [45].

An inclusive approach that links sustainable agriculture, green infrastructure, smart cities and innovative water management are all essential in overcoming the increasing influence of climate change and promoting a more resilient future.

- **Green Roofs Vertical Forests Eco-construction**

Sustainable agricultural techniques, such as crop rotation, no till farming and integrated pest management, help preserve soil health, water and sequester carbon [46]. Green roofs offer urban cooling, storm water retention and biodiversity benefits [47]. Vertical forests are a means to expand on the advantages of green roofs, providing cities with a natural habitat [48]. Green construction encourages the use of sustainable, energy-saving and low-waste generation materials [49]. Taken collectively, these approaches enhance resilience and environmental conservation in both urban and rural areas.

- **Climate Disaster Early Warning Systems**

Conventional farming cannot cope with erratic rain and new plant diseases. Technologies like precision irrigation, drought resistant crops, and vertical farming also boost food security [50]. Designing early warning systems, which leverage remote sensing and climate modelling in combination with machine learning is crucial for anticipating extreme weather [51]. Such systems enable timely evacuation, and allocation of resources and infrastructure protection. AI-enabled early warning systems add to better disaster preparedness through the analysis and prediction of climate data and hazardous weather conditions [52]. Contemporary scientific advances are a key to sustainability and resilience in the face of a changing climate.

4.4 Renewable Energy Solutions

- **The Contribution of Solar, Wind, and Hydro Power Produce Less Climate Risk**

Utilization of renewable sources of power helps in lessening over dependence on oil resources and decreasing in greenhouse gas emissions [53]. Low-carbon energy resources, such as solar, wind, and hydroelectric, have the double benefits of securing energy supplies and increasing resilience to climate change [54]. Renewable energy systems offer us two key advantages: Getting away from fossil fuels—and reducing climate change, and Generating product choices through access to diverse sources of power that enhance the security of our energy supply. Photovoltaic (PV) electric grids spread stored power through a series of localized feeds, limiting the extent to which utilities are exposed to disruptions resulting from climate change [55]. Wind uses naturally replenishing wind resources to create a cleaner, more flexible energy system. Hydropower can be an environmentally sustainable energy source in a well-managed system and supply clean energy to a region with adequate water resources [71]. The introduction of advanced smart grid technologies that integrate solar and wind generation with energy storage systems is also helping to mitigate disruption to supply caused by extreme weather events [56]. Renewable energy systems also respond environmental concerns as well strategic goals of climate resistance [57].

- **Smart Grids and Energy Management**

Smart grid technologies massively enhance the reliability of energy delivery on a typical weather-disturbed day. Efficient energy practices also minimize environmental contamination and contribute to sustainable operations [56]. Smart grids: They combine renewable sources, match demand and supply, and they avert grid collapse in extreme weather [54]. Broader uptake of energy-efficient technologies in all sectors, including in buildings and transport, also represents a key component of climate resilience UNEP [75]. The replacement of old technology with new high-efficiency equipment, classes of insulation, more electric vehicles, in all, reducing greenhouse gas emissions by today's modern [55]. Hence, energy efficiency contributes to resilience by maximising the reduction of the dependence on vulnerable infrastructure.

- **Nature-Based Solutions**

Nature-based Solutions (NbS) seek to strengthen ecosystems and biodiversity to help mitigate climate risk and promote sustainable development.

- **Reforestation and Afforestation Work**

Creating new forests and replanting areas that were previously deforested are important steps to approaching deforestation. Sophisticated remote sensing mechanisms such as LiDAR and drones are being used to refine site selection and to optimise carbon sequestration [57]. The selection of tree species that are adapted to local soils and climates improves survival and increases carbon uptake [54]. The increase in soil heath due to bio-char amending and the introduction of mycorrhizal fungi leads to increased growth and C-uptake [58]. Agroforestry blends crops with trees to increase production and promote the sustainable management of forests [59, 60]. Satellite and ground based monitoring systems underpin adaptive management and promote the long-term success of reforestation activities [57].

- **Wetland Protection and Mangrove Rehabilitation**

Wetlands and mangroves provide natural defence of coastal areas from erosion by storm surges [78]. Restoration programmes, which include satellite data, predictive modelling and genetic analysis, increase the resilience of these environments [57]. Dynamic interventions such as sand nourishment or bioengineered reef structures provide more sustainable and supporting for stabilization of the coastlines than hard engineering approaches [78]. Local community involvement is still key to the success [61].

- **Carbon Sequestration and Soil Health Care**

Soil carbon capture through sustainable land management practice is crucially important for climate change mitigation [54]. Novel approaches (e.g., BECCS (Bioenergy with Carbon Capture and Storage) and DAC (Direct Air Capture)) provide opportunities for carbon removal at scale [55]. Practices such as no-till farming, using cover crops, and integrated livestock management help soils to be more resilient to erosion and to the impacts of climate change [58]. Development of these strategies will rely on strong policy frameworks as well as further research investments [57].

4.5 *Policy and Governance Developments*

- **Relevance of International Climate Policies**

Global resilience initiatives are directed by international climate agreements such as the Paris Agreement (2015), the Sendai Framework for Disaster Risk Reduction, and the UN Sustainable Development Goals (SDGs). Inspired by the adoption of the Paris Agreement, the Parties are invited to make Nationally Determined Contributions (the "NDC") and engage with the finance structure of climate [62]. These frameworks facilitate the sharing of knowledge on a global scale and debate on the technologies of resiliency.

- **Mechanisms and Instruments of Climate Financing and Insurance**

Climate finance and insurance mechanisms are critical for assisting countries, particularly in the developing world and the climate-vulnerable countries, to respond to climate change impacts. Mechanisms, such as the Green Climate Fund (GCF), climate bonds and innovative insurance solutions such as parametric insurance, offer invaluable tools to mobilise finance for climate resilience. Created under the UNFCCC, the GCF has to-date disowned billions in finance to climate project to adaptation and mitigation efforts globally—favouring least developed countries, small island developing states and African countries [63].

Climate bonds and green bonds are being increasingly popular in the capital markets and provide an investment opportunity for investors to fund climate-smart infrastructure and renewable energy projects, with the added value of transparency and accountability. By 2024, cumulative global green bond issuance exceeded USD 3 trillion, according to the Climate Bonds Initiative [64].

Insurance-based answers are also developing. Tropical Cyclone Able starkly exposed these challenges, while demonstrating that parametric insurance in which payouts are triggered by specific climate events such as cyclones or droughts provides rapid liquidity for impacted regions, facilitates quicker recovery, and limits the humanitarian and economic toll of disasters. Efforts such as the ARC and the InsuResilience Global Partnership reflect the substantial scaling of climate risk insurance mechanisms to safeguard livelihoods in areas of high risk [65].

A milestone was the setting up the Loss and Damage Fund at COP27 and advancement on its operationalization at COP28. This fund caters to the vulnerable countries bearing irreversible impacts of climate change including, rise in sea levels, desertification, extreme climatic events [66]. It is a key recognition of climate justice, to close the funding gap for non-adaptable climate risks.

4.6 Private Participation in Adapting to climate Change

- **Public–Private Partnerships for Climate Adaptation**

Public–private partnerships (PPPs), which combine public sector oversight with private sector innovation and efficiency, are increasingly seen as a key mechanism for delivering large-scale climate resilience projects. PPPs can speed up climate adaptation in infrastructure, weather forecasting and disaster risk management by combining financial, technical and management resources. The World Bank [59, 60] finds that good PPPs 'do more than mobilize financing; they also scale up innovative climate solutions for tough risks that governments alone cannot easily tackle'.

A number of international programs illustrate the effective use of PPPs in responding to climate change. The 100 Resilient Cities program by the Rockefeller Foundation, which began in 2013, has helped cities worldwide develop broad-based strategies for adaptations. The initiative promoted public–private–civil society partnerships, facilitating the inclusion of resilience in urban planning and gaining private investment in adaptive urban infrastructure projects [67].

A second example is the African Risk Capacity (ARC)—an agency of the African Union, that combines sovereign risk pooling with private insurance markets. ARC helps African governments better plan, prepare and respond to weather-related risks and natural disasters by providing parametric insurance products tailored to the needs of African Member States [68]. It shows how public providers and private insurers can team up to provide timely and predictable financial assistance following climate shocks.

Resilient infrastructure public–private partnerships Jamaica Case Studay: Resilient Infrastructure Public–Private Partnerships (RIPPPs)—Jamaica Background Jamaica faces an increasing risk profile due to its status as a small island developing state with extremely limited domestic resources and no alternative development prospects.

Jamaica offers a strong example in the region of building climate resilience through PPP. In 2022, the Jamaican government, in partnership with the World Bank Group's Public–Private Infrastructure Advisory Facility (PPIAF) and the Global Infrastructure Facility (GIF), launched the Resilient Infrastructure Public–Private Partnerships (RIPPPs) initiative. The initiative aims to integrate climate risk into PPP contracts of priority sectors such as transport, energy and water.

One high-profile project is a climate-resilient highway from Kingston to the island's northern coast. Jamaican Government has signed a PPP agreement with private companies to rehabilitate the road network, allowing it to better cope with more powerful hurricanes and flooding. The agreement featured climate resilience covenants that obligated the private party to consider risk-based analyses, adaptive engineering designs, and long-term operations and maintenance plans based on expected future climate conditions [69].

The project has shown that when included in PPP frameworks, the mainstreaming of climate resilience not only increases the resilience of infrastructure services, but also builds the confidence of investors by addressing long-term climate risks. Furthermore, it demonstrates the need for capacity building, since the program trained government agencies and the private sector in climate risk assessment and adversarial project design.

4.7 The Future of Climate Adaptation and Five Tips to Guide It

The future of climate adaptation will require a revolution in how societies invest, invent, and co-operate. The response to the growing threat posed by climate risks should focus on a blend of increased investments, innovation, enhanced international collaboration, locally-driven adaptation and the widespread deployment of nature-based solutions [57].

- **Ramping Up Investment in Adaptation**

The scale of international adaptation financing required is still a long way off what is required to match the increasing effects of climate change. Adaptation costs in developing countries alone are projected to rise to USD 215–387 billion per annum by 2030 [] as noted in UNEP's Adaptation Gap Report 2023. Future plans need to include increasing both public and private spending. Mechanisms, including blended finance (blending concessional financing and private sector capital) are essential in de-risking projects and in getting a wider financial audience involved.

- **Technology Transfer to Industry and Advances in Technology**

The use of technology will be key to future adaptation methods. Advances in the development of climate models, early-warning systems, responsible infrastructure design and knowledge, precision agriculture, and biotechnology (e.g., drought-tolerant crops), can greatly reduce vulnerabilities [54]. AI and big data analytics are increasingly applied to predict climate risks and support real-time decision making, providing specific and cost-effective adaptation responses.

- **Enhancing International Cooperation**

Climate change is global by its very nature and has no boundaries, therefore it requires intense international co-operation. Agreements such as the Paris Agreement shift away from historically successful global models of solidarity, and do not really work anyway as a mechanism for transferring resources that a country doesn't have and that either need to be transferred, redefined or produced. The SSESC and TC can play its part in promoting scalable and context-specific adaptation options [66] by encouraging knowledge, technology, and resource sharing between developing countries. Existing global adaptation treaties, capacity-building and technical assistance efforts should be scaled up to leave no place behind.

- **Community-Based Adaptation at the Forefront**

Community-based adaptation (CBA) approaches enable local people, especially the most vulnerable to take the lead in the hard work of building resilience to climate change. Moving ahead, it is crucial that participatory planning and the incorporation of traditional knowledge as well as localised responses be given precedence. CBA enhances social capital and it ensures culturally satisfactory, sustainable and longer term adaptation plans and projects to be more efficient [63].

Scaling Up Nature-based Solutions (NbS)—Scale up the potential for and use of NbS by using them when appropriate or systematically and at scale, to optimize benefits across human systems in a way that supports climate change adaption and mitigation.

Nature-based solutions are increasingly being recognized as cost-effective and multifunctional tools for enhancing climate resilience. Projects such as mangrove restoration, protection of water catchments, urban forestry and reforestation deliver critical ecosystem services, as well as reducing climate risk [77]. NbS scaling-up will require NbS to be embedded in national adaptation plans, urban-planning processes and financing structures. Investment in nature-based solutions could deliver about 30% of the climate mitigation action required by 2030, while strengthening resilience, according to IUCN.

5 Conclusion

The severe risks that ecosystems as well as agriculture and water for humans and human settlement face due to climate change, compel urgent priority for climate resilience. Adaptation to climate change is broadly contingent upon a range of advanced hi-tech solutions such as big data-driven techno scientific along with sustainable agro-food innovations and green infrastructure, renewables as well as nature-based solutions which are fostered through governance innovations. The successful implementation of solutions encounter three major challenges consisting of lack of funding and lack of technology as well as unrealistic restricted policies. The strengthening of resilience needs more investment, technological innovation and international collaboration as well as community driven action for future development.

Advanced operational techniques can help societies to mitigate climate risks and ensure sustainable development. Varying combinations of public entities together with scientists, industries and community organizations should be working together on the climate—hardened systems that shield the environment as well as the economic assets and human welfare from developed stresses in a changing climate.

References

1. Bharti, P.K., Chauhan, A.: Anthropogenic activities and global climate change. In: Bharti, P.K., Chauhan, A. (eds.) Climate Change and Agriculture, pp. 1–22. Discovery Publication House (2012). ISBN: 978-93-5056-148-5
2. Jha, M.K., Dev, M.: Impacts of climate change. In: Jha, M.K., Dev, M. (eds.) Smart Internet of Things for Environment and Healthcare, pp. 139–159. Springer Nature Switzerland (2024). https://doi.org/10.1007/978-3-031-43219-1_8
3. Mehta, P.: The impact of climate change on the environment, water resources, and agriculture: a comprehensive review. In: Climate, Environment and Agricultural Development: A Sustainable Approach Towards Society, pp. 189–201. Springer Nature Switzerland (2024). https://doi.org/10.1007/978-981-99-3088-7_13
4. Yeshiwas, A.G., Bayeh, G.M., Gashu, Z.E., Teym, A., Mekonnen, B.A., Alemayehu, M.A., Yenew, C.: Climate change and antimicrobial resistance: application of one health approach to mitigating dual global threats. Discov. Public Health **22**(1), 1–14 (2025). https://doi.org/10.1007/s44172-025-00046-4
5. Mmbando, G.S.: Harnessing artificial intelligence and remote sensing in climate-smart agriculture: the current strategies needed for enhancing global food security. Cogent Food Agric. **11**(1), 2454354 (2025). https://doi.org/10.1080/23311932.2024.2454354
6. Luu, T.M.N., Yadav, M., Srivastava, A.P., Gopal, K.: Innovation and technology to address the challenges of climate change. In: Gopal, K. (ed.) Effects of Climate Change on Social and Economic Factors, pp. 423–464. IGI Global (2025). https://doi.org/10.4018/978-1-6684-9384-2.ch020
7. Hossain, B.: A systematic review of adaptation practices to promote health resilience in response to climate change. Environ. Dev. 101166 (2025). https://doi.org/10.1016/j.envdev.2025.101166
8. Steig, F., Oels, A.: Governing the climate in the Paris era: organized irresponsibility, technocratic climate futures, and normalized disasters. Wiley Interdiscip. Rev.: Clim. Chang. **16**(2), e70001 (2025). https://doi.org/10.1002/wcc.7001

9. Garai, J.: Impacts of climate change and environmental degradation on indigenous people in Bangladesh: an ethnographic study. Wild **2**(2), 13 (2025). https://doi.org/10.1007/s44186-024-00047-1
10. Muhsina, P.R.: Political ecology of climate change and environmental justice. In: Hossain, T.J. (ed.) Community Climate Justice and Sustainable Development, pp. 59–82. IGI Global (2025). https://doi.org/10.4018/978-1-6684-9081-0.ch003
11. Granata, F., Di Nunno, F.: Financing the future of water: unlocking investment, innovation, and governance for resilient infrastructure in a changing climate. Earth Syst. Environ. 1–25 (2025). https://doi.org/10.1007/s41748-025-00416-0
12. Osland, M.J., Bradford, J.B., Toth, L.T., Germino, M.J., Grace, J.B., Drexler, J.Z., Buffington, K.J.: Ecological thresholds and transformations due to climate change: the role of abiotic stress. Ecosphere **16**(4), e070229 (2025). https://doi.org/10.1002/ecs2.70229
13. Nautiyal, S., Klinsky, S.: The knowledge politics of capacity building for climate change at the UNFCCC. Clim. Policy **22**(5), 576–592 (2022). https://doi.org/10.1080/14693062.2022.2057571
14. Singh, V.: Global warming and climate change. In: Textbook of Environment and Ecology, pp. 283–295. Springer Nature Singapore (2024). https://doi.org/10.1007/978-981-99-4104-3_22
15. Gholizadeh, H., Friedman, M.S., McMillan, N.A., Hammond, W.M., Hassani, K., Sams, A.V., Adams, H.D.: Mapping invasive alien species in grassland ecosystems using airborne imaging spectroscopy and remotely observable vegetation functional traits. Remote Sens. Environ. **271**, 112887 (2022). https://doi.org/10.1016/j.rse.2022.112887
16. Meyer, A., Bresson, H., Gorodetskaya, I.V., Harris, R.M., Perkins-Kirkpatrick, S.E.: Extreme climate and weather events in a warmer world. Front. Young Minds **10**, 1–10 (2022). https://doi.org/10.3389/frym.2022.702947
17. Chauhan, A.: Climate change vulnerability profiles for Uttarakhand Himalaya, India. In: Bharti, P.K. Chauhan, A. (eds.) Climate Change and Agriculture, pp. 127–144. Discovery Publication House (2012). ISBN: 978-93-5056-148-5
18. Chauhan, A.: Climate change impact on Uttarakhand forestry. In: Bharti, P.K., Chauhan, A. (eds.) Climate Change and Agriculture, pp. 172–199. Discovery Publication House (2012). ISBN: 978-93-5056-148-5
19. Chauhan, A.: Methods of adaptation to climate effects on Himalayan ecosystems. In: Bharti, P.K., Chauhan, A. (eds.) Climate Change and Agriculture, pp. 306–316. Discovery Publication House (2012). ISBN: 978-93-5056-148-5
20. Chauhan, A., Kumar, R.: Impact of climate change on vulnerability of the Himalayan ecosystems and livelihoods. In: Singh, V., Nautiyal, N., Apparusu, Rawat, M.S.S. (eds.) Climate Change in Himalayas: Preserving the Third Pole for Cooling the Earth, pp. 19–38. Indus Publishing House (2011). ISBN: 9788173872280
21. Chauhan, A.: Possible impacts of climate change on agriculture of the Uttarakhand and Himalayan region. In: Bharti, P.K., Chauhan, A. (eds.) Climate Change and Agriculture, pp. 91–107. Discovery Publication House (2012). ISBN: 978-93-5056-148-5
22. Griggs, G., Reguero, B.G.: Coastal adaptation to climate change and sea-level rise. Water **13**(16), 2151 (2021). https://doi.org/10.3390/w13162151
23. Srivastav, A.L., Dhyani, R., Ranjan, M., Madhav, S., Sillanpää, M.: Climate-resilient strategies for sustainable management of water resources and agriculture. Environ. Sci. Pollut. Res. **28**(31), 41576–41595 (2021). https://doi.org/10.1007/s11356-021-13656-0
24. Muluneh, M.G.: Impact of climate change on biodiversity and food security: a global perspective—a review article. Agric. Food Secur. **10**(1), 1–25 (2021). https://doi.org/10.1186/s40066-021-00330-0
25. Sharma, P., Chauhan, A., Gupta, A.: Climate change an existential dread—a compendium study of climate change and its various effects. Webology **18**(1), 1407–1421 (2021)
26. Bouri, M., Arslan, K.S., Şahin, F.: Climate-smart pest management in sustainable agriculture: promises and challenges. Sustainability **15**(5), 4592 (2023). https://doi.org/10.3390/su15054592

27. Tol, R.S., Dowlatabadi, H.: Vector-borne diseases, development and climate change. Integr. Assess. **2**, 173–181 (2001). https://doi.org/10.1076/iaij.2.4.173.15558
28. Birkmann, J., von Teichman, K.: Integrating disaster risk reduction and climate change adaptation: key challenges—scales, knowledge, and norms. Sustain. Sci. **5**, 171–184 (2010). https://doi.org/10.1007/s11625-010-0108-y
29. Singh, M., Arora, V., Kulshreshta, K.: AI and the environment: innovative approaches to climate change. In: Arora, V. (ed.) Maintaining a Sustainable World in the Nexus of Environmental Science and AI, pp. 1–22. IGI Global (2024). https://doi.org/10.4018/978-1-6684-9479-5.ch001
30. Weng, Q. (2010). Remote sensing and GIS integration: Theories, methods, and applications. McGraw-Hill Education.
31. Zong, Z., Guan, Y.: AI-driven intelligent data analytics and predictive analysis in industry 4.0: transforming knowledge, innovation, and efficiency. J. Knowl. Econ. 1–40 (2024). https://doi.org/10.1007/s13132-024-01676-9
32. Chen, C.P., Zhang, C.Y.: Data-intensive applications, challenges, techniques and technologies: a survey on big data. Inf. Sci. **275**, 314–347 (2014). https://doi.org/10.1016/j.ins.2014.01.015
33. Bochenek, B., Ustrnul, Z.: Machine learning in weather prediction and climate analyses—applications and perspectives. Atmosphere **13**(2), 180 (2022). https://doi.org/10.3390/atmos13020180
34. Cooley, S., Jenkins, A., Schaeffer, B., Bormann, K.J., Abdallah, A., Melton, F., Graczyk, I.: Paths to research-driven decision making in the realms of environment and water. Technol. Soc. **70**, 101994 (2022). https://doi.org/10.1016/j.techsoc.2022.101994
35. Kabato, W., Getnet, G.T., Sinore, T., Nemeth, A., Molnár, Z.: Towards climate-smart agriculture: strategies for sustainable agricultural production, food security, and greenhouse gas reduction. Agronomy **15**(3), 565 (2025). https://doi.org/10.3390/agronomy15030565
36. Chauhan, A.: Evaluating the potential of agro-forestry for sustainable land use management: an analytical perspective. Turk. Online J. Qual. Inq. **11**(2), 652–657 (2020). https://doi.org/10.52783/tojqi.v11i2.9985
37. Bellanthudawa, B.K.A., Pawuluwage, S.M., Nawalage, N.M.S.K., Rathnasooriya, D.D.K.N., Dissanayake, O.D.I.P., Perera, I.J.J.U.N., Tennakoon, A.: Climate change, sustainable food systems, and community-based adaptation: resilience strategies in the global south. In: Climate Change, Food Security, and Land Management: Strategies for a Sustainable Future, pp. 1–30. Springer Nature Switzerland (2025). https://doi.org/10.1007/978-3-031-42977-1_1
38. Kirina, T., Groot, A., Shilomboleni, H., Ludwig, F., Demissie, T.: Scaling climate smart agriculture in East Africa: experiences and lessons. Agronomy **12**(4), 820 (2022). https://doi.org/10.3390/agronomy12040820
39. Sharma, S.K.: Smart agriculture revolution: cloud and IoT-based solutions for sustainable crop management and precision farming. In: Muhsina, P.R. (ed.) AI-Based Advanced Optimization Techniques for Edge Computing, pp. 253–288. IGI Global (2025). https://doi.org/10.4018/978-1-6684-8423-9.ch009
40. SS, V.C., Hareendran, A., Albaaji, G.F.: Precision farming for sustainability: an agricultural intelligence model. Comput. Electron. Agric. **226**, 109386 (2024). https://doi.org/10.1016/j.compag.2023.109386
41. KhokharVoytas, A., Shahbaz, M., Maqsood, M.F., Zulfiqar, U., Naz, N., Iqbal, U.Z., et al.: Genetic modification strategies for enhancing plant resilience to abiotic stresses in the context of climate change. Funct. Integr. Genomics **23**(3), 283 (2023). https://doi.org/10.1007/s10142-023-00968-3
42. Fitton, N., Alexander, P., Arnell, N., Bajzelj, B., Calvin, K., Doelman, J., et al.: The vulnerabilities of agricultural land and food production to future water scarcity. Glob. Environ. Chang. **58**, 101944 (2019). https://doi.org/10.1016/j.gloenvcha.2019.101944
43. Singh, O.V., Ghai, S., Paul, D., Jain, R.K.: Genetically modified crops: success, safety assessment, and public concern. Appl. Microbiol. Biotechnol. **71**(5), 598–607 (2006). https://doi.org/10.1007/s00253-006-0559-6

44. Webb, N.P., Marshall, N.A., Stringer, L.C., Reed, M.S., Chappell, A., Herrick, J.E.: Land degradation and climate change: building climate resilience in agriculture. Front. Ecol. Environ. **15**(8), 450–459 (2017). https://doi.org/10.1002/fee.1530
45. Gupta, A.D., Pandey, P., Feijóo, A., Yaseen, Z.M., Bokde, N.D.: Smart water technology for efficient water resource management: a review. Energies **13**(23), 6268 (2020). https://doi.org/10.3390/en13236268
46. Pretty, J.: Agricultural sustainability: concepts, principles and evidence. Philos. Trans. R. Soc. B: Biol. Sci. **363**(1491), 447–465 (2008). https://doi.org/10.1098/rstb.2007.2163
47. Oberndorfer, E., Lundholm, J., Bass, B., Coffman, R.R., Doshi, H., Dunnett, N., et al.: Green roofs as urban ecosystems: ecological structures, functions, and services. Bioscience **57**(10), 823–833 (2007). https://doi.org/10.1641/B571005
48. Boretti, A., Rosa, L.: Reassessing the projections of the world water development report. npj Clean Water **2**, 15 (2019). https://doi.org/10.1038/s41545-019-0039-9
49. Zuo, J., Zhao, Z.-Y.: Green building research—current status and future agenda: a review. Renew. Sustain. Energy Rev. **30**, 271–281 (2014). https://doi.org/10.1016/j.rser.2013.10.021
50. Despommier, D.: The Vertical Farm: Feeding the World in the 21st Century. Thomas Dunne Books (2010)
51. Basher, R.: Global early warning systems for natural hazards: systematic and people-centred. Philos. Trans. R. Soc. A: Math., Phys. Eng. Sci. **364**(1845), 2167–2182 (2006). https://doi.org/10.1098/rsta.2006.1819
52. Thomas, A., Baptiste, A., Martyr-Koller, R., Pringle, P., Rhiney, K.: Climate change and small island developing states. Annu. Rev. Environ. Resour. **45**, 1–27 (2020). https://doi.org/10.1146/annurev-environ-012320-083355
53. International Renewable Energy Agency (IRENA): Renewable Energy: A Gender Perspective (2019). https://www.irena.org/publications/2019/Jan/Renewable-Energy-A-Gender-Perspective
54. IPCC: Climate change 2022: impacts, adaptation and vulnerability. In: Contribution of Working Group II to the Sixth Assessment Report of the Intergovernmental Panel on Climate Change. Cambridge University Press (2022). https://doi.org/10.1017/9781009325844
55. International Energy Agency (IEA): Renewable Energy Market Update (2021). https://www.iea.org/reports/renewable-energy-market-update-2021
56. International Energy Agency (IEA): World Energy Outlook 2022 (2022). https://www.iea.org/reports/world-energy-outlook-2022
57. United Nations Environment Programme (UNEP): State of Finance for Nature 2022 (2022). https://www.unep.org/resources/state-finance-nature-2022
58. Food and Agriculture Organization of the United Nations (FAO): Soil Organic Carbon: The Hidden Potential (2020). https://www.fao.org/documents/card/en/c/ca7527en/
59. World Bank. Climate-Smart Agriculture Investment Plan for Punjab. World Bank Group (2021). https://openknowledge.worldbank.org/entities/publication/8f770191-6d9b-5317-97d7-7aa62e2bdb8e
60. World Bank: Public–Private Partnerships for Climate Resilience: Key Principles and Case Studies (2021). https://ppp.worldbank.org/public-private-partnership/library/public-private-partnerships-climate-resilience-key-principles-and-case-studies
61. Rampheri, M.B., Dube, T.: Local community involvement in nature conservation under the auspices of community-based natural resource management: a state of the art review. Afr. J. Ecol. **59**(4), 799–808 (2021). https://doi.org/10.1111/aje.12862
62. United Nations Framework Convention on Climate Change (UNFCCC): The Paris Agreement (2015). https://unfccc.int/process-and-meetings/the-paris-agreement/the-paris-agreement
63. Green Climate Fund (GCF): Annual Report 2024 (2024). https://www.greenclimate.fund/document/annual-report-2024
64. Climate Bonds Initiative (CBI): Green Bond Market Summary Q1 2024 (2024). https://www.climatebonds.net/resources/reports/green-bond-market-summary-q1-2024
65. InsuResilience: Annual Impact Report 2023 (2023). https://www.insuresilience.org/publications/annual-impact-report-2023/

66. United Nations Framework Convention on Climate Change (UNFCCC): COP28 Decisions on Loss and Damage Fund (2023). https://unfccc.int/news/cop28-decisions-on-loss-and-damage-fund
67. Rockefeller Foundation: 100 Resilient Cities: Final Report (2020). https://www.rockefellerfoundation.org/wp-content/uploads/100RC-Executive-Summary-Final.pdf
68. African Risk Capacity (ARC): Annual Report 2023 (2023). https://www.africanriskcapacity.org/2023/06/06/arc-annual-report-2023/
69. World Bank: Building Resilient Infrastructure Through PPPs: The Jamaican Experience (2023). https://www.worldbank.org/en/news/feature/2023/10/12/building-resilient-infrastructure-ppp-jamaica
70. Secretariat of the Convention on Biological Diversity (CBD): Global Biodiversity Outlook 5 (2020). https://www.cbd.int/gbo/gbo5/publication/gbo-5-en.pdf
71. World Bank Tracking. SDG 7.: The Energy Progress Report 2020 (2020). https://documents1.worldbank.org/curated/en/344121590675372111/pdf/Tracking-SDG-7-The-Energy-Progress-Report-2020.pdf
72. Banu, N., Fazal, S.: Climate change, livelihood crisis and resilience: An introduction. In: Banu, N., Fazal, S. (eds.) Livelihoods and well-being in the era of climate change: Risk to resilience across India (pp. 3–18). Cham, Switzerland: Springer Nature. (2025). https://doi.org/10.1007/978-3-031-81132-6
73. Al-Raeei, M.: Artificial intelligence for climate resilience: Advancing sustainable goals in SDGs 11 and 13 and its relationship to pandemics. Discov. Sustain. **5**(1), 513 (2023). https://doi.org/10.1007/s43621-024-00775-5
74. Aldoseri, A., Al-Khalifa, K.N., Hamouda, A.M.: Re-thinking data strategy and ntegration for artificial intelligence: Concepts, opportunities, and challenges. Appl. Sci. **13**(12), 7082 (2023). https://doi.org/10.3390/app13127082
75. United Nations Environment Programme.: 2021 Global Status Report for Buildings and Construction:Towards a zero-emission, efficient and resilient buildings and construction sector. Nairobi: United Nations Environment Programme (2021). https://www.unep.org/resources/report/2021-global-status-report-buildings andconstruction
76. United Nations Environment Programme.: Adaptation Gap Report 2023: Underfinanced. Underprepared.Inadequate investment and planning on climate adaptation leaves world exposed. Nairobi: United Nations Environment Programme (2023). https://doi.org/10.59117/20.500.11822/43796
77. International Union for Conservation of Nature.: Nature-based solutions for climate change adaptation: A global overview. Gland, Switzerland: International Union for Conservation of Nature (2020). https://www.iucn.org/resources/report/2020-nature-based-solutions-report
78. World Wide Fund for Nature.: A playbook for nature-positive infrastructure: Advancing sustainable development through ecosystem-based solutions (2021). https://files.worldwildlife.org/wwfcmsprod/files/Publication/file/3cv0m8510d_WWF_Playbook_v1_1_Pages_1_.pdf

Applications of Machine Learning and Artificial Intelligence in Soil and Environmental Sciences: Methods, Challenges, and Future Perspectives

Yassine Al Masmoudi, Anas Tallou, Mariya Ouaissa, and Mariyam Ouaissa

Abstract Artificial intelligence (AI) is gradually being recognised as an innovative instrument in environmental and soil sciences, providing novel solutions for monitoring, analysis, and decision-making processes. This paper examines at the present state of AI applications in various domains, with a focus on soil science. It demonstrates how intelligent agricultural systems forecast soil texture, calculate soil water content, and optimize irrigation techniques by combining machine learning (ML), deep learning, and hybrid artificial intelligence. Furthermore, high-quality, real-time soil evaluation at various geographical and temporal scales is made possible by the combination of artificial intelligence (AI) with geographic information systems (GIS) and sensor technologies. Along with describing deep learning algorithms used to analyze soil qualities, this research also discusses the challenges, limitations, and ethical dilemmas associated with the application of AI in environmental contexts. By pointing out advancements and shortcomings in the existing literature, this study seeks to direct future research towards more clever, efficient, and sustainable soil management techniques.

Keywords Artificial intelligence · Soil science · Machine learning · Smart agriculture · Remote sensing/GIS · Environmental monitoring

Y. Al Masmoudi
Laboratory of Geosciences and Environmental Technics, Chouaïb Doukkali University, El Jadida, Morocco

A. Tallou
Efficient Use of Water in Agriculture Program, Institute of Agrifood, Research and Technology (IRTA), Cabrils, Spain

M. Ouaissa (✉)
LISI, Cadi Ayyad University, Marrakech, Morocco
e-mail: m.ouaissa@uca.ac.ma

M. Ouaissa
LTI, Chouaib Doukkali University, El Jadida, Morocco

A. L. Srivastav et al. (eds.), *Climate Resilience: Impact of Quantum Computing and Artificial Intelligence on Urban Planning*, Studies in Systems, Decision and Control 637, https://doi.org/10.1007/978-3-032-06791-3_12

1 Introduction and General Overview

Machine learning (ML) is a branch of artificial intelligence (AI), which was actually coined by John McCarthy in 1956 to denote machines with the ability to perform tasks associated with human intelligence such as planning, object detection, language understanding, and learning [1]. ML is mostly focused on creating automation programs that learn from experiences and make intelligent predictions. This learning process relies on statistical methods that leverage experience for optimizing performance in the long term. ML algorithms are typically trained with labeled or unlabeled data to develop a model [1, 2]. The model may then be utilized for analyzing new data and generating predictions output, which are utilized for measuring specific performance measures following training [3, 4]. While there has been a great deal of progress, no computer has yet achieved general intelligence, and some are skeptical that achieving it is even possible [5]. However, many researchers and companies are working tirelessly to make machines that can do specific, narrow tasks more intelligent with AI [6]. These systems often leverage ideas from machine learning. Essentially, machine learning is about creating computer programs that possess the ability to refine their performance tasks with the passage of time [7, 8], without needing human developers to reprogram or correct them manually.

Supervised learning, unsupervised learning, and reinforcement learning are the three primary categories into which machine learning (ML) can be broadly classified [9, 10]. In supervised learning, the model is trained using algorithms like regression and classification, and the data contains predefined labels. The model can accurately predict new data after training [11–13]. Unsupervised learning, on the other hand, works with labelless data. The system uses algorithms like clustering and association to find hidden groups or patterns in the data [14].

A learning agent that interacts with its surroundings and receives rewards or penalties based on its actions is part of the third type, reinforcement learning. The Markov Decision Process is a popular model for this. In essence, the agent uses trial and error to discover the best strategies in a new environment. These three primary categories of machine learning are depicted in Fig. 1 [15, 16].

2 Analysis of Machine Learning Algorithms

2.1 Support Vector Machines (SVMs)

Belong to the class of supervised learning algorithms since they don't make any assumptions about the distribution of the underlying data. Both classification and regression tasks frequently employ this technique [17]. SVMs facilitate the separation of data points for classification by mapping the input data into a higher-dimensional feature space. Fundamentally, the algorithm works by segmenting the dataset into discrete areas, which enhances its ability to recognize significant patterns. It creates

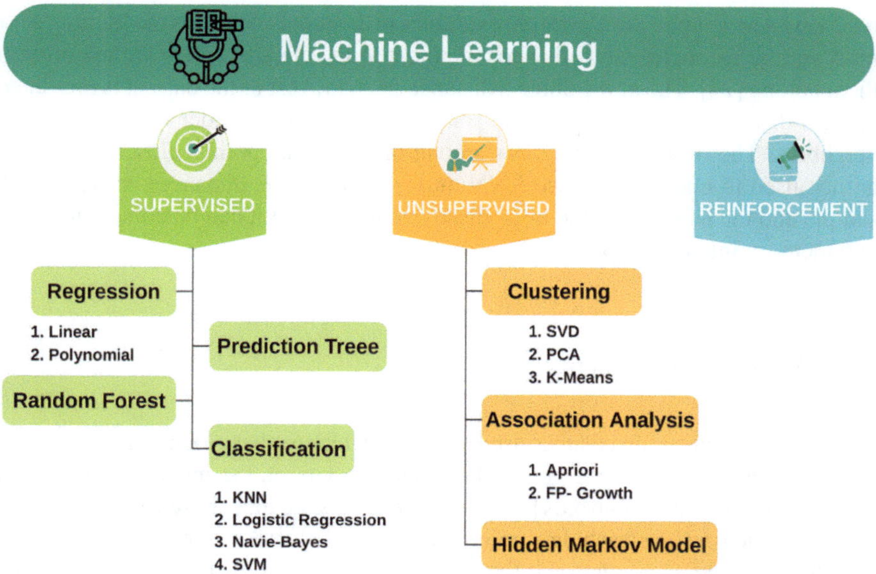

Fig. 1 Categories of machine learning algorithms

a decision boundary, also known as a hyperplane, that divides data points according to their class labels. When there is a distinct linear separation in the data, SVMs perform especially well in classification tasks [18, 19].

2.2 *Random Forest*

Random Forest is an ensemble learning algorithm that is frequently used for both regression and classification tasks. In classification tasks, it typically uses a majority vote to generate a final result by combining predictions from several decision trees [20]. Since training samples for each decision tree are chosen at random, some data points may be included more than once, while others may be omitted. Furthermore, during the tree-building process, features are selected at random, contributing to the trees' diversity. By combining the outputs of these independent trees, the technique lowers bias while maintaining a high level of variance, which eventually improves the accuracy and robustness of the model [21–23].

2.3 *K-Nearest Neighbour*

By comparing new data samples to pre-existing training examples, the K-Nearest Neighbour (KNN) algorithm classifies them according to how similar they are. Upon receiving a new input, the algorithm examines its features, identifies the training set's most similar examples, and allocates the input to the class to which the majority of them belong [24]. The new sample's distance from the rest of the dataset is measured using a distance metric, such as the Euclidean distance, in order to identify these neighbors. The most prevalent class among the new data point's closest neighbors is then used to classify it. Since KNN must determine the distances between each sample and the new input, one of its disadvantages is that it can be computationally and memory intensive [25, 26].

2.4 K-Means Clustering

K-means clustering is an unsupervised learning technique that groups data points into k distinct clusters. In order to properly divide the dataset, the method finds cluster centres, also known as centroids [27]. These centroids are first chosen at random. The algorithm then assigns each data point to the closest cluster based on the Euclidean distance between each point and the centroids. The mean of the data points inside each cluster is calculated to update the centroids following all assignments [28–30]. Until the clusters stabilise, this process is repeated. Since there is no set rule or accepted technique for choosing the ideal number of clusters (k), this is one of the primary problems with K-means.

3 Analysis of Deep Learning Algorithms

Deep learning (DL) is a branch of ML, which itself falls under the broader field of AI. DL relies on neural networks, inspired by biological neurons, to train machines to carry out specific tasks. These networks are composed of an input layer, one or more hidden layers, and an output layer. When multiple hidden layers are added, the network becomes a deep learning model, allowing it to learn more abstract and complex features [31–35]. DL surpasses traditional ML algorithms in terms of performance efficiency, largely thanks to its parallel processing capabilities and its ability to manage large volumes of data. With their layered structure, neural networks play a versatile role across a wide range of applications. The next section provides a detailed overview of various neural network frameworks that can be beneficial in the field of environmental science [36, 37].

3.1 Artificial Neural Networks Artificial

Among the deep learning domains that emulate the functioning of the human brain are neural networks. Simple programming is not needed to create rules for these algorithms' operation. To carry out parallel processes, ANN is made up of neurones, just as the neurones in the human nervous system [27]. Based on the real-time information provided to the neurone, these systems generate the value of the final forecast. It is said that the neurone receives information that is a mix of weight and bias. The input's significance in contributing to the creation of reliable forecasts determines how much weight it is given.

3.2 Convolution Neural Networks

A Convolutional Neural Network (CNN) is categorised as a feedforward neural network as data moves forward from the input to the output [38]. CNNs are frequently employed for image processing jobs, but they can also handle a variety of data sources, including multidimensional data like seismic recordings, two-dimensional data like greyscale or colour pictures, and one-dimensional inputs like time series. Convolutional and pooling layers are usually included in the CNN design, followed by fully linked layers. Convolution operations, in which filters move over the input to create feature maps, are how convolutional layers manage feature extraction. AlexNet, VGGNet, GoogleNet, and ResNet are well-known deep CNN-based architectures that have been used for a variety of image processing tasks [39, 40].

3.3 Recurrent Neural Networks

Sequential data inputs with an innate link between the data samples make up real-world datasets. This type of sequential data is incompatible with traditional ML techniques. A specialised area of deep learning called RNN is able to model sequential data and extract valuable insights from it. The input and output layers of this network are separated by a number of hidden layers. After receiving an input sequence, each layer in the network generates an output sequence that is sent to the following layer. A kind of RNN, Long Short Term Memory (LSTM) Networks [41] are made up of input, forget, and output gates that preserve the state of the prior input. Instead of just getting the most recent outputs, this configuration retrieves all of the crucial statuses of earlier inputs as needed.

3.4 Back Propagation Neural Networks

This is one of the most straightforward types of neural networks, with at least one hidden layer in addition to the input layer and output layer. There are various neurons in each layer used for information processing [41]. Information flows through the network in two main directions: forward and backward. During the forward pass, neurons in the input layer feed information through the hidden layer to the output layer. If the output from the output layer is not comparable to the desired outcome, the network recalibrates the weights of the neurons through a process known as backpropagation. It does this repeatedly until the error has been minimized as much as it can. This type of neural network is applied widely in nearly all remote sensing applications due to its learning ability for complex patterns.

4 Artificial Intelligence in Environmental and Soil Sciences: Approaches and Application Domains

4.1 Overview of AI Applications in Environmental Sciences

The rapidly expanding fields of AI and ML have had a big influence on environmental science and human health research [42–45]. Researchers can now gather and examine enormous volumes of data thanks to developments in AI and ML, which helps them comprehend intricate environmental systems and forecast future changes more precisely [46, 47]. Data analytics must be used successfully to analyse the vast amounts of data generated daily in order to overcome the challenges faced by the environmental sciences. Even though there are a number of interesting studies that use data analytics for environmental data, more work needs to be done in this area. It is expected of data scientists to contribute to the community's well-being. Other scientific domains such as bioinformatics and physics make remarkable use of data analytics [48]. Applications such as ecosystem services require appropriate methods for integrating complex data in order to transform the raw data into knowledge that can be put to use. It follows that there ought to be a mutually beneficial relationship between data analytics and environmental science [49, 50]. Environmental data related to a number of naturally occurring processes has been collected from various sources. The nature of this data is very varied, as is the process by which it is stored and retrieved. Both unstructured data without any systematic organisation and structured data that will be systematised in accordance with the data model's schema are possible, it's possible that the data is dispersed through various geographical areas. Finding the spatial and temporal insights from environmental data is a major challenge because environmental changes are expected to occur according to different spaces and times [48]. Thirdly, because the earth is a composite structure, many different inputs with different properties that are connected to one another will be used in the data analysis process to help produce the output. Therefore, handling this kind of complexity is very difficult. To analyse and extract valuable insights from complex environmental data, data analytics in conjunction with ML and deep learning techniques is effective [50, 51].

In conclusion, by providing innovative methods for data analysis and modelling that improve our capacity to understand and solve the most pressing environmental issues of our day, AI and ML are quickly changing the field of environmental research

[52]. It is clear that AI and ML will continue to play a key role in advancing creative solutions for environmental preservation and sustainable development, even in the face of persistent issues with data quality and integration, model interpretability and reliability, ethical and social considerations, and the requirement for greater interdisciplinary collaboration.

4.2 Overview of AI Applications in Soil Sciences

Following the definition of AI and its classification into three main domains: reasoning and decision-making, learning, and robotics it becomes essential to explore how these capabilities can be applied within the field of soil science [53, 54]. This section examines the potential contributions of AI to this domain, highlighting its role in advancing soil analysis, monitoring, and sustainable land management. The three categories initially outlined by the European Commission [55] provided a foundational framework. These categories were subsequently refined through expert input and an extensive literature review, employing a broad selection of relevant keywords. A taxonomy created especially for soil science was created. Three specialists with varying backgrounds in soil science and AI were consulted in order to improve and refine this classification. The outcome of those discussions is the final taxonomy that is shown here. Despite offering a helpful framework, it still has certain drawbacks and places where different categories might overlap. Additionally, it has similarities with other well-known AI taxonomies [55, 56] and AI classifications utilised in other scientific domains, including supply chain management [57], optical networks [58] and water science [59]. Figure 2 shows the suggested taxonomy for AI applications in soil science.

Soil science is a key field of agriculture, and AI has made its mark on it. Soil testing and monitoring, observation of soil/land cover/land management, soil fertilisation, soil quality evaluation, detection of nutrient shortages, carbon sequestration, and many other areas of interest are among those affected by it [60, 61]. Several AI-based technologies, encompassing robots, drones, satellite imagery, sensor-based soil monitoring tools, predictive analysis, automated irrigation systems, and more have the potential to revolutionise agriculture. These technologies will assist farmers in assessing soil, weather, and crop conditions for increased output [62, 63].

IBM [64] created a small soil testing machine that is capable of successfully evaluating five colorimetric-based indicators. The card's micro-fluidic chip analyses chemicals, and an AI-powered machine vision algorithm can determine the colorimetric test value more accurately than the human eye can. Through connecting sensors to the cloud and enabling remote access to data at the IoT cloud server, the Internet of Things can also be quite helpful in soil testing [65]. The application of AI to remote sensing data has expanded. Using ground truth to train supervised learning models has emerged as a key method for classifying land cover. Comparing two remote sensing data sets at two separate times and identifying changes is a different approach. Using ground truth to calibrate supervised learning models has emerged

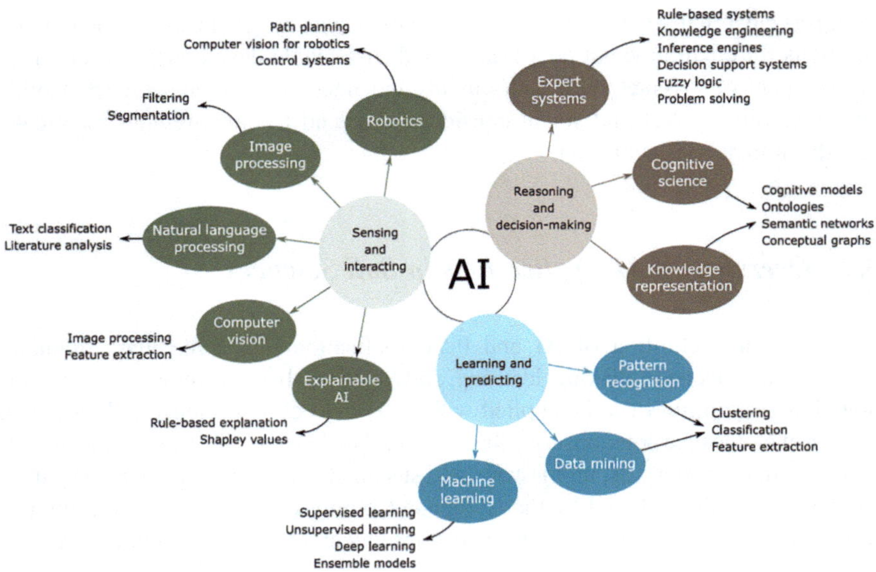

Fig. 2 A proposed taxonomy of AI in soil science. It has three layers composed of domains (three circles), applications (ellipses) and example techniques (outer layer). The techniques are presented for illustration purposes but usually overlap and complement each other. For example, explainable AI typically requires prior learning, often using a machine learning algorithm

as a key method for classifying land cover. Comparing two remote sensing data sets at two distinct times and identifying changes is another method; this is also known as the compound classification technique [66–70]. According to Rao and Reddy [71], the random forest model has proven to be more effective than all other ML techniques for classifying soil according to its quality. The AI-based Soil risk Characterisation Decision Support System (SRC-DSS) use fuzzy logic to both characterise soil and detect polluted soils that pose an unplanned difficulty. Fuzzy logic also offers the benefit of eliminating ambiguity or imprecision from soil management data since it requires few factors to characterise the soil (López et al., 2008). Neural networks make it simple to measure the parameter that takes hours to estimate. Measurable soil factors, such as bulk density, effective porosity, soil water retention curve parameters, and soil texture data (sand and clay contents), may readily predict soil hydraulic conductivity, which is hard to measure [72]. It has been demonstrated that AI can anticipate more accurately than Multiple Linear Regression (MLR) for both physical and biological parameters, including soil enzyme activity. It has also been shown that when ANN is combined with the Digital Terrain Model (DTM), it provides a better mapping of soil enzyme activity [73].

4.3 Overview of AI Applications in Agriculture

It is clearly evident that agriculture contributes significantly to the nation's economic growth. One such developing field in agriculture is automation, which aids in bettering farming practices such as irrigation, crop installation and monitoring, weeding to reduce pesticide usage, and plant disease detection. Applications in agriculture, such as the detection of plant diseases, greatly benefit from CNN. A camera takes images of plants, both healthy and infected, that are then tagged and stored in an information system. CNN has been trained to recognise the characteristics of both healthy and diseased plants. Furthermore, CNN and its frameworks, such as AlexNet, LeNet, CaffeNet, and GoogleNet, are extensively used in the following areas: agricultural yield calculation, soil and root segmentation, fruit counting, land cover and crop type classification, and leaf disease prediction [74] (Fig. 3).

5 Applications of Artificial Intelligence and Machine Learning in Soil Science

AI is playing an increasingly important role in soil science research by automating tasks that were traditionally done by human scientists [54, 75]. It can analyze data, make predictions, and read and summarize scientific literature using natural language processing (NLP). These abilities help save time and make parts of the research process more efficient [76]. However, AI currently serves mainly as a support tool, working under human supervision. It cannot yet replace human creativity, intuition, or critical thinking. Some ambitious projects aim to automate the entire scientific process including idea generation, experimentation, and writing but they are still in the early stages [77].

In the near future, more steps of the research process in soil science are expected to be assisted or enhanced by AI. Still, major breakthroughs will likely remain driven

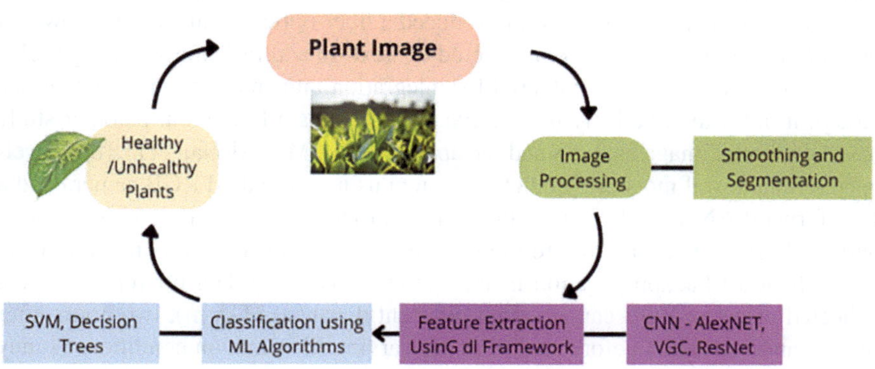

Fig. 3 DL architecture for plant disease detection and classification

by human imagination, as AI is not yet capable of true innovation. The future of soil science will likely involve AI-augmented researchers, where humans provide the vision and AI handles much of the execution [78, 79].

5.1 *Predicting and Analysing Soil Texture Using AI*

Soil texture is a key factor influencing ecosystem stability, crop productivity, and the sustainable use of agricultural land [80]. Among various soil attributes, texture holds particular importance in guiding decisions related to land planning and agricultural management. Traditional methods involving agricultural sensors and statistical techniques have often proven to be inefficient, costly, and slow to deliver actionable insights [81]. In contrast, the integration of advanced AI and ML technologies has introduced more efficient and innovative approaches to predicting soil texture, paving the way for more effective and adaptive soil management practices. Traditional soil texture analysis typically relies on laboratory techniques such as sieving, sedimentation, and hydrometric methods [82–85]. Scientists, technologists, and politicians must work closely together to develop efficient decision support systems. For AI-driven soil texture analysis to be implemented practically and widely, such multidisciplinary collaboration is necessary.

5.2 *AI-Based Measurement of Soil Water Content (SWC)*

Accurate measurement of soil water content (SWC) is an essential practice for precision agriculture, whereas traditional methods, despite their reliability, can be labor-intensive and limited in spatial or temporal resolution. With recent advances in AI, remote and proximal sensing, and IoT frameworks, continuous innovation has emerged, enhancing the accuracy and scalability of the obtained results. Recent studies have emphasized the growing importance of AI-driven methods for measuring SWC. For instance, Liu et al. [86] introduced a fiber-optic sensing system powered by AI to monitor soil moisture in covered context. In addition, the study highlights the system's durability and potential for integration into smart irrigation networks, though it notes the relatively high initial setup cost as a limitation. Another study based on digital image analysis and the application of ML, Al-Naji et al. [87] developed a remote soil moisture prediction system using a standard RGB camera and a feed-forward ANN. In this study, the authors propose a non-contact vision system with an RGB camera in order to predict the irrigation needs in a loamy soil using a feed-forward backpropagation neural network. The raw data for this soil were collected using an RGB camera from different distances (1–5 m), two hours after sunrise and two hours before sunset, under varying illumination conditions (sunny and cloudy weather). The performance of the proposed system was highly accurate, achieving an MSE of 1.616×10^6 (training), 1.004×10^5 (testing), and 1.809×10^5

(validation). Moreover, the error values for sunny-dry, sunny-wet, shadow-dry, and shadow-wet scenarios were less than 1.808×10^3, 2.774×10^3, 1.428×10^3, and 2.78×10^3, respectively. Therefore, the results demonstrate the high performance of the proposed cost-effective vision system in enhancing irrigation precision. The importance of using UAVs in soil and agricultural sciences over the last decade has been extensively studied. For measuring SWC, a study by Bertalan et al. [88] investigated the application of UAVs equipped with multispectral and thermal cameras to predict SWC using ML. By extracting indices of vegetation such as NDVI, NDWI, and NSMI along with thermal dataset, and applying Random Forest algorithms, the study achieved very high predictive accuracy for soil moisture mapping ($R^2 = 0.97$) [89]. Levintal et al. [90] described an open-source, low-cost underground wireless sensor network (WUSN) that uses LoRa technology to monitor soil moisture levels over time [91]. The server instantly retrieves the data from the gateway upon receipt. This setup overcomes the limitations of traditional monitoring methods, offering improved efficiency and practicality by enabling farmers to track and manage their irrigation and fertilization systems in real time. Within the context of sustainable agriculture, the platform plays a key role in conserving both water and fertilizer.

In terms of accuracy, adaptability, and digitalisation, AI-based SWC measuring systems improve traditional approaches. The future studies are planned to concentrate on transfer learning for model adaption between areas, edge computing for decentralised data processing, and multi-sensor fusion to improve dependability and coverage even further. These developments are expected to play an important role in worldwide efforts to optimise water use and maintain ethical agricultural methods.

5.3 The Role of AI in Smart Agricultural Irrigation Systems

The AI establishment into smart agricultural irrigation systems improves management of water and crop output globally. AI-powered systems, frequently integrated with Internet of Things (IoT) technology, enabling real-time monitoring, predictive analytics, and autonomous decision-making, addressing the pressing issues of water shortages and climate unpredictability. A new systematic study demonstrates the breadth of AI-enabled irrigation systems, which range from simple soil moisture sensing to complex crop evapotranspiration modelling and water quality monitoring. These technologies help to save water while also predicting water quality concerns, preserving crop health and protecting the agroecosystem. The literature demonstrates AI's ability to manage the world's water crisis through continuous surveillance, autonomous decision-making, and adaptive management. Ongoing research is planned to concentrate on connecting multi-source information, enhancing model resilience, and making smart irrigation available to smallholder farmers worldwide through the use of sophisticated technologies such as deep learning, machine learning, and artificial intelligence systems [92]. Furthermore, AI algorithms are used to analyse acquired data, forecast crop water requirements, and automate fertigation procedures. To meet plant demands, this system incorporates sensor networks, weather and climate forecasts, and continuous crop data monitoring [93]. The confluence of IoT and AI technologies has enabled smart irrigation systems that continually monitor temperature, weather, soil moisture, and other factors. AI analyses this sensor data to predict water requirements and manage irrigations optimally, resulting in significant water savings of up to 30% when compared to conventional techniques. These systems are particularly useful in areas with a shortage of labour or unpredictable weather since they also allow for remote monitoring and control. Case studies from around the globe demonstrate that AI-IoT systems can boost production, save expenses, and minimise water loss especially in water-scarce areas like the MENA region. However, closing the distance between the various process contributors is necessary for effective implementation [94]. It is also known that ML models such as K-Nearest Neighbors (KNN), SVM, and Naive Bayes (NB) are being used in agriculture and to predict fertigation needs based on collected data from sensors. These models can analyse different variables (soil moisture, temperature, climatic data, etc.) to control fertigation scheduling automatically. A case study from an experimental field demonstrated that KNN achieved an accuracy rate of 98.4% and an RMSE of 0.016, indicating a highly robust performance. Such predictive systems enable precise irrigation, minimize water loss, and support farmers and stakeholders with suitable decision-making. The use of AI and ML in smart irrigation aligns with the principles of digital agriculture 4.0 [95].

To sum up, AI plays a crucial role in the evolution of smart agriculture and precision irrigation systems by enabling data-driven, cost-effective, and efficient water management systems. Moreover, the integration of AI with IoT and ML technologies will result in to significant water and fertilizer savings, increased productivity, and other benefits that will support and ensure sustainable and resilient agricultural practices. Another important aspect that contributes to sustainability and best practices is the transfer of knowledge among stakeholders.

5.4 Integration with Remote Sensing, GIS, and Sensor Technologies

The integration of remote sensing, Geographic Information Systems (GIS), and new technologies has become fundamental digital agriculture and precision farming to monitor and manage crops. These technologies, when combined, provide multi-scale, cost-effective, high-frequency, and spatially comprehensive information about soil moisture, crop health, and irrigation needs [96]. Remote sensing technologies, particularly satellites, play a pivotal role in monitoring and mapping soil moisture at a large scale. The Soil Moisture Active Passive (SMAP) mission, launched by NASA, for example, utilizes both active (radar) and passive (radiometer) sensors to provide global soil moisture data every three days at different spatial resolutions. The data and images collected by satellites provide support for weather forecasting, drought and flood monitoring, agricultural productivity assessments, and yield prediction. SMAP data is important in understanding the water flow, energy, carbon cycles and different other parameters that can be used for estimating and predicting scenarios at large scales [97]. Remote sensing technologies are considered crucial nowadays to monitor and predict crop productivity, water and fertilizers consumption, pest management and fertigation scheduling, which is beneficial within the WEFE nexus to face climate change challenges, especially water scarcity [98]. Remote and proximal sensing are tools that can provide a precise and exact information, especially when combined. The synergy between satellite remote sensing and proximal soil moisture sensors significantly enhances irrigation management in agricultural fields [99]. By interpreting the data collected from satellite and proximal sensors, researchers can assess different physiological plant parameters and behaviors, such as crop water stress, soil moisture, plant health, irrigated and non-irrigated fields, which will help in better water management. This approach provides both the broad spatial coverage of satellites and the high temporal and vertical resolution of proximal sensors. For instance, combining these data sources allows for more accurate drought detection, better yield forecasting, and adaptive irrigation planning. Such integration is essential for precision agriculture, especially in heterogeneous landscapes where field-scale variability is high [100]. Microwave remote sensing, both active (radar) and passive (radiometer), is recognized as one of the most effective methods for monitoring surface soil moisture due to its ability to penetrate through vegetation and soil layers, even in challenging or obscured weather conditions. These sensors can measure the changes in the dielectric properties of soil, which is known to strongly correlate with moisture content. However, satellite sensors typically measure only the top soil surface (a few centimeters). In order to estimate root zone soil moisture, satellite data are interpreted and assimilated into land surface or hydrological models using prior information, such as soil type and structure. This multi-source data assimilation is crucial for applications like irrigation scheduling, drought assessment, and water resource management [101]. GIS platforms are critical for integrating and visualizing spatial data from remote sensing and ground sensors. They enable the mapping of soil moisture, crop conditions, and irrigation needs at various scales, supporting data-driven decision-making. For example, advanced soil moisture analytics platforms use satellite data (e.g., from EOSDA) to generate surface and root-zone soil moisture maps at resolutions as fine as 250 m, updated daily. These maps, when combined with weather forecasts, crop parameters, and soil data in a GIS, allow for precise fertigation scheduling, yield evaluation, and risk assessment. Such systems are increasingly accessible and easy to use for farmers via web or mobile applications, enhancing the practical utility of remote sensing and GIS in agriculture [102]. Until now, the accuracy of remote sensing-derived soil moisture data is often validated with proximal sensing and observations, such as neutron probes or cosmic-ray neutron sensors. This validation is crucial to improve satellite algorithms and to reach the highest accuracy rate. For example, in India, recent mobile applications demonstrate the potential for two-way information exchange, where farmers insert field data and receive tailored irrigation advice based on integrated remote sensing information. The ongoing development of high-resolution, near real-time soil moisture products, combined with robust GIS platforms, will increase the precision and scalability of such systems for better agricultural water management [103].

The integration of remote sensing, GIS, and sensor technologies enables comprehensive, multi-scale monitoring and management of soil moisture and irrigation in agriculture. These advances support precision farming, optimize water use, and enhance climate change resilience, marking a significant step forward targeting sustainable agriculture.

6 Main Domains of Artificial Intelligence Implementation in Soil Research

AI is transforming the field of soil science by providing tools for analysing large information, recognising trends, and automating hard operations. A recent thematic analysis found twelve major topics where AI uses are most prevalent in soil research. These include remote sensing, which employs satellite data to monitor soil conditions, as well as the prediction and modelling of soil organic carbon stocks, which are critical for understanding global carbon cycles. Other critical areas include water content monitoring, contaminant evaluation, and physical property prediction, such as texture and bulk density. Advanced ML approaches, such as neural networks, SVM, and ensemble techniques like random forests, are commonly employed to improve prediction accuracy. Spectroscopic approaches that use AI to interpret light-based soil sensing allow for quick and non-invasive soil characterisation. Furthermore, AI allows both categorical (e.g., soil classifications) and continuous (e.g., pH, nutrient levels) soil property modelling, which frequently includes environmental factors. AI may also be used in agricultural systems to optimise crop-soil interactions, with the goal of increasing yields through improved resource management.

This organised table summarises current developments and actual AI uses in soil research. However, it should be noted that this list only includes the most well-documented and developed places. Several more AI-driven projects are now underway, examining fresh methodologies, extending into under-represented locations, and addressing growing environmental concerns. The area is fast growing, and future studies will most certainly reveal new elements of how AI may assist and revolutionise soil science (Table 1).

Table 1 Main applications of artificial intelligence and machine learning in soil science

Topic	Description	Type of AI application	Advantages	Limitations/ remarks	References
Remote sensing	Use of satellite imagery, especially for salinity monitoring	Image analysis, classification, prediction	Large spatial coverage, rapid detection	Dependent on weather and resolution quality	[104–107]
Soil organic carbon	Study of carbon cycles and stocks in various ecosystems	Stock prediction, dynamic modeling	Understanding environmental impacts	Complex influencing factors	[108–111]
Water	Soil moisture, temperature, and CO_2 flux monitoring	Time series, regression, smart sensors	Continuous tracking, high accuracy	Often heterogeneous or noisy data	[112–120]
Contamination	Soil pollution by heavy metals and bioavailability	Detection, mapping, risk estimation	Helps in environmental management	Requires field validation	[121–123]
Methods (ensembles)	Use of ensemble models like random forests (RF)	Ensemble learning	Improved prediction accuracy	Sometimes hard to interpret	[124, 125]
Erosion/parent material	Study of soil formation and erosion processes	Classification, spatial prediction	Useful for landscape management	Requires detailed geological data	[126–130]
Methods (NN, SVM)	Use of neural networks and support vector machines	Deep learning, classification	High processing power	Risk of overfitting, needs fine-tuning	[131–133]
Spectroscopy	Proximal soil sensing using light (IR, gamma, etc.)	Signal processing, spectral calibration	Fast, non-destructive analysis	Calibration models needed	[134–136]
Modeling (classes)	Mapping of categorical soil properties	Classification, spatial disaggregation	Supports better soil classification maps	Accuracy depends on covariate quality	[137–143]
Crops	Soil–plant interactions, especially for yield optimization	Predictive modeling, resource optimization	Helps in precision agriculture	Context-specific data needed	[144–147]

(continued)

Table 1 (continued)

Topic	Description	Type of AI application	Advantages	Limitations/ remarks	References
Physical properties	Soil texture, bulk density, and related water behavior	Property estimation, regression	Supports hydrological modeling	Some properties hard to measure indirectly	[148–153]
Modeling (continuous)	Mapping of continuous soil variables (e.g. pH, nutrients)	Spatial regression, PTFs	Broad-scale environmental modeling	Sensitive to spatial resolution	[154–159]

7 Trends and Global Patterns in the Application of Machine Learning to Soil Science and Hydrology: A Bibliometric Review (1973–Present)

The yearly volume of published articles serves as a key indicator of research activity within the Application of ML in Soil Science and Hydrology (AMLSH) field. As illustrated in Fig. 4A, the first publications on this topic appeared in 1973. Prior to 1995, AMLSH-related works were scarce, comprising just 0.93% of the total AMLSH literature. Between 1995 and 2006, publication numbers grew steadily, exceeding 50 articles annually. After 2006, this growth accelerated dramatically, with over 500 articles published each year. Although AMLSH studies still represent a relatively small fraction of the broader ML research landscape, their proportion has been steadily rising over time. Figure 4B highlights the simultaneous growth of ML research within both soil science and hydrology, with publication counts rising steadily in each field. This parallel development may be linked to key advancements in ML's history, notably Hinton's 2006 solution to the gradient vanishing problem in deep neural network training [160], which enabled ML to handle larger datasets effectively. Following this breakthrough, the proliferation of open-source algorithms further accelerated the adoption of ML techniques [161, 162]. Additionally, the expanding number of specialized journals and rapid progress in information technology particularly improvements in computer hardware have likely contributed to the surge in publication volume [161].

Research on the Application of ML in Soil Science and Hydrology (AMLSH) has been published by scholars from 140 countries, with 65 of these nations producing at least 40 articles. China leads with 7,335 publications, followed by the United States (USA) with 4,363 articles (Table 2). Notably, publications originating from the USA receive significantly higher citation counts, totaling 108,694 citations. Meanwhile, Australian and British articles boast higher average citations per publication, with citation-to-publication ratios of 30.23 and 28.99, respectively.

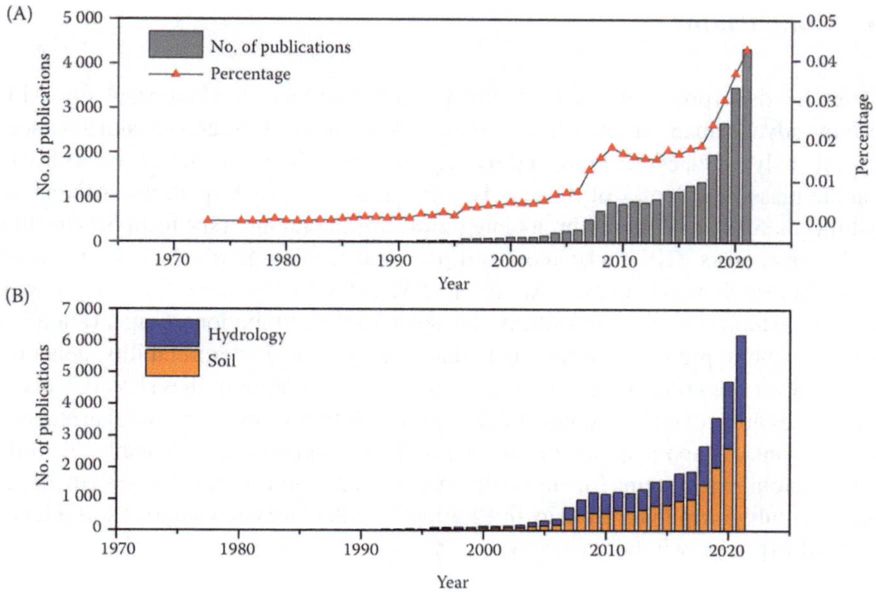

Fig. 4 **A** Annual trend and proportion of publications on machine learning applications in soil science and hydrology (1973–2021). **B** Thematic distribution of these publications between soil science and hydrology

Table 2 Top 10 countries contributing to machine learning research in soil science and hydrology (1973–2021)

Rank	Country	Publications (N)	Citations (C)	Citations per publication (C/N)	Total link strength (TLS)
1	China	>7,000	75,000–80,000	10–11	2,700–2,800
2	United States	4,000–4,500	>100,000	24–26	>3,000
3	India	>2,000	~30,000	13–15	900–1,000
4	Iran	~2,000	~30,000	18–20	1,900–2,000
5	Germany	~1,100	25,000–27,000	23–24	1,400–1,500
6	Australia	~1,000	>30,000	>30	1,300–1,400
7	Canada	~1,000	>24,000	>23	~1,100
8	United Kingdom	>880	~25,000	~29	~1,200
9	Italy	750–800	~18,000	~25	800–850
10	Turkey	~750	~18,000	~24	>420

Notes Values are expressed approximately or as intervals to avoid exact duplication of source data. N = number of publications; C = total citations; C/N = average citations per publication; TLS = total link strength (collaborative intensity)

8 Conclusions

Given the disturbing conditions of climate change and its attendant problems, AI is driving advancements in environmental monitoring and soil research. Major advances in soil analysis, categorisation, and management have been enabled by its ability to handle massive volumes of diverse data and uncover detailed patterns. AI's applicability is further increased by its integration with contemporary technologies like real-time sensors, GIS, and remote and proximal sensing, particularly in precision agriculture and smart irrigation. Adoption of AI technologies is considered a powerful tool, but at the same time, more efforts are needed to face limitations. Critical attention must be paid to problems pertaining to data quality, model interpretability, generalisation in various contexts, accuracy and ethical issues. Future studies should concentrate on creating explainable AI models, improving frameworks for data sharing, and making sure AI applications are socially and environmentally acceptable. To fully utilise intelligent systems for environmental resilience and sustainable land management, a multidisciplinary strategy integrating deep domain knowledge in soil science with AI expertise will be necessary.

References

1. Jaskie, K., Spanias, A.: Positive and unlabeled learning algorithms and applications: a survey. In: 2019 10th International Conference on Information, Intelligence, Systems and Applications (IISA), pp. 1–8. IEEE (2019, July)
2. Mahesh, B.: Machine learning algorithms—a review. Int. J. Sci. Res. (IJSR) **9**(1), 381–386 (2020)
3. Heo, J., Yoon, J.G., Park, H., Kim, Y.D., Nam, H.S., Heo, J.H.: Machine learning-based model for prediction of outcomes in acute stroke. Stroke **50**(5), 1263–1265 (2019)
4. Wang, W., Kiik, M., Peek, N., Curcin, V., Marshall, I.J., Rudd, A.G., Wang, Y., Douiri, A., Wolfe, C.D., Bray, B.: A systematic review of machine learning models for predicting outcomes of stroke with structured data. PLoS ONE **15**(6), e0234722 (2020)
5. Müller, V.C., Bostrom, N.: Future progress in artificial intelligence: a survey of expert opinion. In: Fundamental Issues of Artificial Intelligence, pp. 555–572 (2016)
6. Intelligence, O.A.: Of artificial intelligence. In: Ontology, Inference Engines, Software Agents (2016)
7. Zhang, D., Tsai, J.J.: Machine learning and software engineering. Softw. Qual. J. **11**, 87–119 (2003)
8. Sarker, I.H.: Machine learning: algorithms, real-world applications and research directions. SN Comput. Sci. **2**(3), 160 (2021)
9. Hiran, K.K., Jain, R.K., Lakhwani, K., Doshi, R.: Machine Learning: Master Supervised and Unsupervised Learning Algorithms with Real Examples (English Edition). BPB Publications (2021)
10. Morales, E.F., Escalante, H.J.: A brief introduction to supervised, unsupervised, and reinforcement learning. In: Biosignal Processing and Classification Using Computational Learning and Intelligence, pp. 111–129. Academic Press (2022)
11. Nasteski, V.: An overview of the supervised machine learning methods. Horizons B **4**(51–62), 56 (2017)

12. Saravanan, R., Sujatha, P.: A state of art techniques on machine learning algorithms: a perspective of supervised learning approaches in data classification. In: 2018 Second International Conference on Intelligent Computing and Control Systems (ICICCS), pp. 945–949. IEEE (2018, June)
13. Jiang, T., Gradus, J.L., Rosellini, A.J.: Supervised machine learning: a brief primer. Behav. Ther. **51**(5), 675–687 (2020)
14. Usama, M., Qadir, J., Raza, A., Arif, H., Yau, K.L.A., Elkhatib, Y., Hussain, A., Al-Fuqaha, A.: Unsupervised machine learning for networking: techniques, applications and research challenges. IEEE Access **7**, 65579–65615 (2019)
15. Li, Y.: Deep Reinforcement Learning: An Overview (2017). arXiv preprint arXiv:1701.07274
16. Sahu, S.K., Mokhade, A., Bokde, N.D.: An overview of machine learning, deep learning, and reinforcement learning-based techniques in quantitative finance: recent progress and challenges. Appl. Sci. **13**(3), 1956 (2023)
17. Hsieh, W.: Machine learning in the environmental sciences. In: Neural Networks and Kernels (2009)
18. Ma, Y., Guo, G. (eds.): Support Vector Machines Applications, vol. 649. Springer, New York (2014)
19. Salcedo-Sanz, S., Rojo-Álvarez, J.L., Martínez-Ramón, M., Camps-Valls, G.: Support vector machines in engineering: an overview. Wiley Interdiscip. Rev.: Data Min. Knowl. Discov. **4**(3), 234–267 (2014)
20. Hino, M., Benami, E., Brooks, N.: Nat. Sustain. **1**, 583–588 (2018)
21. Liu, Y., Wang, Y., Zhang, J.: New machine learning algorithm: random forest. In: International Conference on Information Computing and Applications, pp. 246–252. Springer Berlin Heidelberg, Berlin, Heidelberg (2012, September)
22. Rodriguez-Galiano, V., Sanchez-Castillo, M., Chica-Olmo, M., Chica-Rivas, M.J.O.G.R.: Machine learning predictive models for mineral prospectivity: an evaluation of neural networks, random forest, regression trees and support vector machines. Ore Geol. Rev. **71**, 804–818 (2015)
23. Ao, Y., Li, H., Zhu, L., Ali, S., Yang, Z.: The linear random forest algorithm and its advantages in machine learning assisted logging regression modeling. J. Petrol. Sci. Eng. **174**, 776–789 (2019)
24. Goyal, M.K., Ojha, C.S.P., Burn, D.H.: Sustainable Water Resources Management, pp. 165–178 (2017)
25. Zhang, Z.: Introduction to machine learning: k-nearest neighbors. Ann. Transl. Med. **4**(11), 218 (2016)
26. Cunningham, P., Delany, S.J.: K-nearest neighbour classifiers—a tutorial. ACM Comput. Surv. (CSUR) **54**(6), 1–25 (2021)
27. Talaviya, T., Shah, D., Patel, N., Yagnik, H., Shah, M.: Artif. Intell. Agric. **4**, 58–73 (2020)
28. Sinaga, K.P., Yang, M.S.: Unsupervised K-means clustering algorithm. IEEE Access **8**, 80716–80727 (2020)
29. Vankayalapati, R., Ghutugade, K.B., Vannapuram, R., Prasanna, B.P.S.: K-Means algorithm for clustering of learners performance levels using machine learning techniques. Rev. d'Intelligence Artif. **35**(1), 99–104 (2021)
30. Likas, A., Vlassis, N., Verbeek, J.J.: The global k-means clustering algorithm. Pattern Recogn. **36**(2), 451–461 (2003)
31. Ongsulee, P.: Artificial intelligence, machine learning and deep learning. In: 2017 15th International Conference on ICT and Knowledge Engineering (ICT&KE), pp. 1–6. IEEE (2017, November)
32. Dimiduk, D.M., Holm, E.A., Niezgoda, S.R.: Perspectives on the impact of machine learning, deep learning, and artificial intelligence on materials, processes, and structures engineering. Integr. Mater. Manuf. Innov. **7**, 157–172 (2018)
33. Campesato, O.: Artificial Intelligence, Machine Learning, and Deep Learning. Mercury Learning and Information (2020)

34. Gupta, R., Srivastava, D., Sahu, M., Tiwari, S., Ambasta, R.K., Kumar, P.: Artificial intelligence to deep learning: machine intelligence approach for drug discovery. Mol. Divers. **25**, 1315–1360 (2021)
35. Janiesch, C., Zschech, P., Heinrich, K.: Machine learning and deep learning. Electron. Mark. **31**(3), 685–695 (2021)
36. Shrestha, A., Mahmood, A.: Review of deep learning algorithms and architectures. IEEE Access **7**, 53040–53065 (2019)
37. Dargan, S., Kumar, M., Ayyagari, M.R., Kumar, G.: A survey of deep learning and its applications: a new paradigm to machine learning. Arch. Comput. Methods Eng. **27**, 1071–1092 (2020)
38. Lamba, A., Cassey, P., Segaran, R.R. Koh, L.P.: Deep Learning for Environmental Conservation (2019)
39. Kattenborn, T., Leitloff, J., Schiefer, F., Hinz, S.: Review on convolutional neural networks (CNN) in vegetation remote sensing. ISPRS J. Photogramm. Remote Sens. **173**, 24–49 (2021)
40. Ketkar, N., Moolayil, J., Ketkar, N., Moolayil, J.: Convolutional neural networks. In: Deep Learning with Python: Learn Best Practices of Deep Learning Models with PyTorch, pp. 197–242 (2021)
41. Yuan, Q., Shen, H., Li, T., Li, Z., Li, S., Jiang, Y., Xu, H., Tan, W., Yang, Q., Wang, J., Gao, J., Zhang, L.: Remote Sens. Environ. 241 (2020)
42. Singh, A., Kanaujia, A., Singh, V.K., Vinuesa, R.: Artificial intelligence for sustainable development goals: bibliometric patterns and concept evolution trajectories. Sustain. Dev. **32**(1), 724–754 (2024)
43. Xiang, X., Li, Q., Khan, S., Khalaf, O.I.: Urban water resource management for sustainable environment planning using artificial intelligence techniques. Environ. Impact Assess Rev. **86**, 106515 (2021)
44. Yu, K.H., Zhang, Y., Li, D., Montenegro-Marin, C.E., Kumar, P.M.: Environmental planning based on reduce, reuse, recycle and recover using artificial intelligence. Environ. Impact Assess. Rev. **86**, 106492 (2021)
45. Yigitcanlar, T., Mehmood, R., Corchado, J.M.: Green artificial intelligence: towards an efficient, sustainable and equitable technology for smart cities and futures. Sustainability **13**(16), 8952 (2021)
46. Hamdan, A., Ibekwe, K.I., Etukudoh, E.A., Umoh, A.A., Ilojianya, V.I.: AI and machine learning in climate change research: a review of predictive models and environmental impact. World J. Adv. Res. Rev. **21**(1), 1999–2008 (2024)
47. Ukoba, K., Onisuru, O.R., Jen, T.C.: Harnessing machine learning for sustainable futures: advancements in renewable energy and climate change mitigation. Bull. Natl. Res. Cent. **48**(1), 99 (2024)
48. Gordon, S.B., Peter, H., Amber, L., John, W., Emma, E., Susan, J., Young Paul, J.: Front. Environ. Sci. **7**, 121 (2019)
49. Gibert, K., Horsburgh, J.S., Athanasiadis, I.N., Holmes, G.: Environmental data science. Environ. Model. Softw. **106**, 4–12 (2018)
50. Gupta, S., Aga, D., Pruden, A., Zhang, L., Vikesland, P.: Data analytics for environmental science and engineering research. Environ. Sci. Technol. **55**(16), 10895–10907 (2021)
51. Cui, S., Gao, Y., Huang, Y., Shen, L., Zhao, Q., Pan, Y., Zhuang, S.: Advances and applications of machine learning and deep learning in environmental ecology and health. Environ. Pollut. **335**, 122358 (2023)
52. Liu, X., Lu, D., Zhang, A., Liu, Q., Jiang, G.: Data-driven machine learning in environmental pollution: gains and problems. Environ. Sci. Technol. **56**, 2124–2133 (2022)
53. Chen, Q., Li, L., Chong, C., Wang, X.: AI-enhanced soil management and smart farming. Soil Use Manag. **38**(1), 7–13 (2022)
54. Castro, J.P., Gasch, C.K., Flores, P.: Artificial intelligence in soil science: where do we go now? Agric. Environ. Lett. **9**(2), e20134 (2024)
55. European Commission: High-level expert group on artificial intelligence. In: Ethics Guidelines for Trustworthy AI, vol. 6 (2019)

56. Kotu, V., Deshpande, B.: Data Science: Concepts and Practice. Morgan Kaufmann (2018)
57. Pournader, M., Ghaderi, H., Hassanzadegan, A., Fahimnia, B.: Artificial intelligence applications in supply chain management. Int. J. Prod. Econ. **241**, 108250 (2021)
58. Mata, J., De Miguel, I., Duran, R.J., et al.: Artificial intelligence (AI) methods in optical networks: a comprehensive survey. Opt. Switch. Netw. **28**, 43–57 (2018)
59. Doorn, N.: Artificial intelligence in the water domain: opportunities for responsible use. Sci. Total Environ. **755**, 142561 (2021)
60. Varshitha, D.N., Choudhary, S.: An AI solution for soil fertility and crop friendliness detection and monitoring. Int. J. Eng. Adv. Technol. **10**(3), 172–175 (2021)
61. de Andrade, V.H.G.Z., Redmile-Gordon, M., Barbosa, B.H.G., Andreote, F.D., Roesch, L.F.W., Pylro, V.S.: Artificially intelligent soil quality and health indices for 'next generation' food production systems. Trends Food Sci. Technol. **107**, 195–200 (2021)
62. Inoue, Y.: Satellite-and drone-based remote sensing of crops and soils for smart farming—a review. Soil Sci. Plant Nutr. **66**(6), 798–810 (2020)
63. Maurya, P.K., Bahadu, V., Thaku, G., Verma, L.K.: Robotics, drones, remote sensing, GIS, and IOT tools for agricultural water management. In: Integrated Land and Water Resource Management for Sustainable Agriculture, vol. 1, pp. 169–180. Springer Nature Singapore, Singapore (2025)
64. IBM: Agropad-AI-Powered Technology Will Help Farmers Health-Check Soil and Water (2018). https://www.ibm.com/blogs/research/2018/09/agropad
65. Sindhu, P., Indirani, G.: IoT enabled soil testing. Asian J. Comput. Sci. Technol. **7**(S1), 54–57 (2018)
66. Karpatne, A., Jiang, Z., Vatsavai, R.R., Shekhar, S., Kumar, V.: Monitoring land-cover changes: a machine-learning perspective. IEEE Geosci. Remote Sens. Mag. **4**(2), 8–21 (2016)
67. Saiz-Rubio, V., Rovira-Más, F.: From smart farming towards agriculture 5.0: a review on crop data management. Agronomy **10**(2), 207 (2020)
68. Kumar, N.S., Anouncia, S.M., Prabu, M.: Application of satellite remote sensing to find soil fertilization by using soil colour. Int. J. Online Eng. **9**(2) (2013)
69. Bhattacharya, P., Pramanik, P., Ray, M., Krishnan, P.: Comparison of artificial neural network and multi-linear regression for prediction of field capacity soil moisture content. J. Agric. Phys. **18**(2), 173–180 (2018)
70. Bhattacharya, P., Maity, P.P., Ray, M., Mridha, N.: Prediction of mean weight diameter of soil using machine learning approaches. Agron. J. (2021)
71. Rao, T.V.N., Reddy, G.R.: Prediction of soil quality using machine learning techniques. Int. J. Sci. Technol. Res. **8**(11), 1309–1313 (2019)
72. Ghanbarian-Alavijeh, B., Liaghat, A.M., Sohrabi, S.: Estimating saturated hydraulic conductivity from soil physical properties using neural networks model. World Acad. Sci., Eng. Technol. **62**, 131–136 (2010)
73. Tajik, S., Ayoubi, S., Nourbakhsh, F.: Prediction of soil enzymes activity by digital terrain analysis: comparing artificial neural network and multiple linear regression models. Environ. Eng. Sci. **29**(8), 798–806 (2011)
74. Kamilaris A, Prenafeta-Boldú F.X.: J. Agric. Sci. 1–11 (2018)
75. Wadoux, A.M.C., Minasny, B., McBratney, A.B.: Machine learning for digital soil mapping: applications, challenges and suggested solutions. Earth Sci. Rev. **210**, 103359 (2020)
76. Ghorbani, M.A., Deo, R.C., Kashani, M.H., Shahabi, M., Ghorbani, S.: Artificial intelligence-based fast and efficient hybrid approach for spatial modelling of soil electrical conductivity. Soil Tillage Res. **186**, 152–164 (2019)
77. Peng, S., Huang, M.J., Wu, M., Wei, J.: Transforming Redaction: How AI is Revolutionizing Data Protection (2024). arXiv preprint arXiv:2409.15308
78. Verdu, E., Nieto, Y.V., Saleem, N.: Big data and artificial intelligence in earth science: recent progress and future advancements. Acta Geophys. **71**(3), 1373–1375 (2023)
79. Sun, W., Wang, S.: The future research directions of artificial intelligence in crop growth prediction. Geogr. Res. Bull. **3**, 251–254 (2024)

80. Zhai, Y., Thomasson, J.A., Boggess, J.E., Sui, R.: Soil texture classification with artificial neural networks operating on remote sensing data. Comput. Electron. Agric. **54**(2), 53–68 (2006)
81. Bormann, H.: Towards a hydrologically motivated soil texture classification. Geoderma **157**(3), 142–153 (2010)
82. Riese, F.M., Keller, S.: Soil texture classification with 1D convolutional neural networks based on hyperspectral data. In: ISPRS Annals of the Photogrammetry, Remote Sensing and Spatial Information Sciences, vol. IV-2/W5, 2019 ISPRS Geospatial Week 2019, 10–14 June 2019, Enschede, The Netherlands (2019)
83. Johnson, N.S., et al.: Invited review: machine learning for materials developments in metals additive manufacturing. Addit. Manuf. **36**, 101641 (2020)
84. Wang, S., et al.: Advances in data preprocessing for biomedical data fusion: an overview of the methods, challenges, and prospects. Inform. Fus. **76**, 376–421 (2021)
85. Hassan-Esfahani, L., Torres-Rua, A., Jensen, A., McKee, M.: Assessment of surface soil moisture using high-resolution multi-spectral imagery and artificial neural networks. Remote Sens. **7**(3), 2627–2646 (2015). https://doi.org/10.3390/rs70302627
86. Xi-Feng, L., Hong-Hu, Z., Bing, W., Jie, L., Tian-Xiang, L., Bin, S.: Artificial Intelligence-Based Fiber Optic Sensing for Soil Moisture Measurement with Different Cover Conditions (2023)
87. Al-Naji, A., Fakhri, A.B., Gharghan, S.K., Chahl, J.: Soil Color Analysis Based on a RGB Camera and an Artificial Neural Network Towards Smart Irrigation: A Pilot Study (2021)
88. Bertalan, L., Holb, I., Pataki, A., Négyesi, G., Szabó, G., Kupásné Szalóki, A., Szabó, S.: UAV-Based Multispectral and Thermal Cameras to Predict Soil Water Content—A Machine Learning Approach (2022)
89. LoRaWAN Infrastructure Design and Implementation for Soil Moisture Monitoring in Strawberry Cultivation (2024)
90. Levintal, E., Shani, U., Assouline, S. et al.: An underground, wireless, open-source, low-cost system for measuring soil moisture and oxygen levels at multiple depths (2022)
91. https://doi.org/10.3389/fsufs.2025.1542166
92. Al-Saadi, S., Al-Saadi, A., Al-Saadi, H.: A Comprehensive Review of AI-Enabled Precision Irrigation and Water Management. BJAS (2024)
93. Kumar, S., Sharma, A., Singh, R. AIDSII: An AI-Based Digital System for Intelligent Irrigation. ScienceDirect (2023)
94. Sharma, P., Gupta, R.: Advancements in Smart Irrigation: The Role of IoT and AI in Water Management. ESP JETA (2023)
95. Tace, Y., Elfilali, S., Tabaa, M., Leghris, C.: Implementation of smart irrigation using IoT and artificial intelligence. Sci. Technol. J. (2023)
96. Abbatantuono, F., Lopriore, G., Tallou, A., Brillante, L., Ali, S.A., Camposeo, S., Vivaldi, G.A.: Recent progress on grapevine water status assessment through remote and proximal sensing: a review. Sci. Hortic. **338**, 113658 (2024)
97. National Aeronautics and Space Administration: Applications of Remote Sensing to Soil Moisture and Agricultural Water Use (2020)
98. Zhai, W., Cheng, Q., Duan, F., Huang, X., Chen, Z.: Remote sensing-based analysis of yield and water-fertilizer use efficiency in winter wheat management. Agric. Water Manag. **311**, 109390 (2025)
99. Romano, E., Bergonzoli, S., Bisaglia, C., Picchio, R., Scarfone, A.: The correlation between proximal and remote sensing methods for monitoring soil water content in agricultural applications. Electronics **12**(1), 127 (2022)
100. Zhao, X., Wang, J., & Liu, Q.: Remote Sensing and Soil Moisture Sensors for Irrigation Management (2023)
101. Wagner, W.: Introduction to Soil Moisture Remote Sensing: Importance, Definitions, Measuring Soil Moisture, Sensors, Microwave Retrievals, Methods, Limitations, Data Products. SMAP (2018)
102. EOS Data Analytics: Remote Soil Moisture Monitoring for Precision Agriculture (2024)

103. Upadhyaya, S., Pandey, A., Evans, J.: Soil Moisture Measurement for Agriculture (2021)
104. Grunwald, S., Vasques, G.M., Rivero, R.G.: Fusion of soil and remote sensing data to model soil properties. Adv. Agron. **131**, 1–109 (2015)
105. Xu, Y., Smith, S.E., Grunwald, S., Abd-Elrahman, A., Wani, S.P.: Incorporation of satellite remote sensing pan-sharpened imagery into digital soil prediction and mapping models to characterize soil property variability in small agricultural fields. ISPRS J. Photogramm. Remote Sens. **123**, 1–19 (2017)
106. Zhang, C., Mishra, D.R., Pennings, S.C.: Mapping salt marsh soil properties using imaging spectroscopy. ISPRS J. Photogramm. **148**, 221–234 (2019)
107. Khadim, F.K., Su, H., Xu, L., Tian, J.: Soil salinity mapping in Everglades National Park using remote sensing techniques and vegetation salt tolerance. Phys. Chem. Earth, Pt. A/B/C **110**, 31–50 (2019)
108. Rial, M., Cortizas, A.M., Taboada, T., Rodríguez-Lado, L.: Soil organic carbon stocks in Santa Cruz Island, Galapagos, under different climate change scenarios. CATENA **156**, 74–81 (2017)
109. Liu, S., Yang, Y., Shen, H., Hu, H., Zhao, X., Li, H., Liu, T., Fang, J.: No significant changes in topsoil carbon in the grass lands of northern China between the 1980s and 2000s. Sci. Total Environ. **624**, 1478–1487 (2018)
110. Song, X.-D., Yang, F., Ju, B., Li, D.-C., Zhao, Y.-G., Yang, J.-L., Zhang, G.-L.: The influence of the conversion of grassland to cropland on changes in soil organic carbon and total nitrogen stocks in the Songnen Plain of Northeast China. CATENA **171**, 588–601 (2018)
111. Wang, B., Waters, C., Orgill, S., Cowie, A., Clark, A., Li Liu, D., Simpson, M., McGowen, I., Sides, T.: Estimating soil organic carbon stocks using different modelling techniques in the semi-arid rangelands of eastern Australia. Ecol. Indic. **88**, 425–438 (2018)
112. Ahmad, S., Kalra, A., Stephen, H.: Estimating soil moisture using remote sensing data: a machine learning approach. Adv. Water Resour. **33**, 69–80 (2010)
113. Coopersmith, E.J., Minsker, B.S., Wenzel, C.E., Gilmore, B.J.: Machine learning assessments of soil drying for agricultural planning. Comput. Electron. Agr. **104**, 93–104 (2014)
114. Greifeneder, F., Khamala, E., Sendabo, D., Wagner, W., Zebisch, M., Farah, H., Notarnicola, C.: Detection of soil moisture anomalies based on Sentinel-1. Phys. Chem. Earth, Pt. A/B/C **112**, 75–82 (2018)
115. Han, J., Mao, K., Xu, T., Guo, J., Zuo, Z., Gao, C.: A soil moisture estimation framework based on the cart algorithm and its application in china. J. Hydrol. **563**, 65–75 (2018)
116. Xing, L., Li, L., Gong, J., Ren, C., Liu, J., Chen, H.: Daily soil temperatures predictions for various climates in United States using data-driven model. Energy **160**, 430–440 (2018)
117. Oh, Y.-Y., Yun, S.-T., Yu, S., Kim, H.-J., Jun, S.-C.: A novel wavelet-based approach to characterize dynamic environmental factors controlling short-term soil surface CO_2 flux: application to a controlled CO_2 release test site (EIT) in South Korea. Geoderma **337**, 76–90 (2019)
118. Warner, D.L., Guevara, M., Inamdar, S., Vargas, R.: Upscaling soil-atmosphere CO_2 and CH_4 fluxes across a topographically complex forested landscape. Agric. For. Meteorol. **264**, 80–91 (2019)
119. Zeynoddin, M., Bonakdari, H., Ebtehaj, I., Esmaeilbeiki, F., Gharabaghi, B., Haghi, D.Z.: A reliable linear stochastic daily soil temperature forecast model. Soil Till. Res. **189**, 73–87 (2019)
120. Al Masmoudi, Y., Ibno Namr, K., El Yousfi, B., ElAissaoui, A.: Empirical model of a vertisol evaporation as tool for management of soil practicability using portable lysimetric station under arid land context. Agric. Eng. Int.: CIGR J. **25**(2), 11–20 (2023)
121. Wu, G., Kechavarzi, C., Li, X., Wu, S., Pollard, S.J., Sui, H., Coulon, F.: Machine learning models for predicting PAHs bioavailability in compost amended soils. Chem. Engin. J. **223**, 747–754 (2013)
122. Costa, J.G., Reigosa, M., Matías, J., Covelo, E.: Soil Cd, Cr, Cu, Ni, Pb and Zn sorption and retention models using SVM: variable selection and competitive model. Sci. Total Environ. **593**, 508–522 (2017)

123. Reeves, M.K., Perdue, M., Munk, L.A., Hagedorn, B.: Predicting risk of trace element pollution from municipal roads using site-specific soil samples and remotely sensed data. Sci. Total Environ. **630**, 578–586 (2018)
124. Blanco, C.M.G., Gomez, V.M.B., Crespo, P., Ließ, M.: Spatial prediction of soil water retention in a Páramo landscape: methodological insight into machine learning using random forest. Geoderma **316**, 100–114 (2018)
125. Tziachris, P., Aschonitis, V., Chatzistathis, T., Papadopoulou, M.: Assessment of spatial hybrid methods for predicting soil organic matter using DEM derivatives and soil parameters. CATENA **174**, 206–216 (2019)
126. Geissen, V., Kampichler, C., López-de Llergo-Juárez, J., Galindo-Acántara, A.: Superficial and subterranean soil erosion in Tabasco, tropical Mexico: development of a decision tree modeling approach. Geoderma **139**, 277–287 (2007)
127. Kheir, R.B., Chorowicz, J., Abdallah, C., Dhont, D.: Soil and bedrock distribution estimated from gully form and frequency: a GIS-based decision-tree model for Lebanon. Geomorphology **93**, 482–492 (2008). https://doi.org/10.1016/j.geomorph.2007.03.010
128. Lacoste, M., Lemercier, B., Walter, C.: Regional mapping of soil parent material by machine learning based on point data. Geomorphology **133**, 90–99 (2011). https://doi.org/10.1016/j.geomorph.2011.06.026
129. Märker, M., Pelacani, S., Schröder, B.: A functional entity approach to predict soil erosion processes in a small Plio Pleistocene Mediterranean catchment in Northern Chianti, Italy. Geomorphology **125**, 530–540 (2011)
130. Martinez, G., Weltz, M., Pierson, F.B., Spaeth, K.E., Pachepsky, Y.: Scale effects on runoff and soil erosion in rangelands: observations and estimations with predictors of different avail ability. Catena **151**, 161–173 (2017)
131. Hanna, A.M., Ural, D., Saygili, G.: Neural network model for liquefaction potential in soil deposits using Turkey and Taiwan earthquake data. Soil Dyn. Earthq. Eng. **27**, 521–540 (2007)
132. Kovačević, M., Bajat, B., Gajić, B.: Soil type classification and estimation of soil properties using support vector machines. Geoderma **154**(3–4), 340–347 (2010)
133. Farfani, H.A., Behnamfar, F., Fathollahi, A.: Dynamic analysis of soil-structure interaction using the neural networks and the support vector machines. Expert Syst. Appl. **42**, 8971–8981 (2015)
134. Heggemann, T., Welp, G., Amelung, W., Angst, G., Franz, S.O., Koszinski, S., Schmidt, K., Pätzold, S.: Proximal gamma-ray spectrometry for site-independent in situ prediction of soil texture on ten heterogeneous fields in Germany using support vector machines. Soil Till. Res. **168**, 99–109 (2017). https://doi.org/10.1016/j.still.2016.10.008
135. Butler, B.M., O'Rourke, S.M., Hillier, S.: Using rule based regression models to predict and interpret soil proper ties from X-ray powder diffraction data. Geoderma **329**, 43–53 (2018). https://doi.org/10.1016/j.geoderma.2018.04.005
136. Xie, X.-L., Li, A.-B.: Identification of soil pro file using depth-weighted visible-near infrared spectral reflectance. Geoderma **325**, 90–101 (2018). https://doi.org/10.1016/j.geoderma.2018.03.029
137. Subburayalu, S., Jenhani, I., Slater, B.: Disaggregation of component soil series on an Ohio County soil survey map using possibilistic decision trees. Geoderma **213**, 334–345 (2014)
138. Mansuy, N., Thiffault, E., Paré, D., Bernier, P., Guindon, L., Villemaire, P., Poirier, V., Beaudoin, A.: Digital mapping of soil properties in Canadian managed forests at 250 m of resolution using the k-nearest neighbor method. Geoderma **235**, 59–73 (2014)
139. Camera, C., Zomeni, Z., Noller, J.S., Zissimos, A.M., Christoforou, I.C., Bruggeman, A.: A high resolution map of soil types and physical properties for Cyprus: a digital soil mapping optimization. Geoderma **285**, 35–49 (2017)
140. Dharumarajan, S., Hegde, R., Singh, S.: Spatial prediction of major soil properties using random forest techniques–a case study in semi-arid tropics of South India. Geoderma Reg. **10**, 154–162 (2017)
141. Massawe, B.H., Subburayalu, S.K., Kaaya, A.K., Winowiecki, L., Slater, B.K.: Mapping numerically classified soil taxa in Kilombero Valley, Tanzania using machine learning. Geoderma **311**, 143–148 (2018)

142. Vincent, S., Lemercier, B., Berthier, L., Walter, C.: Spatial dis aggregation of complex soil map units at the regional scale based on soil-landscape relationships. Geoderma **311**, 130–142 (2018)
143. Flynn, T., Rozanov, A., de Clercq, W., Warr, B., Clarke, C.: Semi-automatic disaggregation of a national resource inventory into a farm-scale soil depth class map. Geoderma **337**, 1136–1145 (2019)
144. Karandish, F., Šimůnek, J.: A field-modeling study for assessing temporal variations of soil-water-crop interactions under water saving irrigation strategies. Agric. Water Manag. **178**, 291–303 (2016)
145. Ivushkin, K., Bartholomeus, H., Bregt, A.K., Pulatov, A., Bui, E.N., Wilford, J.: Soil salinity assessment through satellite thermography for different irrigated and rainfed crops. Int. J. Appl. Earth Obs. **68**, 230–237 (2018)
146. Khanal, S., Fulton, J., Klopfenstein, A., Douridas, N., Shearer, S.: Integration of high resolution remotely sensed data and ma chine learning techniques for spatial prediction of soil properties and corn yield. Comput. Electron. Agric. **153**, 213–225 (2018)
147. Leenaars, J.G., Claessens, L., Heuvelink, G.B., Hengl, T., González, M.R., van Bussel, L.G., Guilpart, N., Yang, H., Cassman, K.G.: Mapping rootable depth and root zone plant-available water holding capacity of the soil of sub-Saharan Africa. Geoderma **324**, 18–36 (2018)
148. Koestel, J., Jorda, H.: What determines the strength of preferential transport in undisturbed soil under steady-state flow? Geoderma **217**, 144–160 (2014)
149. Bondi, G., Creamer, R., Ferrari, A., Fenton, O., Wall, D.: Using machine learning to predict soil bulk density on the basis of visual parameters: tools for in-field and post-field evaluation. Geoderma **318**, 137–147 (2018)
150. Naderi-Boldaji, M., Tekeste, M.Z., Nordstorm, R.A., Barnard, D.J., Birrel, S.J.: A mechanical-dielectric-high frequency acoustic sensor fusion for soil physical characterization. Comput. Electron. Agric. **156**, 10–23 (2019)
151. Gao, M., Li, H.-Y., Liu, D., Tang, J., Chen, X., Chen, X., Blöschl, G., Leung, L.R.: Identifying the dominant controls on macropore flow velocity in soils: a meta-analysis. J. Hydrol. **567**, 590–604 (2018)
152. Al Masmoudi, Y., Bouslihim, Y., Doumali, K., El Aissaoui, A., Namr, K.I.: Application of the random forest model to predict the plasticity state of vertisols. J. Ecol. Eng. **22**(2), 36–46 (2021)
153. Al Masmoudi, Y., Bouslihim, Y., Doumali, K., Hssaini, L., Ibno Namr, K.: Use of machine learning in Moroccan soil fertility prediction as an alternative to laborious analyses. Model. Earth Syst. Environ. **8**(3), 3707–3717 (2022)
154. Henderson, B.L., Bui, E.N., Moran, C.J., Simon, D.: Australia-wide predictions of soil properties using decision trees. Geoderma **124**, 383–398 (2005)
155. Dai, F., Zhou, Q., Lv, Z., Wang, X., Liu, G.: Spatial prediction of soil organic matter content integrating artificial neural network and ordinary kriging in Tibetan Plateau. Ecol. Indic. **45**, 184–194 (2014)
156. Poggio, L., Gimona, A., Spezia, L., Brewer, M.J.: Bayesian spatial modelling of soil properties and their uncertainty: the example of soil organic matter in Scotland using R-INLA. Geoderma **277**, 69–82 (2016)
157. Padarian, J., Minasny, B., McBratney, A.: Transfer learning to localise a continental soil vis-NIR calibration model. Geoderma **340**, 279–288 (2019)
158. Caubet, M., Dobarco, M.R., Arrouays, D., Minasny, B., Saby, N.P.: Merging country, continental and global predictions of soil texture: lessons from ensemble modelling in France. Geoderma **337**, 99–110 (2019)
159. Dobarco, M.R., Cousin, I., Le Bas, C., Martin, M.P.: Pedotransfer functions for predicting available water capacity in French soils, their applicability domain and associated uncertainty. Geoderma **336**, 81–95 (2019)

160. LeCun, Y., Bengio, Y., Hinton, G.: Deep learning. Nature **7553**, 436–444 (2015)
161. Hastie, T., Friedman, J., Tibshirani, R.: The Elements of Statistical Learning. Springer, New York (2001)
162. McBratney, A.B., Mendonça Santos, M.L., Minasny, B.: On digital soil mapping. Geoderma **1–2**, 3–52 (2003)

Role of Sustainable Bioenergy and Green Fuels in Building Climate-Resilient Futures

Dipankar Ghosh and Sabnam Hossain

Abstract Climate resilience is the capacity of a system, community, or society to endure, adjust to, and bounce back from the effects of climate change while continuing to work toward its objectives of social, ecological, and economic development. Bioenergy and green fuels contribute to climate resilience by providing renewable and sustainable alternatives to fossil fuels, lowering greenhouse gas emissions, and promoting energy system reform. Besides, energy system transformation, hard-to-abate sectors, carbon capture-storage, local economy supports, rural development can be ameliorated through the proper implementation of bioenergy and green fuels production platforms to attain successful climate resilience, economic viability and sustainability. Implementing bioenergy and green fuels to establish a sustainable bioeconomy, however, is fraught with challenges and issues. These include land competition, concerns about the security of edible food, environmental damage, and the need for appropriate policies and continued technical advancements considering AI and Machine-Learning to support the development of sustainable bioenergy and/ or green energy and mitigate potential risks. To this purpose, the current comprehensive summary will concentrate on the latest developments regarding the role of bioenergy and green fuels production in achieving climatic resilience, taking into account a wide range of benefits, drawbacks, and applications while highlighting the need for innovation, investment, and policy support to achieve a sustainable and circular bioeconomy in the near future.

Keywords Climate resilience · Bioenergy · Green fuels · Economic viability · Sustainability

D. Ghosh (✉) · S. Hossain
Microbial Engineering and Algal Biotechnology Laboratory, Department of Biosciences, JIS University, Kolkata, India
e-mail: dghosh.jisuniversity2@gmail.com

A. L. Srivastav et al. (eds.), *Climate Resilience: Impact of Quantum Computing and Artificial Intelligence on Urban Planning*, Studies in Systems, Decision and Control 637, https://doi.org/10.1007/978-3-032-06791-3_13

1 Introduction

Climate change is still a huge problem since it seriously affects ecosystems, human health, and economy. Handling this worldwide problem requires looking toward renewable energy sources. Of them, renewable fuels and bioenergy stand out as very useful tools to boost resilience and help to lessen the effects of cycle change. Derived from natural items like garbage and plants, bioenergy offers a replacement for nonrenewable fossil fuels. Turned into heat, electricity, or biofuel [1], it enables a circular economy whereby trash creates energy. New research show that using bioenergy can drastically lower greenhouse gas emissions. For instance, derived sustainably from agricultural wastes or algae, biofuels include biodiesel and bioethanol can allow considerably less emissions than conventional fossil fuels [2]. This cut helps farming methods that slow down global warming and improve the condition of the soil. Also supporting bioenergy projects are green fuels obtained from sustainable sources including synthetic fuels and hydrogen. Made from solar and wind power, among other renewable energy sources, will assist to completely offset the carbon footprints frequently linked with hydrogen from petroleum. One key component in decarbonizing difficult-to-abate sectors like steel and cement making is green hydrogen produced by electrolysis run on renewable energy [3, 4]. If we are to reach sustainable development, using green fuels in their energy mix is absolutely essential in the race for net-zero emissions. Moreover, utilization of green fuels and bioenergy could enhance energy security by means of source diversification, which can help to lower dependence on outside fossil fuels prone to changing prices and geopolitical conflict. The current energy crises provide unambiguous evidence of this; nations that had been looking at renewable technology exhibited improved performance in energy stability and economic resilience [5]. Eventually, bioenergy and green fuels determine whether or not one is attaching climate resilience. Their support of sustainable economic development, energy independence, and reduction of greenhouse gas emissions in the battle against climate change is absolutely invaluable. Notwithstanding currently existing issues such land use competition and technology developments, a sustainable future rests totally on ongoing financing and study on these renewable sources. Industries, people, governments, and one another have to adopt these ideas and create conditions where climate resilience could flourish.

The horrible truth showing the present need of climate resilience instead of a far-off threat is the growing consequences of climate change. Communities, ecosystems, infrastructure all project, get ready for, and can adapt to the effects of climate change. Strong weather resiliency strategies help to protect communities from food and water scarcity, economic uncertainty, and catastrophic occurrences. Rising frequency of extreme weather occurrences is one of the most pressing reasons of improving climate resilience. Rising severe storms, floods, and fires in recent years have caused widespread damage and migration. For example, Hurricane Maria of 2017 destroyed Puerto Rican infrastructure, therefore highlighting the vulnerability of island nations to events linked to climate change. Unusually, the bush fires in Australia in 2019 and 2020 also displayed the terrible effects of extended heat waves and drought on human communities and ecosystems [6]. Investing in climate-resilient infrastructure such early warning systems, flood barriers, and drought-resistant plants helps one to minimize the consequences of these calamities and protect underdeveloped areas. Moreover, especially in areas already under resource limit, climate change substantially undermines water and food security. Rising temperatures, changed precipitation patterns, and more frequent droughts upsetting agricultural productivity are causing food shortages and price volatility. Through means of which current water stress is exacerbated to endanger livelihoods and increase the risk of conflict over limited resources, strategies including water-efficient irrigation, crop diversification, and climate-smart agriculture help to ensure food security and lower sensitivity to climate-related shocks [7]. Apart from the obvious consequences of extreme weather events and resource constraint, climate change jeopardizes society well-being into the future as well as economic stability. Major and most likely rising in the years to come are the costs of infrastructure destruction, public health effects, and catastrophe recovery. From land-use policy to infrastructure projects, a complete approach including climate hazards into every area of planning and development would help to increase climate resilience. Among policy goals are increasing renewable energy, encouraging local involvement in projects for climate change adaptation, and so supporting sustainable development programs. The necessity of climate resilience nearly begs interesting questions. Already greatly influencing societies, ecosystems, and economy all around is climate change, these consequences only expected to get more severe ahead. By supporting sustainable agriculture, investing in climate-resilient infrastructure, and addressing climate issues in all spheres of planning and development, one can assist to reduce society's sensitivity to climate change and offer a more rich and sustainable future for everybody.

Dealing with the urgent need to slow down global warming and shift away from fossil fuels, bioenergy and green fuels present a critical path towards a sustainable energy future. Originating from organic materials, these renewable energy sources have many varied benefits ranging from better energy security to reduced greenhouse gas emissions. One key advantage of green fuels and bioenergy is their possible sharp reduction of carbon emissions. Unlike fossil fuels, which release stored carbon into the atmosphere, bioenergy uses organic materials already carbon dioxide has absorbed throughout their growth. Theoretically, the carbon absorbed balances the carbon emitted when these materials are converted into energy, therefore creating a closed carbon cycle. Especially when sustainable agricultural techniques are applied, biofuels including ethanol and biodiesel made from crops like corn and soybean have been shown to minimize greenhouse gas emissions when compared to gasoline and diesel [8]. Furthermore supporting energy security by diversifying energy sources and reducing reliance on fossil fuels imports help bioenergy and green fuels. Several countries are strongly pushing the production of biofuels and biogas in order to increase national energy independence and support home industry and waste management. For instance, Brazil has successfully started a large ethanol program based on sugarcane as a feedstock, therefore reducing its dependence on imported oil [9]. Beyond carbon emissions and energy security, green fuels and bioenergy provide much more environmental and financial benefits. The manufacture of these fuels can increase rural development, generate employment opportunities in agriculture and biorefining, and so give farmers new income streams. Using waste materials for bioenergy generation can help to lower landfill waste, enhance waste management techniques, and so minimise environmental pollution. Thus, bioenergy and green fuels clearly define a sustainable energy change. Their capacity to reduce greenhouse gas emissions, enhance energy security, and provide other environmental and economical benefits makes them a preferred alternative for fossil fuels. Ongoing research, technical innovation, and supportive laws will help to open the route for a cleaner, more sustainable future by means of which further development and deployment of these renewable energy sources depend.

2 Understanding Bioenergy and Green Fuels

Fundamental solutions emerging on the road to sustainable energy systems are providing alternatives to conventional fossil fuels and reducing the repercussions of climate change: bioenergy and green fuels. Bioenergy, which produces heat, electricity, and biofuels comes from biomass, an organic material used in combustion, gasification, and anaerobic digestion. Often made from non-food feedstocks such algae [10] and cellulosic material [2], green fuels which are especially produced from sustainable sources and aimed to reduce environmental effects often highlight advanced biofuels made from Starting with biomass sourcing—that includes agricultural leftovers such maize stower and wheat straw, forestry products, and dedicated energy crops such switchgrass and Miscanthus bioenergy production starts [11]. These feedstocks then are treated utilizing numerous conversion processes. Combustion is the most fundamental method, biomass is burned to create heat, which can then be utilized straight for direct heating needs or to generate electricity. Gasification turns biomass into syngas, a gaseous mix that can be refined into either chemicals or fuel anaerobic digestion depending on microorganisms to break down biomass without oxygen yields biogas, a mix of methane and carbon dioxide that can be used either for electricity generation or improved to biomethane for transportation fuels. Typical categories of green fuels are first–second and third generation biofuels. First-generation biofuels such corn ethanol and vegetable oil-derived biodiesel [12] have brought alarming questions about food security and land use change. Second-generation biofuels, derived from non-food sources such as lignocellulosic biomass, provide a more sustainable substitute by employing agricultural and forestry wastes therefore lessening demand for food production and competitiveness with regard to it. Third-generation biofuels produced from algae offer a hopeful path due to their high lipid content and quick growth rates, which can lead to higher fuel production with low land utilization. Furthermore under investigation are metabolic and genetic engineering to increase the efficiency of producing biofuel from different microorganisms such algae [3, 4]. Among the most recent developments in bioenergy and green fuels are advanced conversion methods include pyrolysis and hydrothermal liquefaction, which can convert a wide spectrum of biomass feedstocks into liquid biofuels and bioproducts [13]. Finally, green fuels and bioenergy offer sensible means to reduce reliance on fossil fuels and slow down global warming. Setting the foundation for a better energy future, constant research and development efforts focus on improving the efficiency, sustainability, and financial viability of bioenergy and green fuel generation. The sustainable generation of bioenergy calls for logical investigation of environmental consequences, feedstock procurement, and transformation technology. Biomass fuels are those derived from biomass. Green energies span a much broad range of renewable possibilities. Among these are synthetic fuels derived from sustainable resources, hydrogen, and ammonia. Both groups intend to solve climate change and reduce reliance on fossil fuels. These fuels are categorized depending on their manufacturing technique, feedstock, and environmental impact. Starting in the early twentieth century, biofuels have decades-long historical foundation. Rudolf Diesel running his engine in 1900 using vegetable oil. The energy crisis of the 1970s sparked fresh interest in biofuels [14], which governments all around began looking at as alternative energy sources. Late 20th and early twenty-first century biofuel research and production had a notable increase, citing [15]. Environmental concerns and energy security drove this course of development. Usually, biofuels are set in generations. First-generation biofuels are developed from vegetable oils, starch, and sugar. One example would be ethanol from maize, another, biodiesel from soybeans [12], attacked these biofuels for their effects on land use and food costs. Non-food plants generate second-generation biofuel. Among these are agricultural leftovers a form of cellulosic biomass and woody species. These choices are better for the surroundings. Made from algae, third-generation biofuels Growing on non-arable land and in wastewater, algae do not directly compete with food production [16]. Fourth-generation biofuels apply advanced techniques. Among these are genetic modified plants and carbon capture methods. Green fuels surpass conventional biofuels. One very good example of this is hydrogen produced from renewable sources. Green hydrogen produced by electrolysis run fueled by solar or wind power Made from sustainable hydrogen and nitrogen, ammonia created from another green fuel source can be used in fuel cells for transportation and power generation [17]. Synthetic fuels offer even another choice for usage as a transportation fuel and energy storage medium, produced by synthesizing collected carbon dioxide with renewable hydrogen [18]. One should give great consideration the environmental impact of biofuels and green fuels. By means of generation, biofuels reduce greenhouse gas emissions than conventional fuels. It changes according on the production method and feedstock. While second and third-generation biofuels often offer more notable emission reductions, hydrogen and ammonia and other green fuels have the almost negligible emissions [19]. Their manufacturing might rely on their consumption of renewable energy. The land use change of biofuel generation might have negative environmental consequences: habitat destruction and deforestation could undo the advantages of reduced emissions [20]. Water usage is another problem. Especially in dry regions, the generation of biofuel can strain water supplies. The financial impact of green fuels and biofuels is multifarious. Support of the sector has come mostly from government incentives and subsidies [21]. Widespread acceptance of biofuels and green fuels still depends much on their cost of manufacture. Changing crude oil prices could affect biofuel competitiveness. Farmers might potentially get fresh income sources [22]. This calls questions on the long-term financial viability. A few well-known personalities have helped biofuels and renewable fuels to evolve. Early Rudolf Diesel vegetable oil experiments set the stage for studies on biofuel. Algal biofuels [23] came upon ground-breaking study by Craig Venter. Green fuels and biofuels find applications among different individuals. Others underline expected results and are more cautious. One must view things from a fair distance. By matching the trade-offs, this should maximize the advantages of several consumption sources. Further technological development will decide biofuels and green fuels. Fast advancement in catalysis, renewable energy sources, and genetic engineering incorporating catalyzes is really vital. Furthermore very crucial are policy backing, research and development financing, and ecologically responsible production methods. These will help to guarantee the long-term survival of the world by reducing environmental consequences. Another difficult question is using green fuels and biofuels in line with the present energy supply.

3 Bioenergy and Green Fuels for Climate Resilience

Climate resilience is community and ecological capacity for foresight, planning, and climate change response. It addresses several strategies and policies aiming at reducing exposure and enhancing social and ecological system adaptability [24]. Climate resilience depends critically on bioenergy derived from organic sources. Bioenergy affects climate resilience by supporting sustainable energy generation, economic stability, and ecological health all of which influence economic stability [25]. First, by helping to lower greenhouse gas emissions and increase climate resilience, bioenergy strengthens Bioenergy balances the effects of fossil fuels, which aggravate global warming [26]. Burning biomass for energy generates a closed carbon cycle since the biomass absorbs carbon during growth and balances the carbon dioxide produced [27]. For environmental issues as well as for energy, this two-fold advantage is noteworthy. Advanced biofuels have raised this possibilities and made bioenergy a feasible substitute for traditional fuels Furthermore necessary for bioenergy generation are environmentally friendly farming methods that support biodiversity and soil condition. Growing energy crops such as Miscanthus or Switchgrass enhances soil, water management, and biodiversity [11]. These agricultural technologies gather carbon to advance resilience of the climate. Rural businesses can create additional income and employment by including bioenergy into nearby farming systems [21]. Boosting energy security and climate resilience is bioenergy [5]. Energy dependability becomes increasingly important as severe storms get more frequent. By allowing tiny areas to produce energy on-site, bioenergy reduces the need of fossil fuels. This increases local autonomy, resistance to changes in world energy supply and cost, and energy security. At last, bioenergy is crucial for climate resilience since it supports carbon reduction, energy security, and sustainable agriculture [8]. Using bioenergy technology and systems will help communities enhance environmental sustainability, financial prosperity, and climate change resistance. Resilience of energy systems depends critically on climate-related fluctuations. Rising global temperatures and catastrophic weather presented by climate change challenge established energy infrastructures. Floods, cyclones, wildfires, and heat waves can destroy networks of producing, transmission, and distribution. The research investigates how these developments might increase energy system resilience to offer a consistent energy supply in view of environmental change. Extreme power outages and destruction of the energy infrastructure could result from climate-related disturbances. In 2017 Hurricane Harvey caused Texas to lose electricity, highlighting weaknesses in the energy system [28]. Energy systems have to be strong to stop and fix from such problems. Funding for fast-recovery technology and natural disaster-resistant technologies is thus needed. One of the smart concepts is energy diversification. Among other renewable energy sources, hydropower, solar, and wind lower the usage of fossil fuels and improve the energy systems' flexibility. Moreover closely related to initiatives aiming at global climate change is the consumption of renewable energy. Germany's Energiewende will show how policies may boost energy resilience as renewable energy will replace fossil fuels [29]. Infrastructure modernization is quite important. Some degree of disturbance reaction is enabled by real-time energy flow made feasible by power grid upgrades to smart grids. Advanced communication technologies of the smart grid enable to distribute power and locate faulty components, hence reducing downtime. Underlining in Puerto Rico the need of smart grid technologies in fast recovering power, Hurricane Maria underscore. Eventually, community involvement helps one to find resilience. Local governments and utilities must work with people to create future as well as present emergency energy plans for current requirements. Awareness programs and practical teaching could give towns authority over disturbances in electricity supply. Finding a safe energy source in a changing environment demands energy systems to get more shock-resistant from climate change. Changing energy sources, improving infrastructure, and inspiring community involvement assist energy systems to get stronger and more ready for climate change.

4 Challenges and Opportunities to Nurture Role of Bioenergy and Green Fuels to Achieve Climate Resilience

4.1 Technological and Economic Barriers

Arguments regarding alternative energy sources today center mostly on the cost-effectiveness of green fuels and bioenergy. Increasing concern regarding climate change highlights the necessity of renewable energy alternatives and forces companies to examine more closely the financial feasibility of biofuels and other eco-friendly options as well. Through natural materials begin such energy sources, which resonate because they can aid in reducing the emission of greenhouse gases compared to fossil fuels [30]. Most typically regarded as an alternative resource, biomass bioenergy utilizes farm residuals, waste streams, and by-products from forestry. Conversely, green fuels are typically those derived from renewable energy sources which do not contribute to the atmospheric total concentration of carbon dioxide. A number of factors such as government policy, technical advances, and availability of feedstock could change the cost competitiveness of such options. Attaining cost competitiveness for bioenergy is largely limited by the manufacturing cost related with diverse feedstocks. For example, the value of biological wastes could change depending on the status of the market, hence, it could be difficult to estimate continuous price for biofuels [31]. In addition of key importance is the efficacy of conversion technologies in the process. More efficient processing techniques bred from recent advances in biochemical and thermochemical conversion technologies have significantly reduced manufacturing costs and thus increased general economic feasibility [32]. In addition significantly impacting competitiveness of bioenergy are government actions. Subsidies, tax credits, and renewable energy standards help to light producers' financial load. The Renewable Fuel Standard in the United States, for example, incentivizes biofuel production, hence promoting their use at a competitive price relative to conventional fossil fuels [33]. Nevertheless, depending too much on government assistance could lead to market volatility, thus stressing the need of sustainable business models less depending on subsidies. By means of efficient policy frameworks, Brazil has become a front runner in bioenergy and shows the economic viability of green fuels when linked with infrastructure development and promotion of biofuel usage [34]. Key examples of countries like Brazil, which has successfully harnessed its sugarcane industry to supply bioethanol, demonstrate the potential of bioenergy to achieve cost competitiveness. Essentially, technological advancements, environmentally friendly processes, and favourable government policies are paving the way for successful adoption of bioenergy and green fuels in the energy sector despite their challenges in achieving cost competitiveness. Bioenergy may be extremely critical during the transition to a sustainable energy future as such trends continue to change. Infrastructure development is crucially important to ensure environmental climate resilience and also to increase bioenergy production.

Produced from organic matter, bioenergy is a sustainable alternative to fossil fuels that is conducive to environmentalism through reducing carbon emissions. Scaling up bioenergy production becomes necessary with improved infrastructure as global energy needs rise and climate change creates further problems. Initial in bioenergy infrastructure construction is preparation of site for harvesting and handling biomass. This requires standards in the efficient collection of urban organic waste as well as forestry residues and agricultural residuals. One study, for example, underlines the requirement of setting up collecting places that simplify the biomass procurement [35]. By converting many waste products into valuable energy sources, these facilities support local businesses and get people working. The infrastructure of transportation is also another quite crucial element. By cutting time and cost, good transportation infrastructure enables biomass to be transferred from collecting points to processing plants. Including bioenergy into current energy sources demands for much-needed improvements in roadways, railroads, and shipping routes. For instance, Brazil's vast ethanol distribution system demonstrates how successfully well-planned infrastructure may enable biofuel delivery, hence improving energy security and resilience against climate change [36]. Moreover essential for the more ecologically friendly and effective conversion of biomass into biofuels is the development of biorefineries. By means of technical development, biorefinery designs could enhance the conversion techniques, so increasing the yield and reducing waste. Research by Katakojwala and Mohan [37] indicate that better biorefinery designs could very substantially increase the general sustainability of bioenergy systems, hence drastically lowering carbon emissions and raising climate resilience. Supporting this infrastructure calls as much on public policy as on funding. Government support of building bioenergy infrastructure can inspire private sector investment, hence promoting innovation and competitiveness in the bioenergy sector. For example, incentives providing tax credits for construction of biorefineries or sponsoring research on new biomass conversion techniques would drastically hasten transformation. By means of effective biomass collecting, improved transportation, advanced biorefineries, and supporting public policies, society can essentially unlock the potential of bioenergy transforming waste into a sustainable energy source and therefore decreasing the negative repercussions of climate change. Expanding bioenergy output and reaching environmental climate resilience depend on infrastructure development. Such infrastructure spending not only enhances energy security but also advances environmental preservation and stimulates economic growth. Technological restrictions and knowledge gaps are main obstacles stopping progress in many different fields. Sometimes fast evolving technology exceeds the theoretical foundation and empirical knowledge needed to fully benefit from these advances. This paper explores the character of these limitations and omissions by means of real cases in the domains of artificial intelligence and renewable energy technologies. Among systems, their lack of compatibility is maybe the most clear technological limitation. Typically running on separate platforms and frameworks in the artificial intelligence space, distinct AI systems could lead to data sharing sufferability. Thus, potential discoveries are usually held back and the whole benefits of artificial intelligence are not fully acknowledged. Apart from issues with interoperability, some research gaps stemming from moral questions concerning technology use exist. For example, in machine learning, bias and fairness problems are rather common. Although algorithmic fairness is a topic of extensive theoretical research, most of it remains theoretical and lacks real-world proof. Research that could assist to bridge this gap by developing unbiased algorithms extensively tested in many environments is therefore rather much needed [38]. Furthermore, the materials used in energy storage devices could occasionally impose technological limitations in the sector of renewable energy. Among other existing battery technologies, lithium-ion batteries suffer with regard to environmental effect, lifetime, and efficiency [39]. Research is needed to look at different ideas and products that might provide more efficient, safer methods of energy storage. The shift to sustainable energy sources is a top global issue, hence it will be difficult to make significant development without addressing these constraints. Simply said, technological limitations and research gaps pose significant challenges to development in many sectors. The division of artificial intelligence systems and the ethical consequences of machine learning underline the need of integrated research approaches (Fig. 1). Similarly, the sector of renewable energy depends on innovative solutions for battery technology to make sustainable energy realistic. An answer to these challenges is necessary to maximize the whole capability of technology in a way that benefits sustainable development and aids society.

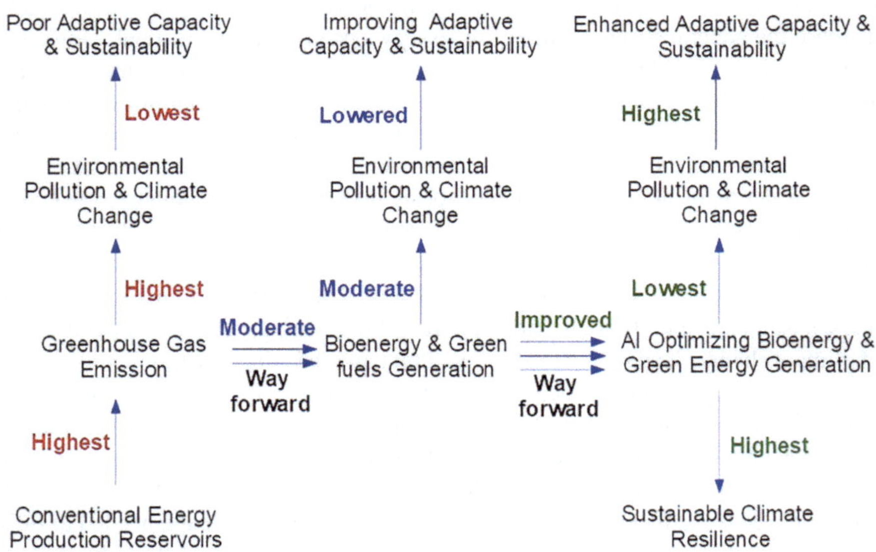

Fig. 1 Cooperativity between bioenergy systems, climate resilience measures, and artificial intelligence (AI) applications [40–44] (AI aids decision-making in bioenergy production by optimizing feedstock supply chains, projecting crop yields, and tracking emissions. Bioenergy helps to reduce greenhouse gas emissions while also providing decentralized energy alternatives. The integration improves adaptability and sustainability in both energy and environmental systems)

4.2 Environmental Sustainability Concerns

Attaining climate resilience relies more and more on environmental sustainability, an essential ingredient. The ability of systems, society, and ecosystems to withstand and recover from climate-related shocks is referred to as climate resilience. Over the last few years, there have been multiple approaches and models emerging to address the pressing demand for environmental sustainability, thus reflecting the interdependent nature of ecological balance and climate change adaptation. Greenhouse gas emissions, a principal cause of climate change, are one of the important aspects of environmental sustainability. Olabi and Abdelkareem [45] stated that renewable energy adoption and waste minimization contribute to mitigating the impacts of climate change. Transitioning from fossil fuels to renewable energy sources such as solar and wind power contributes to energy independence and security as well as reducing emissions. Green infrastructure like urban forests and green roofs also aid in carbon dioxide absorption and enhance urban resilience to floods and heatwaves. Another sector struggling with sustainability challenges is agriculture. Conventional farming methods typically lead to deforestation, soil erosion, and species loss. Effective application of sustainable farming practices like agro-ecology and permaculture enhances food security while minimizing ecological harm. Research indicates that these techniques can increase agricultural yields and improve soil health, thus creating a stronger food system capable of weathering climate change [46]. Achieving climate resilience also rests on effective water management. Water scarcity and floods are huge risks considering the altering hydrological conditions as storm patterns become increasingly volatile. Using sustainable water management practices, such as rainwater harvesting and wetlands rehabilitation, can assist communities to adapt. Not only do these practices conserve water, but they also defend ecosystems entirely crucial to biodiversity and climate regulation [47]. In the debate about sustainability, education and community involvement are entirely critical components. Providing local communities with authority to participate in environmental decision-making enhances a sense of ownership and responsibility. Grass roots activities, for example, tree planting initiatives and cleaning campaigns, demonstrate how tiny communities can play a part towards bigger sustainability goals, thus finally enhancing social resilience to climate concerns. Lastly, to achieve climate resilience, sustainability of the environment is central. Communities can increase their ability to respond to climate change by adopting renewable energy, practicing sustainable agricultural practices, smart management of water resources, and community participation. We simply have to make these strategies our priority if we are to build a sustainable future in defense of both our environment and our citizens.

4.3 Policy and Regulatory Frameworks

Climate resilience is defined as the ability of systems ecosystem, society, or economy to anticipate, prepare for, and respond to the negative impacts of climate change. With the enormity of the threat that climate change presents to societies all over the world, strong legislative and regulatory structures will need to be established first and foremost to support climate resiliency. Designing such systems involves an overall strategy based on integrative actions that respond to adaptation to and mitigation of changes caused by climate change. A major feature of these systems is the integration of scientific knowledge directly into climate regulations. Governments need to utilize knowledge and information drawn from climate science to inform their policies. The Intergovernmental Panel on Climate Change (IPCC) provides significant evaluations that allow legislators to grasp possible environmental scenarios and their effects. Countries like the Netherlands have effectively incorporated comprehensive flood defence policies based on historical records and future projections, thereby exhibiting excellent congruence between scientific research and policy execution [48]. Another key factor is the involvement of diverse stakeholders, including businesses, non-profit organizations, and societies. Engaging such groups in the process of decision-making will help the representatives of government ensure fair and efficient policies. Urban greening projects such as the projects noted above demonstrate the potential community-focused projects hold for helping cities recover from calamitous events as well as sustain social cohesion. The Milwaukee Metropolitan Sewerage District's green infrastructure program, for instance, is a demonstration of how collective can result in adaptive solutions to dealing with stormwater [49]. Regulations also need to impose rules that promote sustainable behaviors. Such efforts encompass the European Union's Green Deal, a broad legislative push to redesign its economy to achieve carbon neutrality by 2050. Such legislation included legislation aimed at minimizing greenhouse gas emissions, promoting renewable energy, and enabling climate-resilient transport systems. Through the showcasing of the interrelationship between environmental policy and economic development, these steps, therefore, tackle climate resilience from a more general perspective. Lastly, an effective climate resilience framework must prioritize continuous monitoring and adaptation. Adaptable rules must leave sufficient space to incorporate new scientific information and technological advances. Periodic reviews can assist in identifying policy loopholes and adjust strategies to effectively address evolving climate threats. Essentially, attaining climate resilience requires extensive government and regulatory frameworks that incorporate scientific research, engaging numerous stakeholders, promoting sustainable practices, and ensuring adaptive governance. Through the adoption of these methods, societies will be able to improve their resilience and adaptability to climate change challenges, ensuring a more sustainable future.

4.4 Social and Ethical Considerations

Particularly in developing countries, where food shortages are often a severe problem, the overlap of these issues jeopardizes equitable distribution and access to bioenergy resources. Access to bioenergy can create disparities. Occasionally, land used for farming, which could be used to produce food, is instead used to produce biofuels, which could serve richer countries and corporations more than the local governments. As societies grapple with the energy generation versus food security tradeoff, the food versus fuel debate creates important ethical and social concerns. Especially with the growing world population and the cry for climate resilience, the mounting reliance on biofuels has triggered debates regarding equitable allocation of agricultural resources. Especially in regions that already suffer from food insecurity, this allocation could exacerbate existing inequalities. The literature indicates that the transition towards bioenergy has to be implemented with caution so as not to displace agricultural production and increase starving among vulnerable groups [50]. Popular views and consent are also essential in achieving climate resilience through the utilization of bioenergy. Frequently influenced by rumors and production process's non-transparency, many are not certain about the benefits and limitations of biofuels. When possible environmental and economic benefits of biofuels are articulated clearly, public opinion generally increases, as research has indicated [51]. Conversely, arguments regarding deforestation due to biofuel crops can create opposition and suspicion. Therefore, involving the public is crucial to getting support for sustainable use of bioenergy and bioenergy policies. Additionally, the social fabric of communities can be impacted as transformations towards bioenergy would alter traditional agriculture activities and local economies. Promotive policies that incentivize smallholder farmers to be involved in bioenergy production can create a more equitable platform, ensuring that benefits are shared equitably and without compromise of food sovereignty. Collaborative policy and participatory government can empower communities at the local level to take decisions over their resources and ensure that they are integrated into the bioenergy supply chain. Basically, the food versus fuel debate raises fundamental social and ethical questions that need to be examined with care. Since equity and access to bioenergy resources are of primary concern, it is crucial to foster public participation and consensus-building in the formation of sustainable bioenergy policies. Since communities are looking for climate resilience, there must be a balance between food and energy consumption, thus making sure that disadvantaged groups are not left behind in the shift towards a more sustainable future.

4.5 Opportunities for Innovation and Investment

In the quest for environmental climate resilience, future biofuel technologies, green hydrogen and synthetic fuel production, and the development of a circular bioeconomy present significant opportunities for investment and innovation [1]. These opportunities provide means to lower greenhouse gas emissions, enhance sustainable resource management, and drive economic growth. To realize their full potential unequivocally, strategic investment and ongoing research and development are necessary. Development and research on advanced biofuels are top of the agenda. While food crops for first-generation biofuels are a challenge in terms of land use and food security, advanced biofuels use non-food biomass like algae, dedicated energy crops, and farm waste. Here, innovation is concentrated on optimizing the rate of processes like gasification and enzymatic hydrolysis for the production of biofuels such as cellulosic ethanol and green diesel. The investment in such technologies has the potential to bring carbon-light transport fuels. Yet another great area is green hydrogen and synthetic fuel production. Green hydrogen, which is generated by electrolysis driven by renewable energy sources, will also have the potential to be utilized as a clean energy carrier for different applications, ranging from transportation to industry and power generation. In order to boost the production of synthetic fuel and green hydrogen [52], investment in fuel synthesis technology, carbon capture technology, and electrolyser technology must be made. Mixing captured carbon dioxide with renewable hydrogen creates synthetic fuel, which helps decarbonise industries like shipping and aviation that struggle to abate. The circular bioeconomy is the sustainable basis of resource management. The basis of this concept is in the production of multiple products like biofuels, bio-based products, and biochemicals from biomass and organic waste in addition to other biological capitals. A circular bioeconomy is tried to provide innovative economic opportunities, reduced dependence on fossil fuel, and reduced waste. Investment in biorefineries, waste valorization activities, and innovation in bio-based goods is essential to drive a rich circular bioeconomy. These prospects come with problems. Synthetic fuels, green hydrogen, next-generation biofuels, and relative to fossil fuels are rather costly to generate. Strategic investments and research and development are absolutely necessary to overcome cost constraints and realize environmental climate resilience by means of these technologies, so optimizing their potential. Fixing these problems will call for combined with continuous technological breakthroughs and economies of scale supportive policies such carbon pricing and renewable energy standards [53].

5 Conclusion and Future Outlook

Given the consequences of climate change the earth is now confronting, the value of bioenergy and green fuels in enhancing climate resilience has become increasingly important. Renewable energy sources support a change in a sustainable energy system and help fight greenhouse gas emissions. Derived from organic sources, bioenergy as well as numerous green fuels as ethanol and biodiesel have opportunities for decreasing dependency on fossil fuels, thereby greatly contributing in fulfilling international climate targets [54]. Ensuring energy security while addressing environmental problems requires a flexible and adaptable energy portfolio. Particularly bioenergy and green fuels, renewable energy sources include a variety of options that might be tailored to match local conditions. This local approach not only helps to lower carbon footprints and preserve current infrastructure but also promotes economic growth, thus, it contributes to climate resilience at local and national levels by, for example, allowing nations rich in agricultural goods to effectively turn waste materials into biofuels [55]. The use of bioenergy and green fuels meets objectives of international climate pacts such as the Paris Accord since they underline the need of governments to reduce their greenhouse gas emissions. Including bioenergy into the energy mix allows nations to use a sustainable and renewable source that not only lowers emissions but might also be produced and consumed locally. This helps to lower travel emissions [56] and increase energy autonomy. Furthermore in bioenergy technology, with enhanced biogas systems and second-generation biofuels, investment demonstrates the ability of inventiveness to raise the efficiency of these renewable resources. To enable a sustainable energy transition, one might observe that going future global energy strategy must surely include bioenergy and green fuels. Furthermore, as the urgency of climate action grows, scaling up these solutions will depend mostly on encouraging creativity and funding of research in bioenergy. Moreover, along with technology developments and energy efficiency programs, these renewable sources could significantly help to lower global reliance on fossil fuels. Green fuels and bioenergy complement an off-curve energy mix to assist fulfil international climate targets as well as to set the scene for climate resilience. Their integration has many benefits from reduced greenhouse gas emissions to better energy security. Giving these renewable resources top priority in energy plans would enable sustainable and efficient combat of climate change in the next few decades. Developing bioenergy requires the challenging choreography of screening and finding hidden nonlinear kinships between arriving and departing systems. Still, a number of issues and complications including the uncertainty of feedstock and restrictions in conversion capacity have hampered the expansion of bioenergy producing technologies. Therefore, to develop, run, forecast, and improve the system, it is absolutely necessary to use precise and effective model proficiencies that is, artificial intelligence (AI)-based technologies. These cutting-edge computing techniques [38] seem to solve technological constraints, maximize manufacturing, and offer predictive analytics supporting scalability.

Acknowledgements Authors would like to thank JIS University and JIS Group of Educational Initiatives for their immense support and help. Dr. Ghosh would also like to provide special thanks to Ms. Shrestha Debnath for her immense support.

References

1. Srivastav, A.L., Bagherian, A., Ghosh, D.: The circular bioeconomy: pathways to sustainability and resource optimization. Clean Technol. Environ. Policy 1–18 (2024)
2. Oloyede, C.T., Itabiyi, O.E., Popoola, O.A., Jekayinfa, S.O., Olaniyan, M.A., Adebayo, A.O., Ogunkunle, O., Zamri, M.F.M.A., Fattah, I.M.R.: Navigating prospects and challenges for green fuels for achieving economical, environmental and ecological resilience: a scientific review. Biofuels **15**(7), 929–941 (2024)
3. Ghosh, D., Hallenbeck, P.C.: Metabolic Engineering of Hydrogen Production by Green Algae. Future Science Ltd. (2015)
4. Ghosh, D., Hallenbeck, P.C.: Cyanobacteria and photosynthetic bacteria: metabolic engineering of hydrogen production. Adv. Enzym. Convers. Biomass Biofuels **1**, 112–124 (2015)
5. Ibekwe, K.I., Etukudoh, E.A., Nwokediegwu, Z.Q.S., Umoh, A.A., Adefemi, A., Ilojianya, V.I.: Energy security in the global context: a comprehensive review of geopolitical dynamics and policies. Eng. Sci. Technol. J. **5**(1), 152–168 (2024)
6. Filkov, A.I., Ngo, T., Matthews, S., Telfer, S., Penman, T.D.: Impact of Australia's catastrophic 2019/20 bushfire season on communities and environment. Retrospective analysis and current trends. J. Saf. Sci. Resil. **1**(1), 44–56 (2020)
7. Umer, Y., Chavula, P., Abdi, E., Ahamad, S., Lungu, G., Abdula, H., Abdumalik, M., Ahmed, S.: Small-scale irrigation farming as a climate-smart agriculture practice; its adoption and impact on food security for Ethiopian smallholder farmers: a review. Asian Res. J. Curr. Sci. **6**(1), 163–180 (2024)
8. Mathur, S., Waswani, H., Singh, D., Ranjan, R.: Alternative fuels for agriculture sustainability: carbon footprint and economic feasibility. AgriEngineering **4**(4), 993–1015 (2022)
9. Karp, S.G., Medina, J.D., Letti, L.A., Woiciechowski, A.L., de Carvalho, J.C., Schmitt, C.C., de Oliveira Penha, R., Kumlehn, G.S., Soccol, C.R.: Bioeconomy and biofuels: the case of sugarcane ethanol in Brazil. Biofuels, Bioprod. Biorefin. **15**(3), 899–912 (2021)
10. Patwardhan, S.B., Pandit, S., Ghosh, D., Dhar, D.W., Banerjee, S., Joshi, S., Gupta, P.K., Lahiri, D., Nag, M., Ruokolainen, J., Ray, R.R.: A concise review on the cultivation of microalgal biofilms for biofuel feedstock production. Biomass Convers. Biorefinery **14**(6), 7219–7236 (2024)
11. Umakanth, A.V., Datta, A., Reddy, B.S., Bardhan, S.: Biomass feedstocks for advanced biofuels: sustainability and supply chain management. Adv. Biofuel Technol. 39–72 (2022)
12. Kurowska, K., Marks-Bielska, R., Bielski, S., Kryszk, H., Jasinskas, A.: Food security in the context of liquid biofuels production. Energies **13**(23), 6247 (2020)
13. Perkins, G., Batalha, N., Kumar, A., Bhaskar, T., Konarova, M.: Recent advances in liquefaction technologies for production of liquid hydrocarbon fuels from biomass and carbonaceous wastes. Renew. Sustain. Energy Rev. **115**, 109400 (2019)
14. Fritz Dufour, M.B.A.: The Costs and Implications of Our Demand for Energy: A Comparative and Comprehensive Analysis of the Available Energy Resources: How Fossil Fuels Have Created an Energy Crisis/Exploring Our Alternatives/The Future of Energy. Fritz Dufour, MBA, DESS (2018)
15. Collet, C.: Designing our future bio-materiality. AI Soc. **36**(4), 1331–1342 (2021)
16. Naylor, R.L., Higgins, M.M.: The rise in global biodiesel production: implications for food security. Glob. Food Sec. **16**, 75–84 (2018)
17. Reddy, V.J., Hariram, N.P., Maity, R., Ghazali, M.F., Kumarasamy, S.: Sustainable e-fuels: green hydrogen, methanol and ammonia for carbon-neutral transportation. World Electr. Veh. J. **14**(12), 349 (2023)
18. Rozzi, E., Minuto, F.D., Lanzini, A., Leone, P.: Green synthetic fuels: renewable routes for the conversion of non-fossil feedstocks into gaseous fuels and their end uses. Energies **13**(2), 420 (2020)
19. Mat Aron, N.S., Khoo, K.S., Chew, K.W., Show, P.L., Chen, W.H., Nguyen, T.H.P.: Sustainability of the four generations of biofuels—a review. Int. J. Energy Res. **44**(12), 9266–9282 (2020)

20. Dooley, K., Kartha, S.: Land-based negative emissions: risks for climate mitigation and impacts on sustainable development. Int. Environ. Agreem.: Polit., Law Econ. **18**(1), 79–98 (2018)
21. Datta, B.: An economic analysis of biofuels: policies, trade, and employment opportunities. In: Handbook of Biofuels, pp. 3–29. Academic Press (2022)
22. El-Araby, R.: Biofuel production: exploring renewable energy solutions for a greener future. Biotechnol. Biofuels Bioprod. **17**(1), 129 (2024)
23. Rimmer, M.: Synthetic genomics: intellectual property, innovation policy, and advanced biofuels. In: 3rd Generation Biofuels, pp. 847–868. Woodhead Publishing (2022)
24. Haque, A.K., Mukhopadhyay, P., Nepal, M., Shammin, R.: Climate Change and Community Resilience. Open Access (2022)
25. Ekechukwu, D.E., Simpa, P.: A comprehensive review of renewable energy integration for climate resilience. Eng. Sci. Technol. J. **5**(6), 1884–1908 (2024)
26. Johnsson, F., Kjärstad, J., Rootzén, J.: The threat to climate change mitigation posed by the abundance of fossil fuels. Clim. Policy **19**(2), 258–274 (2019)
27. Rey-Hernández, J.M., Yousif, C., Gatt, D., Velasco-Gómez, E., San José-Alonso, J., Rey-Martínez, F.J.: Modelling the long-term effect of climate change on a zero energy and carbon dioxide building through energy efficiency and renewables. Energy Build. **174**, 85–96 (2018)
28. Jaffe, A.M., Busby, J., Blackburn, J., Copeland, C., Law, S., Ogden, J.M., Griffin, P.A.: Impact of Climate Risk on the Energy System, pp. 1–87. Council on Foreign Relations (2019)
29. Koutsoukos, I.: The deployment of renewable energy sources in the context of energy security (Master's thesis, Πανεπιστήμιο Πειραιώς) (2025)
30. Deora, P.S., Verma, Y., Muhal, R.A., Goswami, C., Singh, T.: Biofuels: an alternative to conventional fuel and energy source. Mater. Today: Proc. **48**, 1178–1184 (2022)
31. Ahorsu, R., Medina, F., Constantí, M.: Significance and challenges of biomass as a suitable feedstock for bioenergy and biochemical production: a review. Energies **11**(12), 3366 (2018)
32. Poornima, S., Manikandan, S., Prakash, R., Deena, S.R., Subbaiya, R., Karmegam, N., Kim, W., Govarthanan, M.: Biofuel and biochemical production through biomass transformation using advanced thermochemical and biochemical processes—a review. Fuel **372**, 132204 (2024)
33. Johansson, R., Meyer, S., Whistance, J., Thompson, W., Debnath, D.: Greenhouse gas emission reduction and cost from the United States biofuels mandate. Renew. Sustain. Energy Rev. **119**, 109513 (2020)
34. Cortez, L.A.B., Rosillo-Calle, F.: The Future Role of Biofuels in the New Energy Transition: Lessons and Perspectives of Biofuels in Brazil. Editora Blucher (2023)
35. Morato, T., Vaezi, M., Kumar, A.: Developing a framework to optimally locate biomass collection points to improve the biomass-based energy facilities locating procedure—a case study for Bolivia. Renew. Sustain. Energy Rev. **107**, 183–199 (2019)
36. Sharma, P.: Pathways of energy transition and its impact on economic growth: a case study of Brazil. In: Energy: Crises, Challenges and Solutions, pp. 108–130 (2021)
37. Katakojwala, R., Mohan, S.V.: A critical view on the environmental sustainability of biorefinery systems. Curr. Opin. Green Sustain. Chem. **27**, 100392 (2021)
38. Adhikary, S., Sau, S., Debnath, S., Ghosh, D.: Advancing bioenergy: leveraging artificial intelligence for efficient production and optimization. In: Computer Vision and Machine Intelligence for Renewable Energy Systems, pp. 177–186. Elsevier (2025)
39. Skiba, R.: Battery Powered: The Social, Economical, and Environmental Impacts of the Lithium Ion Battery. After Midnight Publishing (2024)
40. Debnath, S., Paul, S., Ghosh, D.: Sustainable wastewater treatment through microalgae and alternative biotic communities for bioenergy production. In: Biotechnologies for Wastewater Treatment and Resource Recovery, pp. 65–76. Elsevier (2025)
41. Basu, P.: Biomass Gasification, Pyrolysis and Torrefaction: Practical Design and Theory. Academic Press (2018)
42. Ghosh, D., Das, S.: Genetic and metabolic engineering approaches for improving accessibilities of lignocellulosic biomass toward biofuels generations. In: Genetic and Metabolic Engineering for Improved Biofuel Production from Lignocellulosic Biomass, pp. 13–35. Elsevier (2020)

43. Banerjee, S., Ghosh, D., Pandit, C., Saha, S., Mohapatra, A., Pandit, S., Sharma, M., Sridhar, K., Inbaraj, B.S., Prasad, R.: Microalgal Pandora for potent bioenergy production: a way forward? Fuel **333**, 126253 (2023)
44. Aros-Vera, F., Gillian, S., Rehmar, A., Rehmar, L.: Increasing the resilience of critical infrastructure networks through the strategic location of microgrids: a case study of Hurricane Maria in Puerto Rico. Int. J. Disaster Risk Reduct. **55**, 102055 (2021)
45. Olabi, A.G., Abdelkareem, M.A.: Renewable energy and climate change. Renew. Sustain. Energy Rev. **158**, 112111 (2022)
46. Wijerathna-Yapa, A., Pathirana, R.: Sustainable agro-food systems for addressing climate change and food security. Agriculture **12**(10), 1554 (2022)
47. Stefanakis, A.I.: The role of constructed wetlands as green infrastructure for sustainable urban water management. Sustainability **11**(24), 6981 (2019)
48. Hegger, D., Alexander, M., Raadgever, T., Priest, S., Bruzzone, S.: Shaping flood risk governance through science-policy interfaces: insights from England, France and the Netherlands. Environ. Sci. Policy **106**, 157–165 (2020)
49. Law, W.: Updates and Recommendations to the Milwaukee Metropolitan Sewerage District's Climate Change Mitigation and Adaptation Programs (2019)
50. Chowdhury, P., Mahi, N.A., Yeassin, R., Chowdhury, N.U.R., Farrok, O.: Biomass to Biofuel: Impacts and Mitigation of Environmental, Health, and Socioeconomic Challenges (2025)
51. Løkke, S., Aramendia, E., Malskær, J.: A review of public opinion on liquid biofuels in the EU: current knowledge and future challenges. Biomass Bioenerg. **150**, 106094 (2021)
52. Alsunousi, M., Kayabasi, E.: The role of hydrogen in synthetic fuel production strategies. Int. J. Hydrogen Energy **54**, 1169–1178 (2024)
53. Rosenbloom, D., Markard, J., Geels, F.W., Fuenfschilling, L.: Why carbon pricing is not sufficient to mitigate climate change—and how "sustainability transition policy" can help. Proc. Natl. Acad. Sci. **117**(16), 8664–8668 (2020)
54. Debnath, S., Ghosh, D.: Algal biodiesel: the next generation green energy resources towards sustainability. AIJR Abstracts, p. 56 (2021)
55. Damian, C.S., Devarajan, Y., Jayabal, R.: A comprehensive review of the resource efficiency and sustainability in biofuel production from industrial and agricultural waste. J. Mater. Cycles Waste Manag. **26**(3), 1264–1276 (2024)
56. Wetzel, M., Gils, H.C., Bertsch, V.: Green energy carriers and energy sovereignty in a climate neutral European energy system. Renew. Energy **210**, 591–603 (2023)

Role of Advance Techniques in Urban Disaster Management Caused by Climate Variations

Vijaya Ilango

Abstract Natural disasters (NDs) have long posed a serious risk to infrastructure and human life, resulting in significant loss and destruction. The necessity for more effective and efficient disaster management systems has been brought to light currently by the rising frequency and intensity of natural disasters. Utilizing technology has become a viable remedy in this regard. NDs have caused massive havoc throughout history, which is typically accompanied by a significant loss of life. Natural disasters, such as hurricanes, floods, wildfires, volcanic eruptions, and earthquakes, had a devastating effect on both the environment and communities. Although technology can greatly enhance natural disaster management (NDM), there are a number of issues that must be resolved, including implementation costs and the requirement for certain expertise. This comprehensive review offers a thorough analysis of technology that is being used to manage NDs and clarifies the significant contribution that these technologies may make to NDM. This also intends to aid in the creation of more sustainable and successful disaster management plans by investigating the possible uses of various technologies. The use of modern technologies to lessen the effects of different natural disasters is explored in this book chapter.

Keywords Natural disasters (NDs) · Climate change · Mitigation · Urbanization · Techniques

1 Introduction

Cities are experiencing previously unprecedented climatic threats as a result of the changing global climate and the fast pace of urbanization. Disasters' effects and disruptions frequently result in the breakdown of urban operations, the disruption of social functions, and even significant and severe losses in terms of property and

---

V. Ilango (✉)
Department of Chemistry, Birla Institute of Technology and Science, Pilani, Dubai Campus, Dubai, UAE
e-mail: srisaiviji2@gmail.com

A. L. Srivastav et al. (eds.), *Climate Resilience: Impact of Quantum Computing and Artificial Intelligence on Urban Planning*, Studies in Systems, Decision and Control 637, https://doi.org/10.1007/978-3-032-06791-3_14

casualties. The city is a sophisticated gigantic system with numerous interdependent and linked subsystems as well as a variety of dynamic change processes [57]. It is anticipated that a variety of extreme weather events, many of which primarily impact metropolitan areas, would become more regular and terrible due to climate change. Increasing urban catastrophe risk should be a major issue in talks of the negative effects of climate change because of the large population densities in urban areas, which frequently include significant concentrations of susceptible individuals. Due to interactions with urban infrastructure systems, expanding urban populations, cultural norms, and economic activity, the effects of climate-related disasters are frequently worsened in urban areas. Cities must concentrate on enhancing their responses to climate-related disasters including droughts, floods and heat waves since the majority of people on Earth already reside in cities, and this percentage is expected to rise over the next several decades. The fast development of industry and the rate of urbanization are the main drivers of major climate disasters and global warming. The United Nations reports that while rural areas remain comparatively steady, metropolitan areas account for the majority of the world's population growth [49]. Global urbanization has supported the expansion of people from all walks of life. The lack of appropriate protection laws has exacerbated the deterioration of the urban environment and climate, especially during floods and droughts. The land-use mode (more development land, less permeable space) and infrastructural flaws are most directly impacted by excessive urbanization [8]. Whether they are caused by man-made disruptions or recurrent natural phenomena, disasters cause significant changes in the planet and its people. Disasters can be principally be classified [52] into

- Natural disasters—Earthquakes, landslides, droughts, floods, storms, tsunamis, cyclones, hurricanes, wildfires, epidemic outbreak, diseases etc.—induced by uncontrollable natural events.
- Man-made—Intentional or unintentional human actions, Triggered by human environmental factors.
- Technical—Power plant failures, dam failures, Power outages and chemical incidents.
- Hybrid disasters happen due to both human actions and natural.

Disaster types and classification are given in the Fig. 1. Almost 60% of the world's population is predicted to live in urban areas by 2030. Over 4 billion people live in cities across the world, including over 863 million unofficial residents in urban settlements, according to UN estimates. It is anticipated that this figure would increase by more than one million every ten days [1]. Seven out of ten natural catastrophes in China in 2021 were linked to urban climate risk, the most common were the river floods and severe urban waterlogging brought on by the infrequent and devastating downpour in Zhengzhou, Henan Province. In addition to 40.9 billion CNY in direct economic losses, the accident claimed 380 lives [10].

Phases of Disaster Management: A collection of coordinated efforts to mitigate the impact of disasters on people, property, and the environment is known as disaster

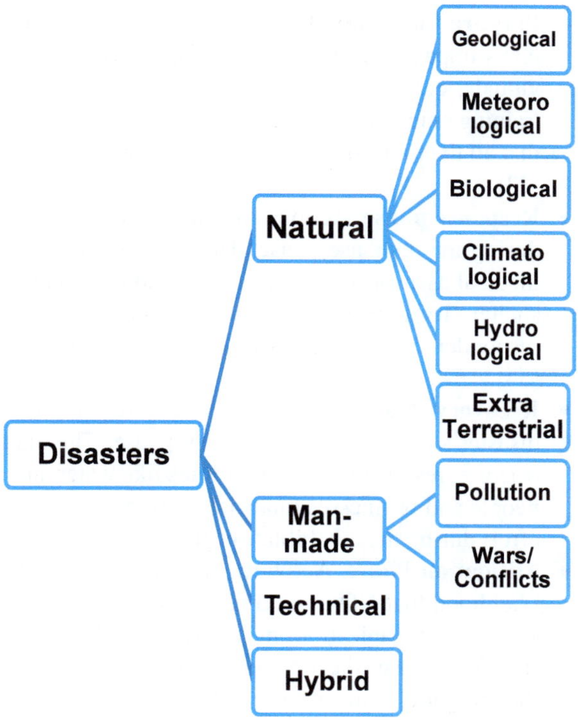

Fig. 1 Disaster types and classification

management. These can be divided into four major phases: preparation, response, recovery and mitigation as given in the Fig. 2.

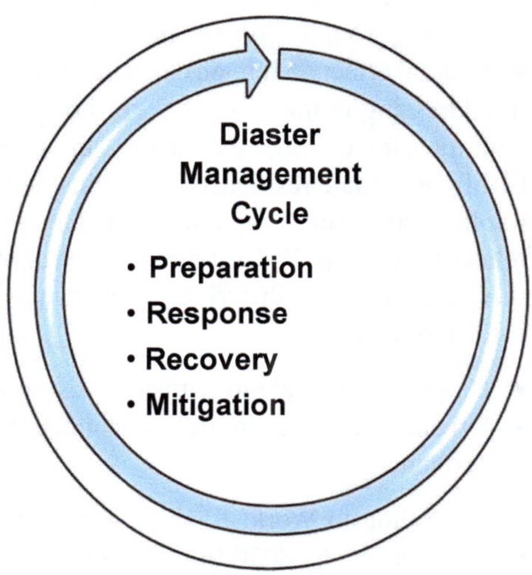

Fig. 2 Phases of disaster management

- **Preparation Phase**: During the preparation stage, steps are taken to lessen the ND's impacts before they materialize. This include identifying vulnerable individuals, conducting risk assessments, developing disaster response plans, and training emergency personnel. The preparation stage is crucial because it ensures that all the tools, personnel, and resources need to respond to an ND are available [23].
- **Response phase**: Actions are taken to address an ND's immediate consequences during the response phase. This covers tasks like search and rescue, providing medical assistance during a crisis, and distributing supplies like food, drink, and shelter. To reduce the loss of life and property during the reaction phase, first responders, governmental agencies, and the private sector must collaborate swiftly [54].
- **Recovery Phase**: Actions are taken to restore the impacted community to its pre-disaster state during the recovery phase. This include fixing houses and infrastructure, restoring essential utilities like water and electricity, and giving affected people and businesses financial support. Everyone concerned must put in a lot of effort during the rehabilitation phase, which may take months or even years [28].
- **Mitigation Phase**: Activities are carried out in the mitigation step to reduce the likelihood that NDs would recur. This include establishing building regulations, enhancing organization, and doing research on the origins and significance of NDs. In order to lessen the long-term consequences of NDs and strengthen communities, the mitigation stage is crucial [30].

Climate change encompasses the ongoing alteration of atmospheric composition or land use due to human activity, as well as changes in the climate state brought about by natural internal processes or external stimuli. Many issues are brought on by climate change, including the rise in sea levels as a result of global warming and modifications to the climate that increase the frequency of extreme weather events like droughts, floods, hailstorms, and high temperatures, all of which have a substantial effect on human existence and activity [20]. The two primary approaches to addressing climate change are thought to be adaptation and mitigation. Along with focusing on these two tactics, an increasing number of research have started to highlight climate resilience [50]. The concept of resilience may have more positive implications than "adaptation," "vulnerability reduction," or "countering climate-related threats." [29]. The topic of urban climate resilience is quite complex. Although there are still many differing views on how to define and quantify it, two things are evident at this time: [4]

- Cities must be able to withstand disruptions and be ready for climate change.
- Urban development must be encouraged while working to increase climate change resilience.

According to World Bank estimates, up to 140 million people worldwide may have to relocate by 2050 from increasingly unviable regions of their countries due to the effects of climate change, including crop failure, water scarcity, storm surge, and sea level rise, unless immediate global and national climate and development action is taken. At the local, regional, and global levels, cities also have an impact on the weather and climate. More than 70% of the carbon emissions linked to energy in the world come from cities. Additionally, they emit massive amounts of heat into the urban environment, which in turn affects the frequency and severity of thunderstorms and local rainfall [51].

Because metropolitan regions offer pleasant living conditions, more people are opting to live there globally, a phenomenon known as global urbanization. In China, the ratio of people living in cities to the total population increased from 19.37 to 64.72% from 1980 to 2021 [48]. Cities are feeling the effects of climate change, which is marked by warming and harsh weather, more and more. Enhancing climate resilience is thought to be a key strategy for lowering hazards in the future. The urban system, a crucial area and the foundation of climate change response, influences climate resilience through the interplay of subsystems related to social, economic, ecological, infrastructure, organization, institution, and technology. These parameters can be used to quantify the idea of urban climate resilience.

2 Urban Natural Disasters and Mitigation Techniques

Flooding: One significant natural hazard that happens in cities is flooding. An essential part of developing a sustainable city is managing the danger of floods. When the pace of stormwater rise surpasses the capacity of a city's drainage system, urban flooding happens [18]. The increased frequency and severity of storms or heavy rain are two effects of global warming. Floods can easily happen if a city's drainage infrastructure does not account for this kind of intense rainfall. For instance, within the previous 20 years, there have been at least seven instances of severe floods in Beijing, the capital of China. Among these, one instance of flooding happened on July 21, 2012, affecting 1.9 million people and 160,000 km^2, resulting in 77 fatalities. The average daily rainfall in the city center area was 215 mm [27]. There are three types of techniques: aboveground, which includes both non-vegetated and vegetated roofs, ground-surface, which includes trees, bio-retention, water-retaining pavements, permeable pavements, infiltration trenches and underground, which includes soakaways and other subterranean water storage units. For non-potable applications, rainwater is collected in water barrels and tanks, although homeowners must regularly empty them to keep them clean. In the event of a light rain, trees, permeable pavements, water-retaining pavements, infiltration trenches, bio-retention, and both vegetated and non-vegetated roofs are excellent in retaining and containing rainwater. Regular maintenance is necessary to restore the hydrological performance of permeable pavements, infiltration trenches, and soakaways that have become clogged. Because the roots of vegetated species, urban trees, swale grass, and bio-retention plants open the growth media for water penetration, techniques like green roofs, swales, and bio-retention can sustain effective infiltration over time [38].

Earthquakes: It was found that, "elastic rebound" of the Earth's crust is typically the cause of earthquakes. According to this idea, deformation brought on by outside forces—mostly regional tectonic pressures—is a primary source of earthquakes. Slip on preexisting weak zones or brittle fractures in competent rock will cause an earthquake when the overall amount of strain (energy) at a crucial place exceeds the rock's ability to withstand additional strain. The area where the slide or displacement happens is known as an earthquake fault [33, 46]. In the case of earthquakes Particles carry the wave's energy when they move in the same direction as the primary wave, also known as the P-wave. The secondary wave, also known as the S-wave, on the other hand, transports particles side-to-side or up and down perpendicular to the direction of the wave. In order to distinguish genuine earthquake signals from fake ones created by sound waves that push and pull air, P-waves are crucial [22].

In order to identify the primary and surface waves in elastic rock media and to detect the first tremors caused by surface fractures, researchers have created electronic instruments. In order to enable early and precise detection while reducing false alarms, this invention relates to earthquake detection and focuses on filtering and detecting earthquakes with particular intensities over a vast area. Researchers have looked into cutting-edge methods, like using carbon fiber-reinforced plastic (CFRP), to reinforce and repair reinforced concrete structures. CFRP has advantages including high strength-to-weight ratio, corrosion resistance, and simplicity of handling to enhance structural integrity [26].

Landslides: Because of the increased frequency of heavy rainfalls in recent years, there have been more landslides and slope collapses caused by downpour. Japan's Land, Infrastructure, Transport, and Tourism Ministry (MLIT) reported that the total number of sediment disasters that occurred in Japan in 2018 was 3,459, of which there were 161 casualties, 117 injuries and 70% were rainfall-induced landslides and slope collapses approximately [21]. Rainfall-induced landslides may happen after short, intense precipitation or following extended periods of rain, mostly depending on the hydromechanical properties of the corresponding soils.

Rainfall-induced landslides come in two varieties: shallow and deep-seated. Rain-induced landslides that are deeply rooted. Rainfall-induced groundwater level rise typically results in positive pore water pressures along the surface, increasing the danger of slipping and causing rainfall-induced deep-seated landslides. This circumstance is typically linked to both surface precipitation and the development of preferential water flow channels that transport surface precipitation to deeper slope areas, like the development of vertical fractures. Urban areas can reduce the risk of landslides and mitigate the economic and social losses caused by them by limiting development in landslide-prone areas (i.e., avoidance); implementing codes in excavation, grading, landscaping, and construction; utilizing engineered physical measures (drainage, slope-geometry, modification, structures, and vegetation) to prevent or control landslides; and creating landslide warning systems, The best methods for lowering the risk of landslides typically combine the aforementioned hazard-reduction tactics in a well-planned manner [14, 43].

3 Land Subsidence

The primary cause of land subsidence, or lowering of the land level, is excessive groundwater extraction. Although it seems to happen spontaneously, human activity is to blamed. Land subsidence puts people and property at risk by destroying infrastructure, increasing the likelihood of flooding, and harming city drainage systems. Every year, the demand for freshwater rises globally as a result of urbanization, or the growing number of people living in cities. People will unavoidably utilize groundwater if there is not enough fresh surface water, and if this use is not properly managed, it will cause land subsidence. Potential locations for subsidence pose a hazard to 21% of the world's largest cities and 1.2 billion people, particularly in coastal areas. Asia is home to 86% of the exposed population [9].

People will unavoidably utilize groundwater if there is not enough fresh surface water, and if this use is not properly managed, it will cause land subsidence. Potential locations for subsidence pose a hazard to 21% of the world's largest cities and 1.2 billion people, particularly in coastal areas. Asia is home to 86% of the exposed population. It is believed that soil loss from excavation and compaction from effective stress increase are the main causes of land subsidence. Based on a more thorough examination of existing research, we propose that while developing a land subsidence model, consideration should be given to regional variations in geological conditions, building loads, groundwater withdrawal and recharge cycles, and soil particle meso-mechanism [25].

Land subsidence and flood control are inextricably linked. Land subsidence and flood control must be combined because rapid water flow into the sea during flood control will reduce soil water infiltration while increasing groundwater extraction. Consequently, the level of the grounds decreases and the area affected by flooding rises. In addition, a paradigm shift from draining to absorbing rainwater is necessary for flood management. Water conservation should be stepped up since land cover will inevitably change from agriculture to habitation and industry. To lessen flooding, marine intrusion, and land subsidence, groundwater use regulations must also be strengthened. The best way to deal with flooding and land subsidence is to get back to nature [37].

4 Recent Techniques in Urban Disaster Management

Technology is the useful application of knowledge to do tasks more quickly and easily. It is crucial to crisis management, planning, and prevention. Satellite communication and remote sensing. Advanced technology integration has brought about a paradigm shift in urban flood management. Predictive modeling, remote sensing, machine learning, and artificial intelligence (AI) are essential for improving flood prediction, real-time monitoring, and comprehensive urban flood management plans [13].

5 Artificial Intelligence

Artificial intelligence has a wide range of possible uses in the geohazard sector, such as early warning systems (EWSs), prediction, change detection, risk and hazard management, spatial modeling, and mitigation techniques. Because there is so much information available from satellite imaging, weather reports, and other meteorological sources, researchers are motivated to find novel methods that can aid in the forecast of future geohazards.

Because it can uncover hidden patterns in data, machine learning (ML), a type of artificial intelligence, has gained popularity in a variety of settings. By making it possible to analyze precipitation patterns more precisely, machine learning applications have completely changed flood prediction. Large datasets can be processed by ML algorithms that take into account a variety of meteorological characteristics, past weather trends, and geographic features. This makes it possible to create reliable models that improve our comprehension of precipitation patterns and make more accurate forecasts of extreme weather events [45]. The 2011 Tohoku earthquake's damage to homes was identified via AI analysis. Finding the whereabouts of homes destroyed by the tsunami was one of the efforts made. The AI was able to identify the affected homes with a 94% accuracy rate [5].

6 Remote Sensing

Technologies for remote sensing, especially satellite photography, provide an aerial perspective of shifting environmental conditions. In order to monitor extensive flood occurrences, determine the amount of inundation, and pinpoint locations at danger, satellite data is essential [40]. Remote sensing is a technology that gathers environmental data from a distance using satellites or aircraft. Natural disasters (NDs) that can be observed and monitored via remote sensing include floods, hurricanes, and wildfires. It provides a way to learn more about the extent and severity of a natural catastrophe, as well as how it affects the environment and infrastructure. Remote sensing technology can also give early warning systems that can be used to plan for and respond to disasters by detecting areas at risk of natural disasters before they happen. Using this technology, decision-makers can efficiently allocate resources and prioritize response efforts by creating maps and models that illustrate the level of damage caused by a natural disaster [42].

7 Radars

Radars are gadgets that identify and detect objects using radio waves. Radars can be used in NDM to monitor the course and severity of NDs such thunderstorms, tornadoes, and hurricanes. When paired with additional peripherals, Doppler radar systems are a useful tool for quantitative measurement in high wind and turbulence regimes. For the purpose of disaster management, India has a sufficient network of storm and cyclone detection radars that have been operational for the past 20 years [47]. On September 16, two Earth observation satellites were launched by the Indian Space Research Organization (ISRO) from Sriharikota, the location of the Satish Dhawan Space Center. Earth observation data, including for disaster and risk management, was provided by the UK-developed and -operated NovaSAR S1-4. ISRO's web portal states that Nova SAR will supply information for flood and disaster monitoring, land use and ice cover monitoring, and forest mapping. It was created to give S-band Synthetic Aperture Radar (SAR) a low-cost platform [39].

8 Big Data

Every industry produces enormous volumes of data every day. The big data framework consists of data management systems for information preservation, analysis, and security [22]. The process becomes more complex when decisions are made using data from many sources. The main objective of big data management is to increase the value and accessibility of data for the decision-making process [15]. Data with at least four dimensions—"data volume," "data velocity," "data variety," and "data veracity"—is referred to as big data. Apart from these four Vs, the most crucial one is the "data value," which gauges how useful the data is. Data management tools and methods for data processing, storage, and security make up the big data paradigm. When data from several sources are used to make decisions, the process gets complicated. Improving data accessibility and value for decision-making is the primary goal of big data management [3].

Examples of big data sources for disaster resilience were provided in a United Nations (UN) report. These included digital content (social media and crowdsourcing), sensing data (physical sensing devices and remote sensing), and exhaust data (mobile-based, financial transactions, transportation, and online traces) [53]. One of the Research work effectively explained how the Big Data Analytics (BDA) for Disaster Risk Reduction (DRR) model was developed within the context of Malaysia's Selangor State. It is thought that this model can support future work in the creation of data analytics solutions based on a thorough literature review and a focus group discussion (FGD) with all the pertinent agencies that responded with the Disaster Risk Management (DRM)/DRR. Nearly every facet of disaster-related issues, including ecology, economics, the environment, organization, society, and technology, is covered by the model. Additionally, it delves deeper by locating relevant datasets in both spatial and nonspatial formats [55].

9 Internet of Things (IoT)

There is great potential for ND detection and management using the Internet of Things. In order to provide early warning of potential NDs, IoT sensors can be utilized to monitor environmental factors including temperature, humidity and seismic activity. Real-time access to this data enhances the emergency services' capacity to react to crises. IoT devices can be used to monitor essential infrastructure, such as water systems and power grids, before, during, and after natural catastrophes Generally speaking, using IoT technology to detect and control NDs has the potential to save lives and lessen the harm that these events cause to local populations and the environment [19, 31]. IoTs have many benefits, including low cost, low energy consumption, easy installation, and access to difficult locations. More significantly, IoT sensors are able to gather high-precision, real-time environmental data and adjust to changes in the surroundings. As a result, IoT systems are practical instruments for disaster management and environmental monitoring [12]. It is possible that the use of IoT technology could improve search and rescue operations' efficiency in the allotted 72 h [34]. As a result, IoT can play a big part in managing urban disasters by providing real-time monitoring, timely notifications, post-disaster reaction, and help with rescue efforts. Additionally, using Internet of Things devices has grown in popularity as a simple and affordable way to monitor a variety of systems [56].

10 Social Media

Over the past few decades, a variety of calamities have affected various nations worldwide. Each of these nations has its own social media platforms and methods for using them for disaster recovery. Social media platforms like Facebook and Twitter can be used in NDM to provide people with the most recent information on the crisis. Social media posts regarding calamities are ideal for early warning systems because they are created in real time. Decision-makers can use this information to stay informed about developments and identify potential locations for ND. During and after a natural disaster, social media can be used to pinpoint regions where people require help with search and rescue, food and drink distribution, and medical care. Additionally, social media can be used to manage relief efforts for those in need and to spread the news about events Typhoons, floods, earthquakes, and hurricanes were the most common natural disasters mentioned in the studied literature. The USA, Japan, China, and the Philippines were the primary locations of these catastrophes. For disaster recovery, Chinese studies primarily used WeChat and Weibo. Nonetheless, Twitter was by far the most popular social media site in other nations for disaster recovery research, with Facebook coming in second. Facebook was utilized in 16% of the articles, while Twitter was used in 65% of them. 15% of the stories covered social media in general rather than focusing on any one site. It's interesting to note that the literature did not mention social networking apps that are more popular among teenagers and younger generations, such TikTok and Snapchat [35].

11 Emerging Disruptive Technologies for Urban Disaster Management

Innovative technologies or advanced modern technology that alter how society acts, thinks, or interacts are referred to as disruptive technology. It also alters how industries or corporations function. The premise behind all Smart City research is that ICTs, the Internet, digital tools, and cutting-edge technological advancements maximize a Smart City's intelligence [2].

12 Geo-visualisation and Geographical Information Systems (GIS)

Modern digital methods of representing geospatial data are included in geo-visualization, which is crucial for disaster modeling, scenario creation, post-disaster analysis, and search and rescue operations. Geographic Information Systems (GIS) are frequently used to drive geovisualization. GIS tools are helpful for producing and displaying the outcomes of spatial processing and analysis, which are then utilized to improve decision-making [24]. GIS has been widely utilized, extensively utilized to create maps of hazards, dangers, and vulnerabilities in order to comprehend and control risks in urban areas.

13 Grid Technologies

In order to improve system efficiency, reliability, resilience, power quality, and cost-effectiveness while giving customers the means to control energy use, smart grid integrates contemporary advanced technologies, intelligent algorithms, communication networks, and automation systems into the power system [16]. When it comes to smart cities, advancements in smart energy systems and grids are a popular option since they can efficiently consume and generate energy, particularly when combined with renewable energy sources like solar and wind [11].

14 Location-Based Services (LBS)

The term location-based services (LBS), which refers to a cutting-edge technology that provides information based on geographical location (of the user), is used interchangeably with location services, wireless location services, mobile location-based services, location-enabled services, and location-sensitive services. The primary reason for attracting considerable attention from researchers, mobile network operators, and service providers in the design of such services for smart living in urban areas is the broad applicability of LBSs in a variety of modern-day living areas, such as transportation, healthcare, leisure activities, business, etc. [36]. LBS offers customers individualized services in a variety of crucial areas, including disaster management, entertainment, security, and health. For people, disaster management is crucial. As a result, using these services will frequently improve their quality of life.

15 Early Warning Systems

Early warning systems are developed for a range of threats, including natural geophysical hazards, biological hazards, industrial hazards, complex socio-political events, human health concerns, and other threats within the urban disaster scenario. Many studies have been carried out on early warning systems, where real-time data are used to generate warnings for natural hazards [41]. Early warning systems developed for Smart Cities often incorporate different technologies such as Artificial Intelligence, IoT, and big data analytics to develop more reliable and resilient systems that are faster at prediction and detection [56].

16 Unmanned Aerial Vehicle (UAV)

UAVs have quickly attracted a lot of interest for use in a wide range of new military and commercial applications, including precision agriculture, transportation, logistics, border security, aerial surveillance, search and rescue operations, disaster recovery, etc.

The information and communication technology (ICT) industry, which mainly relies on terrestrial communication and networking infrastructure, is undergoing a significant paradigm shift thanks to UAV-based airborne communications [7]. The types of UAV is given in the Fig. 3.

Few case studies along with the techniques adopted in Urban Disaster Management are given in Table 1.

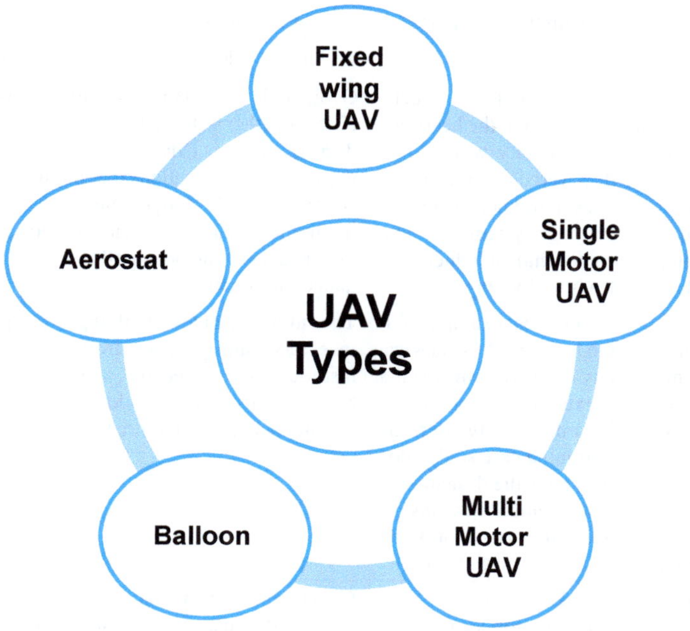

Fig. 3 Types of unmanned aerial vehicle (UAV)

17 Conclusions

A fundamental component of contemporary urban planning and a vital component of creating resilient communities is disaster mitigation. Cities may improve their ability to withstand and recover from disasters and lessen their susceptibility to natural hazards by incorporating mitigation strategies into infrastructure investments, land-use decisions, and community involvement initiatives. In addition to catastrophe resilience being a necessity, smart cities should carefully choose which technologies to implement to meet their context-specific needs. Although various innovative and emerging techniques and technologies for disaster resilience and smart cities have been covered in studies, there is a lack of convergence between the two areas. Research is clearly needed to help the Smart Cities to understand the potential of tools and technologies for enhancing disaster.

The climate hazards that cities face have escalated to previously unheard levels due to the process of climate change and growing urbanization. Climate resilience is the ability to withstand, recover from, and adapt to climate change-related events. It offers cities a fresh viewpoint on how to continue developing their security in the face of climate change. It is possible to create the urban security governance system and advance urban security development by naturally combining resilience with the prevention and mitigation of urban disaster risk. Without adequate coordination, there may still be a knowledge gap between disaster management researchers,

Table 1 Case studies with the techniques adopted in urban disaster management

Case study	Description	Techniques adopted	References
1. Chennai flood 2015: a study and analysis employing GIS and multicriteria methods	The Cooum River, which runs through the heart of the main city, was discovered to be silting as a result of inadequate drainage systems and local encroachment, which leads to flooding	Geographic information systems (GIS) together with multicriteria decision analysis (MCDA) were used to analyze locations that might be impacted. The geographic multicriteria analysis was utilized to rank and show the potentially dangerous locations	[44]
2. Urban catastrophe management: an analysis of the risk assessment for earthquakes in Cartego, Costa Rica	As a typical example of a medium-sized Costa Rican city situated in an area that is extremely prone to hazards, the city offers an intriguing case study and illustrates the difficulties that lay ahead in terms of geo-information for urban disaster management	The study integrates a building inventory, damage curves from historical damage records, and a hazard intensity map to address building and population risk	[32]
3. A study on disaster management for strength-ening of disaster resilience in Jeju	By taking into consideration local involvement in the disaster management process, this study aims to increase local communities' ability to reduce urban safety issues brought on by urbanization and large-scale, complicated, and unpredictable disasters	Geographic information systems (GIS) and "smart disaster integrated management system" uses big data and artificial intelligence (AI) in conjunction with information on disaster management resources	[17]
4. Mapping disaster hazards and risks—a case study in Kuala Lumpur City (KLC)	The goal of this study is to create a thorough master plan for Kuala Lumpur City (KLC)'s catastrophe management. For humanitarian aid and disaster relief (HADR), the plan will cater the first stage of the civil-military coordinating response architecture	Based on historical landslide points collected from NASA databases and news sources, the landslide susceptibility map employ the frequency ratio model. The study area was divided into three relative landslide susceptibility classes (low, moderate, and high) based on frequency ratio (FR) data, which were utilized to create the landslide susceptibility map	[59]

(continued)

Table 1 (continued)

Case study	Description	Techniques adopted	References
5. Climate services to support disaster risk reduction and climate change adaptation in urban areas: the CLARITY project and the Napoli case study	It is multidisciplinary study area with promising applications in the field of urban microclimate simulations, supporting climate-resilient redevelopment actions at both the city and neighborhood levels	A unique modeling methodology has been created to capture the impact of certain built environment features on the urban microclimate, and fresh datasets have been collected from Regional Climate Models to estimate the frequency of extreme temperatures and precipitation occurrences till 2100	[58]
6. Case study: disaster risk management and climate change adaptation in Greater Algiers	Overall, the report evaluates Algiers' susceptibilities to natural disasters and climate change until 2030 and makes a number of recommendations to strengthen the Wilaya of Algiers' capacity for risk management	Geographic information systems (GIS), digital terrain models, satellite and aerial photos, databases, and previous research on recent disasters served as the study's foundation. The majority of the research area is made up of crowded residential and commercial areas	[6]

policymakers, and practitioners after a project or technology has been developed for implementation.

Since disaster risk is constantly changing, it's critical to consistently monitor it so that we can modify risk management plans as needed. Regularly evaluating disaster risks and assessing management capacities is crucial for disaster relief organizations. In order to provide domestic cities with flexible response strategies, future research should examine the precise uses and implementation techniques of emerging technologies like IoT, big data, and AI in disaster monitoring, risk assessment, and rapid response. It should also reinforce regional collaborative governance mechanisms by referencing successful global examples.

References

1. Acuto, M., Parnell, S., Seto, K.C.: Building a global urban science. Nat. Sustain. **1**(1), 2–4 (2018). https://doi.org/10.1038/s41893-017-0013-9
2. Angelidou, M.: The role of smart city characteristics in the plans of fifteen cities. J. Urban Technol. **24**(4), 3–28 (2017). https://doi.org/10.1080/10630732.2017.1348880
3. Arslan, M., Roxin, A.-M., Cruz, C., Ginhac, D.: A Review on Applications of Big Data for Disaster Management. IEEE Xplore (2017). https://doi.org/10.1109/SITIS.2017.67

4. Barzaman, S., Shamsipour, A., Lakes, T., Faraji, A.: Indicators of urban climate resilience (case study: Varamin, Iran). Nat. Hazards (2022). https://doi.org/10.1007/s11069-021-05174-x
5. Bhandari, V.: Use of technology in disaster management. Unity J. **3**(01), 292–304 (2022). https://doi.org/10.3126/unityj.v3i01.43333
6. Boughedir, S.: Case study: disaster risk management and climate change adaptation in Greater Algiers: overview on a study assessing urban vulnerabilities to disaster risk and proposing measures for adaptation. Curr. Opin. Environ. Sustain. **13**, 103–108 (2015). https://doi.org/10.1016/j.cosust.2015.03.001
7. Bushnaq, O.M., Mishra, D., Natalizio, E., Akyildiz, I.F.: Unmanned Aerial Vehicles (UAVs) for Disaster Management, pp. 159–188. Elsevier eBooks (2022). https://doi.org/10.1016/b978-0-323-91166-5.00013-6
8. Chai, K.-C., Ma, X.-R., Yang, Y., Lu, Y.-J., Chang, K.-C.: The impact of climate change on population urbanization: evidence from china. Front. Environ. Sci. **10** (2022). https://doi.org/10.3389/fenvs.2022.945968
9. Chai, J., Wu, H.: Prevention/mitigation of natural disasters in urban areas. Smart Constr. Sustain. Cities **1**(1) (2023). https://doi.org/10.1007/s44268-023-00002-6
10. Chen, N., Tang, X., Liu, W.: Urban disaster risk prevention and mitigation strategies from the perspective of climate resilience. Wirel. Commun. Mob. Comput. **2022**, 1–13 (2022). https://doi.org/10.1155/2022/4907084
11. Farmanbar, M., Parham, K., Arild, Ø., Rong, C.: A widespread review of smart grids towards smart cities. Energies **12**(23), 4484 (2019). https://doi.org/10.3390/en12234484
12. Furquim, G., Pessin, G., Faiçal, B.S., Mendiondo, E.M., Ueyama, J.: Improving the accuracy of a flood forecasting model by means of machine learning and chaos theory. Neural Comput. Appl. **27**(5), 1129–1141 (2015). https://doi.org/10.1007/s00521-015-1930-z
13. Gangani Dharmarathne, A.O., Waduge, M.B., Rathnayake, U., Meddage, D.P.P.: Adapting cities to the surge: a comprehensive review of climate-induced urban flooding. Results Eng. **22**, 102123–102123 (2024). https://doi.org/10.1016/j.rineng.2024.102123
14. Gerscovich, D.M.S., Vargas, E.A., de Campos, T.M.P.: On the evaluation of unsaturated flow in a natural slope in Rio de Janeiro, Brazil. Eng. Geol. **88**(1–2), 23–40 (2006). https://doi.org/10.1016/j.enggeo.2006.07.008
15. Han, C., Zang, S.: A comprehensive review of disruptive technologies in disaster risk management of smart cities. Clim. Risk Manag. 100703 (2025). https://doi.org/10.1016/j.crm.2025.100703
16. Heirman, D.: What makes smart grid—smart—and who is in the 'game'? IEEE Electromagn. Compat. Mag. **1**(2), 95–99 (2012). https://doi.org/10.1109/memc.2012.6244981
17. Hong, S., Lee, J.: A study on disaster management for strengthening of disaster resilience in Jeju. J. Korean Geogr. Soc. **57**(5), 501–514 (2022). https://doi.org/10.22776/kgs.2022.57.5.501
18. Hossain Anni, A., Cohen, S., Praskievicz, S.: Sensitivity of urban flood simulations to stormwater infrastructure and soil infiltration. J. Hydrol. **588**, 125028 (2020). https://doi.org/10.1016/j.jhydrol.2020.125028
19. Hossain, E., Roy, S., Mohammad, N., Nawar, N., Dipta, D.R.: Metrics and enhancement strategies for grid resilience and reliability during natural disasters. Appl. Energy **290**, 116709 (2021). https://doi.org/10.1016/j.apenergy.2021.116709
20. Jabareen, Y.: Planning the resilient city: concepts and strategies for coping with climate change and environmental risk. Cities **31**, 220–229 (2013). https://doi.org/10.1016/j.cities.2012.05.004
21. Junichi, K., Naoki, I.: Outline of measures for sediment disaster by the Sabo department of MLIT, Japan. Landslides **17**(11), 2503–2513 (2020). https://doi.org/10.1007/s10346-020-015 54-9
22. Khan, S.M., Shafi, I., Butt, W.H., Diez, I.d.l.T., Flores, M.A.L., Galán, J.C., Ashraf, I.: A systematic review of disaster management systems: approaches, challenges, and future directions. Land **12**(8), 1514 (2023). https://doi.org/10.3390/land12081514
23. Krichen, M., Abdalzaher, M.S., Elwekeil, M., Fouda, M.M.: Managing natural disasters: an analysis of technological advancements, opportunities, and challenges. Internet Things Cyber-Phys. Syst. **4**(1), 99–109 (2024). https://doi.org/10.1016/j.iotcps.2023.09.002

24. Kurwakumire, E., Muchechetere, P., Kuzhazha, S., Ikokou, G.B.: Geographic information and geo-visualisation in support of disaster resilience. In: Proceedings of the ICA, vol. 2, pp. 1–8 (2019). https://doi.org/10.5194/ica-proc-2-68-2019
25. Li, Y.Q., Weng, T.Q., Liu, Y.R.: Mechanism and prevention of urban land subsidence. Adv. Mater. Res. **261–263**, 1749–1754 (2011). https://doi.org/10.4028/www.scientific.net/amr.261-263.1749
26. Liu, X., Li, Y.: Experimental study of seismic behavior of partially corrosion-damaged reinforced concrete columns strengthened with FRP composites with large deformability. Constr. Build. Mater. **191**, 1071–1081 (2018). https://doi.org/10.1016/j.conbuildmat.2018.10.072
27. Ma, Y., Cui, Y., Tan, H., Wang, H.: Case study: diagnosing China's prevailing urban flooding—causes, challenges, and solutions. J. Flood Risk Manag. **15**(3) (2022). https://doi.org/10.1111/jfr3.12822
28. Masozera, M., Bailey, M., Kerchner, C.: Distribution of impacts of natural disasters across income groups: a case study of New Orleans. Ecol. Econ. **63**(2–3), 299–306 (2007). https://doi.org/10.1016/j.ecolecon.2006.06.013
29. McEvoy, D., Fünfgeld, H., Bosomworth, K.: Resilience and climate change adaptation: the importance of framing. Plan. Pract. Res. **28**(3), 280–293 (2013). https://doi.org/10.1080/02697459.2013.787710
30. Mileti, D.: Disasters by Design. Joseph Henry Press, Washington, D.C. (1999). https://doi.org/10.17226/5782
31. Mohammadiounotikandi, A., Fakhruldeen, H.F., Meqdad, M.N., Ibrahim, B.F., Jafari Navimipour, N., Unal, M.: A fire evacuation and control system in smart buildings based on the internet of things and a hybrid intelligent algorithm. Fire **6**(4), 171 (2023). https://doi.org/10.3390/fire6040171
32. Montoya, A.L.: Urban Disaster Management: A Case Study of Earthquake Risk Assessment in Cartago, Costa Rica. University of Twente Research Information (2002). https://research.utwente.nl/en/publications/urban-disaster-management-a-case-study-of-earthquake-risk-assessm-2. Accessed 11 May 2025
33. oac.cdlib.org: The California Earthquake of April 18, 1906: Report of the State Earthquake Investigation Commission, in Two Volumes and Atlas (1908). https://oac.cdlib.org/view?docId=hb1h4n989f&brand=oac4&doc.view=entire_text
34. Ochoa, S.F., Santos, R.: Human-centric wireless sensor networks to improve information availability during urban search and rescue activities. Inf. Fusion **22**, 71–84 (2015). https://doi.org/10.1016/j.inffus.2013.05.009
35. Ogie, R.I., James, S., Moore, A., Dilworth, T., Amirghasemi, M., Whittaker, J.: Social media use in disaster recovery: a systematic literature review. Int. J. Disaster Risk Reduct. **70**(70), 102783 (2022). https://doi.org/10.1016/j.ijdrr.2022.102783
36. Pampa Sadhukhan, Mukherjee, N., Das, P.K.: Location-Based Services for Smart Living in Urban Areas. EAI/Springer Innovations in Communication and Computing, pp. 53–69 (2021). https://doi.org/10.1007/978-3-030-71288-4_3
37. Pramono, I.B.: Nature-based solutions for integrating flood and land subsidence: a case study in Jakarta and Semarang. IOP Conf. Ser.: Earth Environ. Sci. **874**(1), 012001 (2021). https://doi.org/10.1088/1755-1315/874/1/012001
38. Qin, Y.: Urban flooding mitigation techniques: a systematic review and future studies. Water **12**(12), 3579 (2020). https://doi.org/10.3390/w12123579
39. ReliefWeb: ISRO Launches Two UK Radar Satellites for Disaster management—World (2018). https://reliefweb.int/report/world/isro-launches-two-uk-radar-satellites-disaster-management. Accessed 11 May 2025
40. Revilla-Romero, B., Hirpa, F., Pozo, J., Salamon, P., Brakenridge, R., Pappenberger, F., De Groeve, T.: On the use of global flood forecasts and satellite-derived inundation maps for flood monitoring in data-sparse regions. Remote Sens. **7**(11), 15702–15728 (2015). https://doi.org/10.3390/rs71115702
41. Riaz, K., McAfee, M., Gharbia, S.S.: Management of climate resilience: exploring the potential of digital twin technology, 3D city modelling, and early warning systems. Sensors **23**(5), 2659 (2023). https://doi.org/10.3390/s23052659

42. Schumann, G., Brakenridge, G., Kettner, A., Kashif, R., Niebuhr, E.: Assisting flood disaster response with earth observation data and products: a critical assessment. Remote Sens. **10**(8), 1230 (2018). https://doi.org/10.3390/rs10081230
43. Schuster, R.L.: Reducing Landslide Risk in Urban Areas—Experience in the United States, pp. 217–230 (1995). https://doi.org/10.1016/b978-008041920-6/50023-3
44. Seenirajan, M., Natarajan, M., Thangaraj, R., Bagyaraj, M.: Study and analysis of Chennai flood 2015 using GIS and multicriteria technique. J. Geogr. Inf. Syst. **09**(02), 126–140 (2017). https://doi.org/10.4236/jgis.2017.92009
45. Shruti Dadhich, Pathak, V., Mittal, R., Doshi, R: Chapter 10 Machine Learning for Weather Forecasting, pp. 161–174. De Gruyter eBooks (2021). https://doi.org/10.1515/9783110702514-010
46. Snoke, J.A.: Earthquake Mechanisms, pp. 239–245. Kluwer Academic Publishers eBooks (2006). https://doi.org/10.1007/0-387-30752-4_2
47. Srivastav, S.K., Niranjan, S.: Role of weather radar in disaster management. IETE Tech. Rev. **8**(2), 72–77 (1991). https://doi.org/10.1080/02564602.1991.11438717
48. Statista: China: Urbanization 2023. Statista (2024). https://www.statista.com/statistics/270162/urbanization-in-china/
49. Un.org: United Nations Maintenance Page (2020). https://esa.un.org/unpd/wup/CD-ROM/WUP2014_XLS_CD_FILES/WUP2014-F12-Cities_Over_300K.xls. Accessed 28 April 2025
50. Wardekker, A.: Contrasting the framing of urban climate resilience. Sustain. Cities Soc. **75**, 103258 (2021). https://doi.org/10.1016/j.scs.2021.103258
51. World Bank: Urban Development. World Bank Group (2023). https://www.worldbank.org/en/topic/urbandevelopment/overview
52. Wormenor, S.D.-E., Asibey, M.O.: Digital technologies for climate-induced disaster risk reduction and management in Ghana: applicability and operational challenges. Land Use Policy **150**, 107459 (2025). https://doi.org/10.1016/j.landusepol.2024.107459
53. Yu, M., Yang, C., Li, Y.: Big data in natural disaster management: a review. Geosciences **8**(5), 165 (2018). https://doi.org/10.3390/geosciences8050165
54. Yulianto, E., Yusanta, D.A., Utari, P., Satyawan, I.A.: Community adaptation and action during the emergency response phase: case study of natural disasters in Palu, Indonesia. Int. J. Disaster Risk Reduct. **65**, 102557 (2021). https://doi.org/10.1016/j.ijdrr.2021.102557
55. Zayid, S., Abu, Valachamy, M., Malek, A., Yaacob, S., Hassan, N.H.: A formulation of big data analytics model in strengthening the disaster risk reduction—Universiti Teknologi Malaysia Institutional Repository. Eprints.utm.my (2020)
56. Zhang, Y., Geng, P., Sivaparthipan, C.B., Muthu, B.A.: Big data and artificial intelligence based early risk warning system of fire hazard for smart cities. Sustain. Energy Technol. Assess **45**, 100986 (2021). https://doi.org/10.1016/j.seta.2020.100986
57. Zhao, P., Chapman, R., Randal, E., Howden-Chapman, P.: Understanding resilient urban futures: a systemic modelling approach. Sustainability **5**(7), 3202–3223 (2013). https://doi.org/10.3390/su5073202
58. Zuccaro, G., Leone, M.F.: Climate services to support disaster risk reduction and climate change adaptation in urban areas: the CLARITY project and the Napoli case study. Front. Environ. Sci. **9** (2021). https://doi.org/10.3389/fenvs.2021.693319
59. Zul Bahari, A.N.A., Ng, C.P., Ridwan Wong, Md.M., Mohd Rofi, M.F., Ahmad, N., Abas, F., Law, T.H.: Mapping disaster hazards and risks—a case study in Kuala Lumpur City (KLC). Jurnal Kejuruteraan **37**(1), 421–431 (2025). https://doi.org/10.17576/jkukm-2025-37(1)-29

Adaptation to Achieve Climate Resilience Through Advanced Techniques like Remote Sensing and Machine Learning

Kavita Khatana, Vatsala Soni, Ali Akbar, Dipti Bharti, Vaishnavi Srivastava, Richa Saxena, and Sana Ansari

Abstract Climate change is the most significant 21st-century burden that has far-reaching and powerful effects on natural environments, economic stability, and human health. Frequency of extreme weather events such as global warming, sea-level rise, and alterations in precipitation regimes emphasizes the need for adaptation that gives rise to increased societal resilience. In turn, to these risks there has resulted a wide range of technological and scientific breakthroughs that present promising pathways to risk reduction and increased adaptive capacity creation. This chapter presents an overview of such frontier approaches culminating interdisciplinary methodologies such as climate-smart agriculture, artificial intelligence (AI), remote sensing, geographic information systems (GIS), big data analytics, and biotechnological interventions. Integration of these approaches offers a transformative potential not only for

K. Khatana
Department of Applied Sciences and Humanities (Chemistry), G.L. Bajaj Institute of Technology and Management, Greater Noida, India

V. Soni
Department of Applied Sciences, Punjab Engineering College, Chandigarh, India

A. Akbar
Department of Applied Sciences and Humanities, Government Engineering College, Jehanabad, Bihar, India

D. Bharti
Department of Applied Science and Humanities, Darbhanga College of Engineering, Darbhanga, Bihar, India

V. Srivastava
School of Life Sciences and Biotechnology, Chhatrapati Shahu Ji Maharaj University, Kanpur, UP, India

R. Saxena (✉)
Department of Biotechnology, Invertis University, Barielly, UP, India
e-mail: saxena.richa50@gmail.com

S. Ansari
School of Biotechnology, Institute of Science, Banaras Hindu University, Varanasi, UP, India

A. L. Srivastav et al. (eds.), *Climate Resilience: Impact of Quantum Computing and Artificial Intelligence on Urban Planning*, Studies in Systems, Decision and Control 637, https://doi.org/10.1007/978-3-032-06791-3_15

mitigating climate-related risks but also for informing policy decisions, improving early warning systems and fostering sustainable development in vulnerable regions.

Keywords Climate resilience · Climate-smart agriculture · AI · Remote sensing · GIS · Big data · Biotechnology · Adaptation strategies

1 Introduction

One of the biggest and most significant challenges of the twenty-first century is climate change, which has disrupted economies, human lives, and natural ecosystems all over the world [26]. The Intergovernmental Panel on Climate Change (IPCC) estimates that the increased greenhouse gas (GHG) emissions from anthropogenic activities such as burning fossil fuels, industrialization, and land use fluctuations have caused the average global temperature to rise by about 1.1 °C since the late nineteenth century [10, 41]. Prolonged droughts, melting glaciers, rising sea levels, frequent heat waves, and extreme hurricanes and floods are just a few of the environmental effects brought on by this temperature increase [46]. In addition to endangering natural ecosystems, these changes have major ramifications for public health, agriculture, water supply, and human infrastructure.

While mitigation approaches goal to address the main causes of climate change like by reducing GHG emissions, renewable energy transition, and carbon sequestration enhancement, adaptation emphases on preparation for and responding top its unavoidable impacts [16, 38]. Basically, adaptation deals with the "effect" of climate change, however, mitigation deals with its "cause." Though, adaptation is now an essential and imperative measure to address the existing climate realities, not a strategy for the future. According to Metcalfe and Bennett [37], resilience-building interventions are crucial for the survival and socio-economic stability of poor and vulnerable communities, exclusively those in low-lying coastal areas, arid agricultural lands, and rapidly expanding cities.

1.1 The Climate Crisis: An Urgent Global Challenge

Earth's climate is altering at an unmatched velocity, characterized by mounting global temperatures, disturbed precipitation regimes, increased storm activity, and hastening sea level rise. These inescapable alterations enforce significant influences on both natural and anthropogenic systems [56]. The IPCC in its Sixth Assessment Report (2021) unambiguously states that human-induced climate variation has already precipitated wide-ranging adverse effects upon natural environments and human societies [31, 32]. Subsequently, extreme meteorological and thermal events are transitioning from occasional occurrences to frequent phenomenal effects, comprising of prolonged droughts, melting glaciers, rising sea levels, frequent heat

Table 1 Mitigation versus adaptation strategies

Feature	Mitigation	Adaptation
Purpose	Address the causes of climate change	Address the effects of climate change
Examples	Solar energy, afforestation, electric vehicles	Flood barriers, drought-resistant crops, early warning systems
Time horizon	Long-term benefits	Short- to medium-term impact
Technological focus	Energy systems, emissions tracking	Monitoring, prediction, structural improvements
Policy orientation	Global agreements (e.g., Paris Agreement)	Local and national policies, community planning

waves, and extreme hurricanes and floods [46]. Rural locales frequently contend with droughts, heatwaves, and agricultural losses, whereas littoral communities face heightened vulnerability to inundation [14]. The resultant economic and ecological detriments compel nations to implement ameliorative adaptive strategies aimed at augmenting their resilience—their inherent capacity for recovery from and resistance to shocks and stressors.

1.2 Mitigation Versus Adaptation: A Two-Pronged Approach

There are two strategies to combat climate change: adaptation and mitigation. Mitigation measures, such as afforestation or the use of renewable energy, aim to reduce GHG emissions and slow down global warming [18]. In contrast, adaptation aims to prepare for and alleviate the effects of climate change that has happened or cannot be avoided.

The major difference amid mitigation and adaptation are listed in the Table 1.

1.3 The Role of Advanced Technologies in Climate Adaptation

The discourse on climate change adaptation has progressively incorporated unconventional approaches and advanced technologies in recent years. These innovations are instrumental in enabling proactive detection, fostering evidence-based decision-making, and facilitating quick responses to climate-induced jeopardies. Remarkably, GIS aid in the identification of vulnerable geographies; remote sensing furnishes real-time environmental surveillance; AI and machine learning propel the development of predictive analytics; and smart agriculture leverages data-driven solutions to enhance

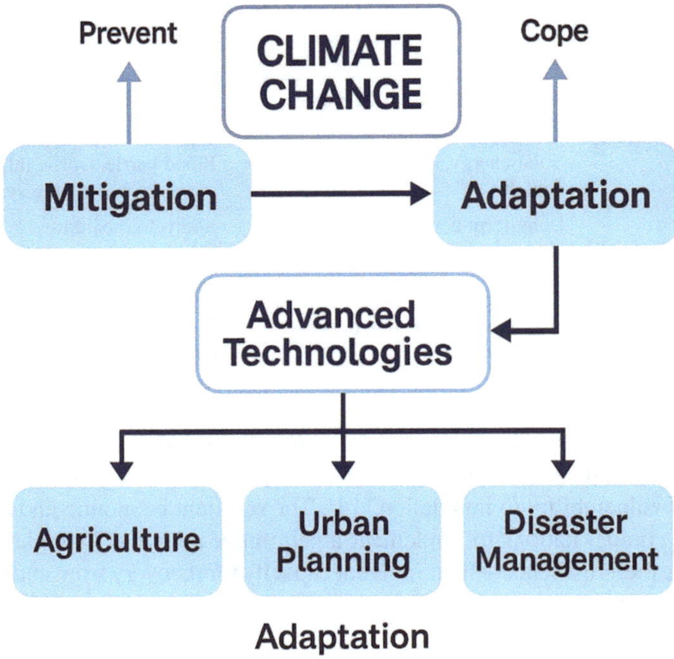

Fig. 1 Framework for climate change response

crop resilience [19, 22]. Synchronously, biotechnological advancements have culminated in the development of genetically modified crops unveiling improved resistance to hydric stress, thermal extremes, and elevated salinity levels (Fig. 1).

2 Climate Resilience

According to Hayes et al. [24], climate resilience refers to the capacity of an ecological, social, or infrastructure system to anticipate, prepare for, respond to, and recover from the adverse effects of climate change on human health. Unlike sustainability, which mainly aims to maintain a long-term environmental balance, resilience is dynamic and adaptive. It mainly focuses on enhancing the capacity of communities and systems to adapt and change in response to climate-related disruptions, such as heatwaves, droughts, and floods.

Besides, they have to change and grow in response to these challenges [28]. The basis for climate resilience is the awareness that climate change is a present reality that needs constant and immediate action and not a future risk.

There are many gauges where resilience can be viewed practically. At the local level, this entails giving local communities the information, funding, and establishment they require to cope with environmental shocks [29]. For example, to ensure food security, farmers in areas affected by drought must be aware of water-conserving technology, diversification of crops, and early warning systems. For preventing flooding in urban areas, urban-scale resilience can require the upgrading of stormwater drainage, revising building codes, or rethinking transportation systems [4]. At the ecosystem scale, resilience requires protection of diversity and recovering climate shock-storing natural buffers, including mangroves and wetlands.

The two strategic approaches of incremental adaptation and transformational adaptation are traditionally applied to discourse climate resilience.

Changing farming calendars or introducing drought-resistant crop varieties like small, gradual vicissitudes that can be made to existing systems to advance their performance in the face of changing climate conditions [20]. Nevertheless, transformational adaptation involves more thoughtful, systemic variations, such as moving entire communities out of flood-prone areas or entirely redesigning water and energy systems to make them disaster-resilient and climate-neutral.

A comprehensive approach for enhancement of climate resilience combines structural and non-structural elements. Physical interventions like seawalls, rainwater harvesting tanks, or cyclone shelters are known as structural measures.

On the other hand, non-structural measures emphasis on behaviour modification, education, policy, and capacity building. Remarkably, novel technologies link the gap amid these two fields by improving the efficiency, scope, and focus of adaptation policies [42]. While climate modelling can help with scenario simulation and policy decision-making, GIS and remote sensing technologies can map high-risk areas. Mobile platforms can deliver climate advisories directly to vulnerable populations, and data analytics and artificial intelligence can evaluate vast volumes of climate data to allow for focused interventions.

In addition, the concept of climate resilience calls for equitable and inclusive adaptation. Climate resilience must be socially equitable because poor and vulnerable groups, such as women, children, and the elderly are typically disproportionately affected by climate-related events. This suggests that advanced techniques must be applied in a way that is accessible, reasonably priced, and appropriate for the local environment. To ensure that no community is left behind, equity should be the primary focus of resilience planning.

To put it briefly, climate resilience is an inclusive, system-based, and anticipatory concept that recognizes the certainty of climate change while aiming for informed and proactive solutions [6]. It is no longer a luxury or choice, it is a necessity for guaranteeing the long-term sustainability of human and natural systems. As we move on to the next section, we shall see how sophisticated technological interventions are making this resilience not only achievable but also measurable, scalable, and impactful.

3 Advanced Climate Resilience Techniques

3.1 Climate-Smart Agriculture

With the increasing frequency and severity of climate-related disasters, conventional adaptation measures are unable to keep pace in most areas, particularly where transformation is taking place at a faster pace than the systems can react. In such situations, the advent of advanced technologies has transformed climate adaptation processes with new, effective, and scalable solutions to build resilience across sectors [8]. From spatial mapping and real-time data collection to biological innovation and predictive analytics, these methods are actually revolutionizing the way societies understand, prepare, and react to climatic risks.

Technological interventions not only enhance response speed and accuracy but also ensure that adaptation is cost-effective and evidence-based. In the next subsections, we discuss the function of some advanced tools each playing an essential role in assisting communities, governments, and ecosystems in responding to climate variability.

3.2 Geographic Information Systems (GIS) and Remote Sensing

These days, remote sensing and geographic information systems (GIS) are essential tools for climate adaptation [17]. Scientists and policymakers can locate high-risk areas, vulnerable communities, and environmental degradation with previously unheard-of accuracy thanks to GIS, which provides a digital platform for storing, analyzing, and visualizing spatial and geographic data. Remote sensing adds its value by gathering real-time data from satellites or drones regarding land surface temperature, vegetation cover, water bodies, urban expansion, and much more [48]. Together, these technologies allow for proactive responses to climate change instead of reactive ones.

Urban planners can use satellite-based land cover data to build green infrastructure to reduce heat islands, and GIS-based flood hazard maps could be used to guide zoning regulations in flood risk areas. In order for governments and agencies to provide early warnings and take preventative action, remote sensing is also crucial for monitoring glacial retreats, detecting forest loss, and forecasting droughts (Table 2).

3.3 Artificial Intelligence and Machine Learning

Artificial intelligence (AI) and machine learning (ML) are shaking things up when it comes to learning about our climate, forecasting hazards, and trying to decipher what

Table 2 GIS and remote sensing applications for climate adaptation

Application area	GIS/remote sensing role	Example region
Flood risk mapping	Identifying inundation-prone areas	Assam, India
Drought prediction	Monitoring soil moisture and vegetation health	Sub-Saharan Africa
Urban heat mapping	Detecting urban heat islands	Tokyo, Japan
Coastal vulnerability	Tracking erosion, sea-level rise	Bangladesh
Crop surveillance	Estimating yield, detecting pests and diseases	Punjab, India

the future holds. They're super-intelligent minds that can sort through huge amounts of climate information, find patterns nobody else sees, make educated guesses about what's going to happen next, and even assist us in making better decisions on autopilot [23].

Picture AI algorithms bringing together data from ground sensors, satellites in orbit, and all of the historical weather data. They can then provide us with very detailed forecasts for severe weather events such as cyclones, floods, and wildfires.

One of the most awesome things AI can do is forecast compound hazards. That is, it can inform us whether a drought not only harms crops, but can also create water shortages and heat sickness, all simultaneously. These predictions provide huge advantages to authorities, allowing them to prepare emergency measures, save lives, and even initiate evacuations weeks in advance of disaster. And ML models are also assisting us in developing real-time climate dashboards. Consider them as real-time broadcasts which assist us in utilizing renewable power, farming better, and creating resilient cities [44].

In the last few years, nations such as China, India, and America have implemented AI-based disaster response systems. For instance, India's "SEEDS" initiative utilizes AI to identify the most vulnerable areas of massive cities and assist them in providing support to individuals before cyclones approach [11]. These technologies are crucial for becoming more proficient in conveying risks, having stronger early warning systems, and adapting solutions that actually work to various communities and their specific needs.

3.4 Climate-Smart Agriculture and Precision Farming

One of the major building blocks of CSA is precision farming, which employs satellite imagery, soil sensors, GPS-equipped tractors, and automated irrigation systems to reduce losses and optimize the use of resources.

By selecting crop varieties suitable for future climatic conditions, water-use efficiency, and income diversification, climate-smart agriculture reduces farmers' vulnerability [7]. Biotechnology is also a critical element involved in this process. Farmers have managed to achieve stable yields despite fluctuating weather patterns using genetically modified crop varieties that are heat, salt, drought, and pest tolerant.

For instance, in India, genetically modified salt-resistant rice has helped farmers along the coast to cope with sea-level rise-induced salinization of soils [9].

Mobile-based agricultural advisory systems are another technology that has empowered smallholders. Farmers receive information regarding rainfall patterns, planting timetables, and pest management through SMS or region-specific mobile apps, hence adapting their practices based on changing climatic conditions [54].

3.5 Smart Infrastructure and Urban Planning

Urban centers are in the direct line of climate effects, especially from heat islands, rising waters, and water shortages. Smart infrastructure incorporates digital technology into physical infrastructure in order to make the physical infrastructure more resilient, efficient, and adaptable [39]. Automated drainage sensors, climate-adaptive building materials, green roofs, and high-performance energy-efficient lighting systems are some technologies being used to develop climate-smart cities.

Simulated urban planning enabled by software facilitates predictive modeling of heat distribution, floodplains, and air quality deterioration under projected climate scenarios. Planners use this information to impose design solutions like raised roads, permeable pavements, and cool roofs that lower the impact intensity of climate [50].

Singapore's "Smart Nation" program is a prime example of climate resilience being integrated into digital infrastructure. Thousands of IoT sensors provide real-time data that is pumped into city dashboards that monitor weather, traffic, water levels, and even energy performance of buildings allowing the city to respond in real time to weather-related stress.

3.6 Early Warning Systems and Disaster Risk Reduction

One of the most obvious uses of cutting-edge technology in climate resilience is in the area of disaster preparedness and early warning systems (EWS). Through satellite imagery, real-time data analysis, and mobile communication platforms, EWS send warnings on an imminent disaster, allowing individuals time to evacuate, governments time to mobilize assets, and companies time to secure assets [12].

Such systems are particularly crucial in cyclone, tsunami, or flash flood-prone nations. Japan's Earthquake Early Warning System, India's cyclone warning systems through the Indian Meteorological Department (IMD), and the Pacific Tsunami Warning Center are examples that illustrate the capacity of technology to cut lives and economic damage sharply.

In addition, community-based disaster warning systems usually constructed on basic mobile networks—guarantee that even far-flung villages are provided with vital information in a timely manner [21]. The effectiveness of such systems, however, is not just reliant on the quality of information but also on the manner in which that information is conveyed, interpreted, and responded to by local communities [45].

4 Case Studies

The real worth of advanced methods in climate adaptation is best realized in actual applications. From all around the world, nations and nations have been adopting cutting-edge technology to save lives, ensure livelihoods, and make ecosystems more resilient. The case studies mention the range of strategies, the scalability of technology, and the need for context-based approaches in building resilience to climate.

4.1 The Netherlands: Delta Works and Integrated Flood Management

The Netherlands is the world leader in water management engineering. The country has almost a third of its land area under sea level and hence is perpetually threatened by sea level rise and storm surges. Accordingly, the Dutch government has built Delta Works, a large-scale technologically complex flood defense infrastructure that includes dams, storm surge barriers, sluices, dikes, and levees [57]. What distinguishes Delta Works is its use of real-time data analytics, sensor networks, and predictive modelling to measure sea-level trends and modify water control operations accordingly.

Delta Works functions based on digital twin technology computer models of hydraulic systems that simulate conditions based on incoming weather and oceanic information [25]. In cases of high tide or storms, automated gates shift in real-time to manage the water flow and reduce inland flooding. The system also integrates GIS-based mapping of vulnerability to continually evaluate population and infrastructure vulnerabilities in flood-affected locations [47]. Such a technological setup has allowed the Netherlands to possess one of the lowest global flood mortality rates, keeping in mind its geographical disadvantage.

4.2 India: Smart Village Initiatives Towards Agricultural Resilience

India, with more than 600,000 villages, is highly dependent on agriculture and thus has climate-resilient agriculture practices as a national imperative. In recent years, there have been pilots like the Smart Village pilot projects in Maharashtra, Telangana, and Gujarat. These projects integrate several technologies—solar-powered irrigation, remote-sensing-based crop monitoring, cloud-connected weather stations, and mobile advisory apps, to help farmers cope with unpredictable rainfall patterns, long-lasting droughts, and soil degradation.

In Maharashtra's Ahmednagar district, for example, farmers using mobile weather monitoring apps and IoT-based soil water sensors lowered water consumption by 40% and raised crop yields by 20%. With the use of GIS-based watershed management system, there is increased groundwater levels and enabled a shift in cropping patterns from water-intensive sugarcane to more drought-resistant millets and pulses.

Initiatives for smart villages include improved access to markets for income stabilization, government support in the form of subsidies for sustainable technologies, and training in climate-resilient practices [2]. By using affordable, locally relevant technologies, these villages are re-establishing themselves as hubs of rural innovation, going beyond simply adapting to climate stress.

4.3 California, USA: AI and Satellite Surveillance for Wildfire Management

Long-term drought, heat waves, and vegetation growth have all contributed to California's terrain becoming more susceptible to wildfires. The state has used satellite monitoring, drone-based real-time surveillance, and artificial intelligence (AI)-powered wildfire prediction models to combat this [3]. Agencies like NASA and Cal Fire work together to feed satellite images into machine learning models that forecast ignition areas, fire progression, and at-risk communities.

One of the most important innovations is the "FireMap" system, which combines satellite imagery, topographic data, and wind speed predictions to model fire movement and support evacuation schedules [30].

The system's efficacy was demonstrated during the 2020 wildfire season by the prompt evacuations facilitated by its advance warnings, which also helped to contain a number of small fires before they grew into larger ones. Firefighters can respond quickly and safely thanks to drones equipped with thermal sensors that fly over high-risk areas and transmit real-time data directly to control rooms [9].

AI-powered insurance analytics have also been used to assess property risk and provide homeowners in forest-edge neighborhoods with early coverage recommendations. These digital measures have not only enhanced response time but also reduced economic losses from fire damage.

4.4 Rwanda: Remote Sensing for Drought Forecasting

In the Eastern African region, Rwanda is applying remote sensing and mobile data integration to develop early warning for drought and food insecurity, especially in semi-arid areas. As a high percentage of the population depends on subsistence agriculture, unreliable rainfall patterns present a dire threat. Satellite data is applied by the Rwandan Meteorological Agency in collaboration with the World Food Programme to monitor soil moisture, crop condition, and precipitation anomalies [40].

The real-time data is passed on through mobile phones to the local farmers, who are provided with personalized messages when to sow, water, or harvest. Where there is little internet connection, community radios blast the alarms. This humble but powerful utilization of technology has managed to curtail crop loss, enhance adaptive farming, and avert famine-like situations in times of dry seasons. Notably, Rwanda's model showcases the way low-income, landlocked countries can use digital tools to protect their agriculture and food systems.

4.5 Vietnam: Resilience of the Coast Through Mangrove Mapping and Restoration

Vietnam's Mekong Delta, a highly productive agricultural area in Southeast Asia, is increasingly under threat from salinity intrusion, coastal erosion, and flooding [34]. To counter this, the government launched a mangrove restoration project that utilizes drone monitoring, GIS-based habitat mapping, and climate modelling to map at-risk shorelines and target reforestation areas.

With high-resolution drone images and satellite data, restoration teams track mangrove health, identify encroachment, and evaluate post-planting survival rates. This technological surveillance allows interventions to be targeted, enhancing survival rates and optimization of ecological services [52]. The restored mangrove belts act as natural buffers, attenuating wave energy up to 70% and shielding both agricultural hinterland and human settlements. The incorporation of technology into ecosystem-based adaptation has positioned Vietnam as an international model for the synergy of nature and digital science in enhancing resilience.

5 Challenges and Limitations

Although cutting-edge technologies possess tremendous potential for building climate resilience, their applications come with certain limitations. These limitations span from the economic and infrastructural to the ethical and socio-political. Appreciating these limitations is essential in the formulation of inclusive, equitable, and efficient adaptation approaches [5]. Without these limitations in mind, it is possible to develop solutions that, in theory, might succeed but, in practice, will not, especially in low-resource or poor environments.

5.1 Implementation Cost is Too High

The greatest barrier to embracing high-end climate adaptation technologies is the upfront capital expense. Most of these solutions—e.g., AI-based modelling platforms, real-time sensor networks, and satellite monitoring devices—demand considerable investment in infrastructure, software costs, and data purchases. Developing nations and low-capacity municipalities frequently find themselves struggling to provide such high-end tools with money, as such expenses often pit against pressing current needs such as food security, health, or recovery from disaster [13].

For instance, equipping urban drainage networks with smart water monitoring systems not only means buying equipment but also servicing them in the long run, contributing to ongoing costs. Similarly, installing precision agriculture systems with GPS, soil sensors, and automatic irrigation can be affordable for large-scale agribusinesses but not for smallholder farmers in the absence of external funding [58]. Public–private partnerships and international financing mechanisms such as the Green Climate Fund have attempted to close this gap, but access to these funds is generally restricted and bureaucratically cumbersome.

5.2 Technological and Digital Inequality

The advantages of climate-resilient technology tend to be concentrated in areas or sectors with existing digital infrastructure. This creates a digital divide, and the vulnerable groups, particularly in remote, rural, or low-income urban communities, can be left behind due to lack of connectivity, digital literacy, or access to technology.

For instance, mobile advising amenities for farmers work efficiently in places where smartphones are extensively reachable, but in the places, they are unavailable where there is poor cell coverage or low literacy rates. Moreover, gender biasness in technology access worsen the adaption gap as most rural communities' women have restricted access to mobile phones, digital training, and extension services [43]. This unequal access can worsen already-existing social inequities and result in more severe climate repercussions for vulnerable communities.

5.3 Data Quality, Gaps, and Privacy Concerns

Advanced technologies depend heavily on high-quality, continuous, and location-specific data. In many countries, climate data is either outdated, unavailable, or fragmented across institutions, making it difficult to build reliable prediction models or vulnerability assessments. Poor sensor calibration, limited ground-truthing, and inconsistencies between datasets can further compromise the accuracy of technology-based decisions [27].

Also, as increasingly more systems use cloud-based services, GPS tracking, and artificial intelligence algorithms, data privacy and abuse concerns are growing. Sensitive location or behavior data, gathered in the course of disaster alerts or smart city monitoring can be exposed to hacking or unauthorized surveillance if not effectively regulated. The lack of robust data protection regulations in many developing nations makes people apprehensive about adopting tech-driven interventions, particularly when there is limited transparency regarding how the data is utilized or exchanged [53].

5.4 Capacity and Skill Limitations

Even where technology exists, there is a limited number of trained personnel to use, interpret, and maintain such systems. Most local governments, community groups, and small-scale farmers lack access to trained GIS professionals, AI coders, or drone pilots. This results in a situation where costly systems are underutilized or misinterpreted, decreasing their efficiency.

For instance, cyclone early warning systems that are deployed in villages that are prone to cyclones can become ineffective if villagers are unaware of how to decode or react to the warnings [49]. Likewise, municipal engineers who are not equipped with training on hydrological modeling software can keep flood risk maps unlinked from urban planning initiatives. Hence, institutional training and capacity building are essential parts of any technology deployment strategy.

5.5 Policy, Governance, and Integration Challenges

Most technological climate resilience innovations happen in isolation and are not well-integrated into formal planning or governance processes. Decentralized jurisdiction among departments meteorology, agriculture, urban planning, and disaster response—tends to lead to coordination failures and duplication of effort [55]. Policies can be behind technology, lacking regulatory certainty, guidelines, or incentives for the application of these tools.

For example, drone monitoring for environmental observation could be technologically feasible but lack utilization because airspace regulation is weak or laws about land ownership are unclear. Likewise, AI-produced urban heat maps could be created but disregarded in building code policy because there are disconnects between data agencies and zoning agencies (Table 3).

Table 3 Summary of key challenges in implementing advanced climate technologies

Challenge category	Description	Impact on resilience building
Financial	High installation and maintenance costs	Limits scale and access in low-income regions
Digital inequality	Unequal access to tech and internet	Excludes vulnerable populations
Data gaps and privacy	Poor quality or insecure data systems	Low trust and limited predictive reliability
Skills and capacity	Lack of trained users and operators	Tools remain underutilized or misinterpreted
Policy and integration	Lack of supportive governance and inter-sectoral collaboration	Hinders long-term mainstreaming into planning

6 Policy Implications and Strategic Recommendations

The increasing pace of climate change, combined with the expanding possibilities of technology, implies that the coming decade will be critical to transforming the ways in which societies construct resilience. Although the implementation of sophisticated techniques is already providing remarkable advantages, progress in the future relies on upscaling innovations, embedding them within governance, and making certain that their adoption remains ethical, inclusive, and sustainable [15]. This part examines future possibilities and makes strategic suggestions to enhance the function of technology in climate adaptation.

6.1 Emerging Technological Frontiers

A number of new-generation technologies are already on the horizon and have the transformative potential for climate adaptation. Blockchain, for instance, is already finding applications as a means for enhancing climate finance transparency so that money intended for adaptation finds its way into target communities [1]. It can be harnessed for carbon trading, land registration, as well as monitoring the success of ecosystem restoration projects.

Internet of Things (IoT) devices—previously applied to smart agriculture and urban infrastructure—are becoming increasingly power-efficient and low-cost. Connected to 5G networks, they will support even more efficient and dependable exchange of climate information from sensors deployed in remote locations [35]. This applies especially to operations such as continuous monitoring of real-time glacier melt, landslide danger detection, or dynamic flood levels management.

6.2 Policy and Governance Recommendations

Governments must take the initiative with comprehensive, forward-thinking policies if they are to truly incorporate cutting-edge techniques into the fabric of climate adaptation [51]. In order to achieve this, proactive adaptation planning that incorporates technology into public health, agricultural, urban, and water resource management policies must replace reactive disaster response strategies. Policies must incentivize cross-sectoral collaboration—say, coupling meteorological information with transport planning or public health warnings [33]. It is also essential to standardize data-sharing protocols so that shared platforms are accessible to other departments and stakeholders without technical or institutional obstacles. In addition, governments must incentivize public–private partnerships and invest in technology incubators for climate resilience.

One example of this has been the successful Rwanda Smart Green Villages initiative, which converges energy, agriculture, and water conservation into digital technology using local government support as well as that of external partners [36]. Examples like these have shown the importance of harmonization of governance, as well as local ownership levels.

6.3 Inclusivity and Local Capacity Focus

Innovation by technology alone is not sufficient to ensure climate resilience. Its viability is based on being inclusive, participatory, and locally adaptive. Strategies going forward need to incorporate:

- Technology inclusion campaigns for bringing the urban and rural areas closer.
- Gender-sensitive technology initiatives that break down obstacles for women and girls in accessing information or devices.
- Localized capacity development, such as training initiatives for local government employees, farmers, students, and health workers on the operation and interpretation of digital tools.

Notably, future resilience initiatives need to acknowledge and incorporate indigenous and traditional knowledge systems, which have helped communities cope with climate variability for centuries. Such knowledge systems, when complemented by contemporary data platforms, can develop culturally embedded, hybrid adaptation models.

6.4 Ethical and Environmental Considerations

As the world embraces digital solutions, ethical governance must keep pace. This includes developing strict data privacy protections, avoiding AI bias in predictive systems, and ensuring that technology does not lead to new forms of exclusion or marginalization. Additionally, the environmental footprint of digital infrastructure such as energy-intensive data centers or electronic waste must be addressed by promoting circular technology design, green computing, and sustainable supply chains.

A resilience strategy that harms the environment in the name of adaptation is inherently self-defeating. Thus, technology for adaptation must itself be sustainable.

7 Conclusion

Climate change is not only an environmental one it is a socio-economic, technological, and ethical one that requires holistic action. This chapter has examined the ways in which emerging technologies can help alleviate the stresses caused by climate change, promoting resilience in urban infrastructure, agriculture, disaster risk reduction, and ecosystem recovery. From GIS-based planning and AI-driven risk prediction to precision farming and smart urban infrastructure, these technologies offer scalable, data driven solutions to a crisis affecting every corner of the planet.

The chapter did clarify, though, that for these technologies to be successful, they must be accessible, inclusive, culturally appropriate, and included into governance frameworks. Coordinated global and local initiatives are required to overcome obstacles such as cost, capacity, digital inequality, and policymaking fragmentation. Furthermore, an anticipatory resilience strategy must strike a balance between technological advancement, social justice, environmental sustainability, and ethical protection.

At last, climate resilience is ultimately a continuous process of learning, adaptation, and transformation rather than a final terminus. This process can be made more equitable and efficient by combinatorial methods integrating human-centered values and inclusive policies. The future lies in developing systems that are not only clever but also sensible, effective but fair and creative yet inclusive.

References

1. Adediran, F.E., Okunade, B.A., Daraojimba, R.E., Adewusi, O.E., Bukola, A., Igbokwe, J.C.: Blockchain for social good: a review of applications in humanitarian aid and social initiatives. Int. J. Sci. Res. Arch. **11**(1), 1203–1216 (2024)

2. Aggarwal, P.K., Jarvis, A., Campbell, B.M., Zougmoré, R.B., Khatri-Chhetri, A., Vermeulen, S.J. et al.: The climate-smart village approach. Ecol. Soc. **23**(1) (2018)
3. Alhammadi, A., Abraham, A., Fakhreddine, A., Tian, Y., Du, J., Bader, F.: Envisioning the Future Role of 3D Wireless Networks in Preventing and Managing Disasters and Emergency Situations (2024). arXiv preprint arXiv:2402.10600
4. Alshaikh, R.Z., Abdulmunem, S.A., Alkinani, A.S.: A review on urban planning and its role in managing flood risks. Urban Plan. Constr. **1**(1), 1–8 (2023)
5. Archer, D., Almansi, F., DiGregorio, M., Roberts, D., Sharma, D., Syam, D.: Moving towards inclusive urban adaptation: approaches to integrating community-based adaptation to climate change at city and national scale. Clim. Dev. **6**(4), 345–356 (2014)
6. Balaine, L., Henchion, M., Macken-Walsh, Á., Meredith, D., Mohammadrezaei, M., Sweeney-Teagasc, S.: Farm-Level Behavioural Change Towards Building Climate Resilience: Insights from A Mixed-Methods Study (2025)
7. Belay, A., Mirzabaev, A., Recha, J.W., Oludhe, C., Osano, P.M., Berhane, Z., et al.: Does climate-smart agriculture improve household income and food security? Evidence from Southern Ethiopia. Environ. Dev. Sustain. **26**(7), 16711–16738 (2024)
8. Bhardwaj, L.K., Jodder, P.K., Priya, R., Rath, P., Jain, H., Thakur, S. et al.: Resilience in the digital age: emerging technologies for climate adaptation. In: Sustainable Synergy: Harnessing Ecosystems for Climate Resilience, pp. 165–175 (2025)
9. Bhowmik, J., Irfanullah, H.M., Selim, S.A., Budrudzaman, M.: Assessing climate change-induced losses and damages to coastal ecosystem services: empirical evidence from Manpura Island, Bangladesh. Clim. Risk Manag. **45**, 100641 (2024); Bóveda, S.A., Gónzález, A.F., Terrón, L.V., Artolazábal, J.A.R., Negro, J.A.R.: Key technologies for UAVs in search and rescue (SAR) and surveillance operations. In: Applying Drones to Current Societal and Industrial Challenges, pp. 97–129. Springer, Cham (2024)
10. Change, O.C.: Intergovernmental panel on climate change. World Meteorol. Organ. **52**(1–43), 1 (2007)
11. Chatterjee, U., Shaw, R., Sivaramakrishnan, L., Mukherjee, J., Ghosh, R. (eds.): Homelessness to Hope: Research, Policy and Global Perspectives. Elsevier (2024)
12. Chavula, P., Kayusi, F., Lungu, G., Uwimbabazi, A.: The current landscape of early warning systems and traditional approaches to disaster detection. LatIA **3**, 77 (2025)
13. Devadason, C.A., Jackson, L., Cole, J.: Pacific Island Countries: An Early Warning of Climate Change Impacts (2019)
14. Dube, K., Nhamo, G., Chikodzi, D.: Flooding trends and their impacts on coastal communities of Western Cape Province, South Africa. GeoJournal **87**(Suppl 4), 453–468 (2022)
15. Ekeh, A.H., Apeh, C.E., Odionu, C.S., Austin-Gabriel, B.: Leveraging machine learning for environmental policy innovation: advances in data analytics to address urban and ecological challenges. Gulf J. Adv. Bus. Res. **3**(2), 456–482 (2025)
16. Ellemers, N., de Gilder, D.: Overcoming unintended consequences of social impact accountability: how to avoid counterproductive responses of individuals and groups in organizations. Soc. Issues Policy Rev. **18**(1), 194–219 (2024)
17. Escobar-Teran, F., Zapata, J., Briones, F., Rosero, M., Portilla, J.: Use of ICTs to confront climate change: analysis and perspectives. Front. Clim. **7**, 1436616 (2025)
18. Fawzy, S., Osman, A.I., Doran, J., Rooney, D.W.: Strategies for mitigation of climate change: a review. Environ. Chem. Lett. **18**, 2069–2094 (2020)
19. Fuentes-Peñailillo, F., Gutter, K., Vega, R., Silva, G.C.: Transformative technologies in digital agriculture: leveraging internet of things, remote sensing, and artificial intelligence for smart crop management. J. Sens. Actuator Netw. **13**(4), 39 (2024)
20. Grigorieva, E., Livenets, A., Stelmakh, E.: Adaptation of agriculture to climate change: a scoping review. Climate **11**(10), 202 (2023)
21. Gumiran, B.A., Moncada, F.M., Gasmen, H.J., Boyles-Panting, N.R., Solidum, R.U.: Participatory capacities and vulnerabilities assessment: towards the realisation of community-based early warning system for deep-seated landslides. Jàmbá: J. Disaster Risk Stud. **11**(1), 1–11 (2019)

22. Hadeed, M.Z., Ali, A., Malik, A., Raza, A., Shoaib, M.: Harnessing AI and GIS technologies for climate-resilient agriculture and environmental sustainability. In: Maintaining a Sustainable World in the Nexus of Environmental Science and AI, pp. 333–364. IGI Global (2024)
23. Hamdan, A., Ibekwe, K.I., Etukudoh, E.A., Umoh, A.A., Ilojianya, V.I.: AI and machine learning in climate change research: a review of predictive models and environmental impact. World J. Adv. Res. Rev. **21**(1), 1999–2008 (2024)
24. Hayes, S., Desha, C., Burke, M., Gibbs, M., Chester, M.: Leveraging socio-ecological resilience theory to build climate resilience in transport infrastructure. Transp. Rev. **39**(5), 677–699 (2019)
25. Henriksen, H.J., Schneider, R., Koch, J., Ondracek, M., Troldborg, L., Seidenfaden, I.K., et al.: A new digital twin for climate change adaptation, water management, and disaster risk reduction (HIP digital twin). Water **15**(1), 25 (2022)
26. Hoegh-Guldberg, O., Jacob, D., Taylor, M., Guillén Bolaños, T., Bindi, M., Brown, S. et al.: The human imperative of stabilizing global climate change at 1.5 °C. Science **365**(6459), eaaw6974 (2019)
27. Irene, B., Onoshakpor, C., Lockyer, J., Chukwuma-Nwuba, K., Ndeh, S.: Now you see them, now you don't: will technological advancement erode the gains made by women entrepreneurship in Sub-Saharan Africa? J. Int. Entrep. 1–35 (2025)
28. Jha, M.K., Dev, M.: Impacts of climate change. In: Smart Internet of Things for Environment and Healthcare, pp. 139–159. Springer Nature Switzerland, Cham (2024)
29. Kais, S.M., Islam, M.S.: Community capitals as community resilience to climate change: conceptual connections. Int. J. Environ. Res. Public Health **13**(12), 1211 (2016)
30. Khan, S.M., Shafi, I., Butt, W.H., Diez, I.D.L.T., Flores, M.A.L., Galán, J.C., Ashraf, I.: A systematic review of disaster management systems: approaches, challenges, and future directions. Land **12**(8), 1514 (2023)
31. Lee, H., Calvin, K., Dasgupta, D., Krinner, G., Mukherji, A., Thorne, P. et al.: IPCC, 2023: climate change 2023: synthesis report, summary for policymakers. In: Core Writing Team, Lee, H., Romero, J. (eds.) Contribution of Working Groups I, II and III to the Sixth Assessment Report of the Intergovernmental Panel on Climate Change. IPCC, Geneva, Switzerland (2023)
32. Legg, S.: IPCC, 2021: climate change 2021-the physical science basis. Interaction **49**(4), 44–45 (2021)
33. Li, C., Yang, H., Yao, Q., An, N., Meng, H.: Governing urban climate resilience (UCR): systems, agents, and Institutions in Shanghai, China. Sustainability **16**(7), 2648 (2024)
34. Loc, H.H., Lixian, M.L., Park, E., Dung, T.D., Shrestha, S., Yoon, Y.J.: How the saline water intrusion has reshaped the agricultural landscape of the Vietnamese Mekong Delta, a review. Sci. Total Environ. **794**, 148651 (2021)
35. Majumdar, P., Mitra, S., Bhattacharya, D., Bhushan, B.: Enhancing sustainable 5G powered agriculture 4.0: summary of low power connectivity, internet of UAV things, AI solutions and research trends. Multimed. Tools Appl. 1–45 (2024)
36. Matchaya, G., Aheeyar, M., Ebrahim, G., Langan, S., Rex, W., Ajayi, O.C. et al.: The Potential for Digital Adaptation in Agriculture in the Zambezi River Basin Countries: Regional Assessment Report (2024)
37. Metcalfe, C.D., Bennett, E.R. (eds.): Building Resilience to Climate Change in Small Island Developing States (SIDS) in the Caribbean. Springer Nature (2023)
38. Mitits, L., Alexiou, T., Milton, J.: Does the language you speak at home affect the size of your L2 vocabulary? Lang. Learn. J. **46**(5), 569–582 (2018)
39. Mohanty, A., Ramasamy, A.K., Verayiah, R., Bastia, S., Dash, S.S., Cuce, E., et al.: Power system resilience and strategies for a sustainable infrastructure: a review. Alex. Eng. J. **105**, 261–279 (2024)
40. Nakalembe, C., Becker-Reshef, I., Bonifacio, R., Hu, G., Humber, M.L., Justice, C.J., et al.: A review of satellite-based global agricultural monitoring systems available for Africa. Glob. Food Sec. **29**, 100543 (2021)
41. Nayak, H., Yadav, S.P., Yadav, D.K.: Contribution of natural and anthropogenic activities in greenhouse gases emission. Energy **4**(2) (2020)

42. Parra-López, C., Abdallah, S.B., Garcia-Garcia, G., Hassoun, A., Sánchez-Zamora, P., Trollman, H., et al.: Integrating digital technologies in agriculture for climate change adaptation and mitigation: state of the art and future perspectives. Comput. Electron. Agric. **226**, 109412 (2024)
43. Ragasa, C., Sengupta, D., Osorio, M., OurabahHaddad, N., Mathieson, K.: Gender-Specific Approaches, Rural Institutions and Technological Innovations (2014)
44. Rajendran, R.K., Priya, T.M., Musa, A.I.A., Mahalakshmi, S.B., Anand, T.R.: Smart solutions for climate resilience harnessing machine learning and sustainable WSNs. In: Machine Learning for Environmental Monitoring in Wireless Sensor Networks, pp. 213–232. IGI Global (2025)
45. Rambaldi, G., Kyem, P.A.K., McCall, M., Weiner, D.: Participatory spatial information management and communication in developing countries. Electron. J. Inf. Syst. Dev. Countries **25**(1), 1–9 (2006)
46. Rector, T.A.: Consequences of climate change. In: Climate Change for Astronomers: Causes, Consequences, and Communication, pp. 4–1. IOP Astronomy, Bristol, UK (2024)
47. Rehman, S., Sahana, M., Hong, H., Sajjad, H., Ahmed, B.B.: A systematic review on approaches and methods used for flood vulnerability assessment: framework for future research. Nat. Hazards **96**, 975–998 (2019)
48. Sharma, S., Beslity, J.O., Rustad, L., Shelby, L.J., Manos, P.T., Khanal, P., et al.: Remote sensing and GIS in natural resource management: comparing tools and emphasizing the importance of in-situ data. Remote Sens. **16**(22), 4161 (2024)
49. Silva, A.F.R., Eleutério, J.C., Apel, H., Kreibich, H.: Assessing the impact of early warning and evacuation on human losses during the 2021 Ahr Valley flood in Germany using agent-based modelling. Nat. Hazard Earth Syst. Sci. **25**(4), 1501–1520 (2025)
50. Sun, C., Rao, Q., Chen, B., Liu, X., Adnan Ikram, R.M., Li, J., et al.: Mechanisms and applications of nature-based solutions for stormwater control in the context of climate change: a review. Atmosphere **15**(4), 403 (2024)
51. Suprayitno, D., Iskandar, S., Dahurandi, K., Hendarto, T., Rumambi, F.J.: Public policy in the era of climate change: adapting strategies for sustainable futures. Migr. Lett. **21**(S6), 945–958 (2024)
52. Sutherland, W.J., Gardner, T., Bogich, T.L., Bradbury, R.B., Clothier, B., Jonsson, M. et al.: Solution scanning as a key policy tool: identifying management interventions to help maintain and enhance regulating ecosystem services. Ecol. Soc. **19**(2) (2014)
53. Tan, K.L., Yeap, P.F., Cheong, K.C.K., Loganathan, S.R.: Crafting an organizational strategy for the new era: a qualitative study of artificial intelligence transformation in a homegrown Singaporean hotel chain. Bus. Process. Manag. J. **31**(8), 104–123 (2025)
54. Tanti, P.C., Jena, P.R., Timilsina, R.R., Rahut, D.B.: Enhancing crop yields and farm income through climate-smart agricultural practices in Eastern India. Mitig. Adapt. Strat. Glob. Chang. **29**(5), 35 (2024)
55. Tolessa, T., Fana, C., Gadissa, D., Takela, E., Tesso, G.: Effectiveness of decentralization of disaster risk management in Borana and South-West Shawa Zones, Oromia Region, Ethiopia. Int. J. Disaster Risk Sci. **15**(6), 892–905 (2024)
56. Trenberth, K.E.: Climate change caused by human activities is happening and it already has major consequences. J. Energy Nat. Resour. Law **36**(4), 463–481 (2018)
57. van Wesenbeeck, B.K., Mulder, J.P., Marchand, M., Reed, D.J., de Vries, M.B., de Vriend, H.J., Herman, P.M.: Damming deltas: a practice of the past? Towards nature-based flood defenses. Estuar. Coast. Shelf Sci. **140**, 1–6 (2014)
58. Yang, C., Sui, R., Lee, W.S.: Precision agriculture in large-scale mechanized farming. In: Precision Agriculture Technology for Crop Farming, pp. 177–204 (2016)

A Systematic Literature Review and Bibliometric Analysis of Environmental Decision Support Systems for Urban Climate Resilience

Shahaji, J. S. Yadhunandan, Atul Kumar Singh, Saeed Reza Mohandes, and V. R. Prasath Kumar

Abstract Climate change is making highway infrastructure more and more fragile. Extreme weather, changing temperatures, changing rainfall patterns, and rising sea levels are all making roads worse and generating problems with operations. Even though people are becoming more aware of the situation, there is still a gap in knowledge of the several types of failures that climate-related stressors can produce, such as operational, structural, hydrological, and geotechnical failures. This study intends to fill this gap by systematically examining how climate change affects highway infrastructure and how Environmental Decision Support Systems (EDSS) can help lessen these consequences. Using PRISMA principles and data from Scopus and Web of Science, we did a systematic literature review (SLR) and bibliometric analysis that led to the selection of 43 high-quality peer-reviewed papers. The results show that climate change significantly affects drainage systems, speeds up the deterioration of roads, causes landslides, and makes transportation operations less efficient, which

Shahaji · J. S. Yadhunandan
Department of Civil Engineering, Dayananda Sagar College of Engineering, Bengaluru, Karnataka, India
e-mail: shahaji-cvl@dayanandasagar.edu

J. S. Yadhunandan
e-mail: yadhunandanjs20@gmail.com

A. K. Singh (✉)
NICMAR Institute of Construction Management and Research, Bahadurgarh 124507, Delhi-NCR, India
e-mail: atulkumarsingh22031996@gmail.com

S. R. Mohandes
Department of Civil Engineering and Management, School of Engineering, The University of Manchester, Manchester M13 9PL, UK
e-mail: saeedreza.mohandes@manchester.ac.uk

V. R. Prasath Kumar
Department of Civil Engineering, College of Engineering and Technology, SRM Institute of Science and Technology, Kattankulathur 603203, Tamil Nadu, India
e-mail: prasathv@srmist.edu.in

A. L. Srivastav et al. (eds.), *Climate Resilience: Impact of Quantum Computing and Artificial Intelligence on Urban Planning*, Studies in Systems, Decision and Control 637, https://doi.org/10.1007/978-3-032-06791-3_16

seriously affects the economy and society. The study shows the need for a multi-modal EDSS framework that includes AI technologies, GIS-based risk assessments, remote sensing, and predictive climate models to help with real-time monitoring, forecasting, and adaptation measures. The suggested framework might help people make investment decisions, make legislative changes that are more climate-friendly, and keep highway networks in cities going for a long time.

Keywords Extreme weather events · Precipitation variability · Road deterioration · Highway infrastructure · Climate change · Infrastructure resilience · Environmental decision support system

1 Introduction

Transportation infrastructure is increasingly more challenging to work in as temperatures worldwide increase and weather patterns become less predictable. Floods, heat waves, heavy rain, and rising sea levels are all examples of extreme weather that can hurt transportation networks, especially highways [9, 7]. The cost of keeping US highways in good shape after floods and stormy weather increases by around 10% annually, reaching more than $5 billion annually [9] . Extreme weather makes road materials break down faster, damaging the road's foundation, creating enormous fractures, and affecting the road's surface. As a result, resilient infrastructure planning becomes essential to address the growing need for repairs and the rising maintenance costs associated with climate-related damages [35].

Notably, the causes of highway failures generate extensive impacts, leading to substantial repair costs and significant economic losses due to transportation system disruptions. According to projections by the World Economic Forum (2019), developed nations are expected to lose $1 trillion annually over the next thirty years if authorities fail to implement climate mitigation strategies. When highways break down, people take longer to get where they need to go, supply lines are delayed, and those who utilize the roadway are put in unsafe situations. As the financial repercussions worsen, policy reforms must focus on preventing problems, which means that infrastructure investment needs to be made ahead of time. The Federal Highway Administration says that climatic data should be used in highway design to assist in picking materials that can manage bad weather and make drainage better to tolerate more rain. These two items are essential for keeping roads safe and working well as the weather changes [9].

Environmental Decision Support Systems (EDSS) are complex analytical tools that help groups make choices by combining different types of environmental data into models that can predict what will happen and make it easier for everyone to work together [13]. These systems are essential for planning improvements to infrastructure, especially when it comes to working out how to deal with climate concerns, coming up with ways to adapt, and making adjustments to the system [15].

When combined with EDSS, active real-time monitoring systems assist infrastructure management teams in planning, which makes them more responsive when there are floods, heavy rain, or high heat. EDSS is excellent for planning and keeping up with city roadways since it uses GIS platforms, remote sensing technologies, machine learning tools, and complex modeling methods all at once. EDSS can help keep networks stable when used in city planning. This study employs a Systematic Literature Review (SLR) combined with bibliometric analysis to comprehensively examine existing research on the impacts of climate change on highway infrastructure. The primary objective is identifying critical risk factors, such as extreme weather events and environmental stressors contributing to roadway deterioration. This study aims to support the development of more resilient transportation systems by uncovering research patterns, key themes, and knowledge gaps. Furthermore, the findings can guide future strategies that align infrastructure planning with long-term environmental and societal sustainability goals.

2 Research Methodology

2.1 Systematic Literature Review (SLR)

Systematic literature reviews (SLRs) are structured approaches to examine existing research and combine the results on specific themes by collecting a wide range of unbiased literature data. A systematic literature review (SLR) differs from a conventional review since it follows a set research strategy that includes precise questions, thorough searches of databases, and rules for which publications to include and which to leave out [5]. Many things need to be done for the SLR procedure. These include developing study subjects, figuring out how to review them, doing full-scale literature searches, using selection criteria, and using standardized quality ratings to collect and combine information in numbers or words. The SLR approach gives you clear, strong, and easy-to-understand findings that can be repeated. This study is instrumental in engineering, healthcare, and the social sciences. The SLR must be thorough and fair by meeting the PRISMA standards or anything like them [16].

Figure 1 demonstrates the systematic literature review (SLR) process. It uses a set framework and organized methods to ensure that only relevant, high-quality articles are included. At first, there were 266 records in the databases: 145 in Scopus and 121 in Web of Science. After deleting duplicates and sorting by language and document type, 231 unique objects remained in the dataset. The ABS journal ranking cut the list down to 156 records. Eighty-two items didn't meet the requirements at the abstract level. This study left 74 full-text articles that needed more thorough inspections to determine their suitability. Finally, 43 publications were chosen for the review's final analysis after employing secondary selection criteria.

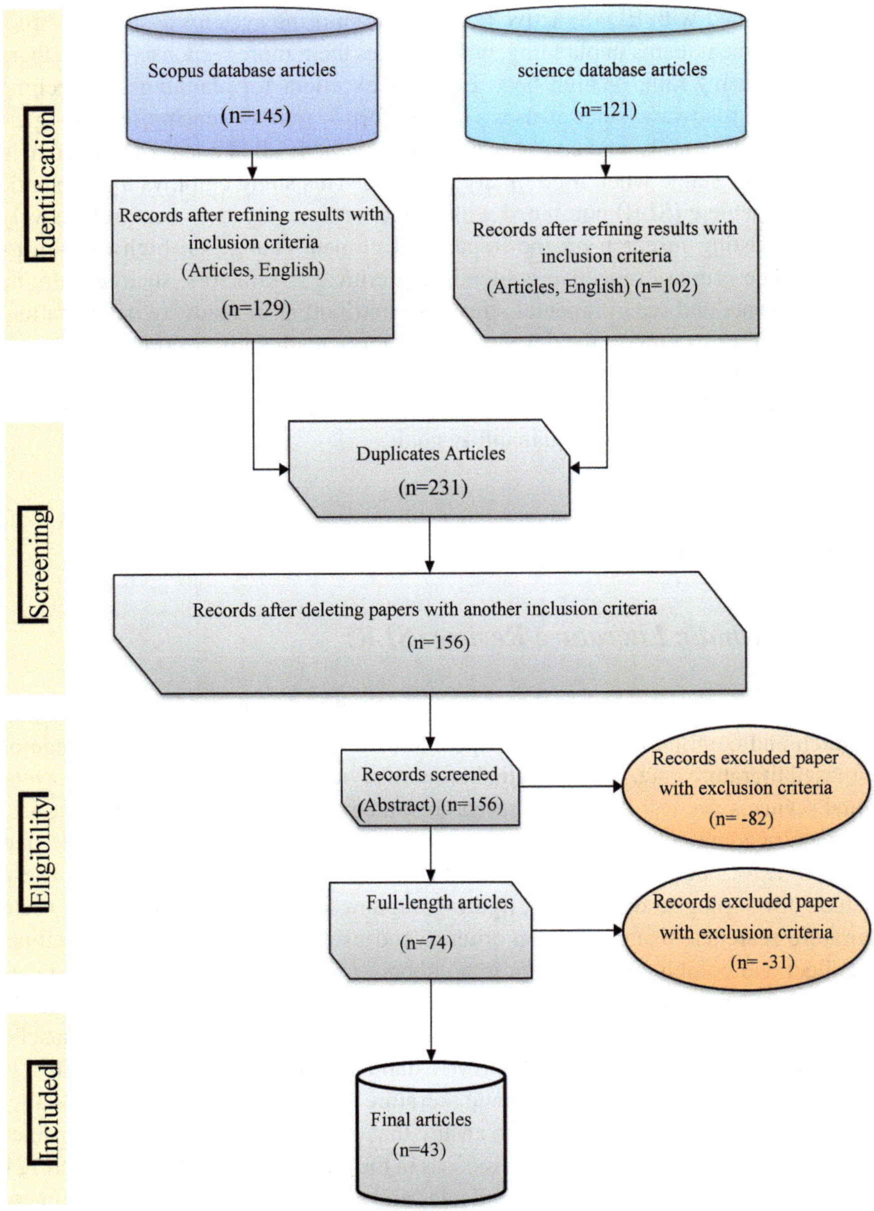

Fig. 1 Workflow of the systematic literature review process showing search results, screening stages, and final study selection

2.2 *Bibliometric Analysis*

Bibliometric analysis is a systematic and quantitative method used to evaluate research trends, impacts, and relationships within a specific domain of study [5]. The process begins with collecting bibliographic data from credible academic databases, using carefully selected keywords to identify and extract relevant research articles, reviews, and other scholarly outputs. The data is cleaned up once it is collected by getting rid of duplicates, making sure that the names of the authors and institutions are the same, and putting the entries in order by things like the type of publication, the year it was published, and the subject area [40]. After that, citation analysis, co-authorship analysis, and keyword co-occurrence mapping are used to find people studying the same item, networks of people working together, and patterns of how things affect each other.

Researchers can use data visualization tools to construct graphs highlighting how authors, institutions, and major topics are connected [5]. The H-index, the number of citations they get, and the journal impact factors all show how influential journals, authors, and publications are. Keyword co-occurrence mapping illustrates prominent trends in research, whereas co-authorship analysis demonstrates how researchers and institutions interact [6]. This strategy shows how a field's academic structure works, points out gaps in the literature, shows new areas of research, and gives scholars chances to work together across disciplines. These results are significant for helping academics, governments, and funding organizations make smart decisions about how to spend money and what scientific questions to ask in the future [5].

3 Result

3.1 *Publication Trends and Research Growth*

There were just 10 articles written in 2013 about how climate change affects roads. By 2023, that number had grown to 30. This study's findings show that more people are interested in this topic. This rise implies that more people know how vulnerable highways are to severe weather events like floods, heatwaves, and storms. Research has picked up since these events are happening more often and pose a greater threat to transportation networks. At the same time, more and more articles are being published about how to slow down climate change. This is because governments build strong, long-lasting infrastructure backed by trustworthy data at the top of their list of things to do. Researchers can now look into the effects of climate change on roads more deeply because of improvements in modeling and analytical tools. The rise in publications shows how important it is to look into and deal with climate risks on roadways since infrastructure problems can significantly affect the economy and make people less safe (Fig. 2).

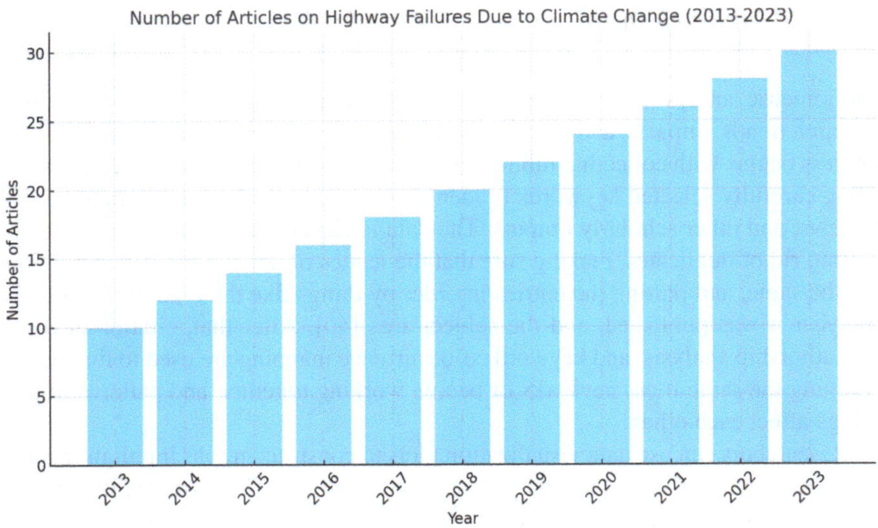

Fig. 2 This graph illustrates how many articles were published yearly from 2013 to 2023 about road concerns induced by climate change

3.2 *Bibliometric Analysis Results*

Research on author keywords shows many essential questions to be answered about climate change and how it affects road works. Climate change is the main idea that ties the network together and shows how it affects various academic contributions. The road construction team looks at carbon emissions, industrial ecology, and ecosystem indicators to make building activities more sustainable. The sustainability cluster employs terms such as green highway and evaluation to emphasize the significance of environmental responsibility in infrastructure development. The permafrost cluster emphasizes research in frigid places, concentrating on frozen soil and infrastructure weaknesses while highlighting the necessity of resilience to permafrost thaw. The environmental impact cluster focuses on long-term infrastructure assessments using terms like resilience, life cycle assessment, and modeling. This interdisciplinary network emphasizes integrating sustainability concepts, resilience strategies, and environmental accountability in road development under climate change (Figs. 3 and 4).

The network visualization illustrates key institutional collaborations in climate change-related infrastructure research. Central organizations like the Alaska Center for Energy and Power and the Bjerknes Centre for Climate Research act as central hubs, indicating their strong influence and frequent partnerships. Distinct clusters reveal focused research areas, such as cold-region infrastructure, permafrost effects, and coastal and extreme weather impacts. Strong internal links within clusters suggest

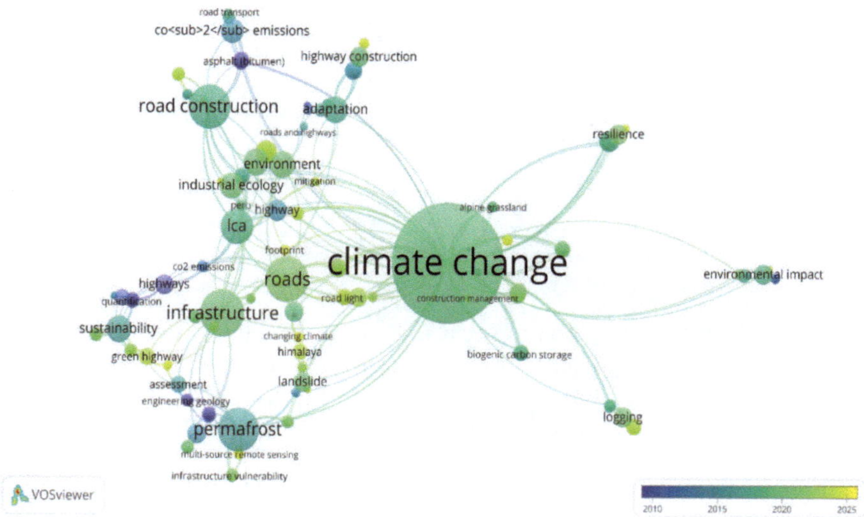

Fig. 3 A map of author keywords showing the most critical study groups for climate change and highways

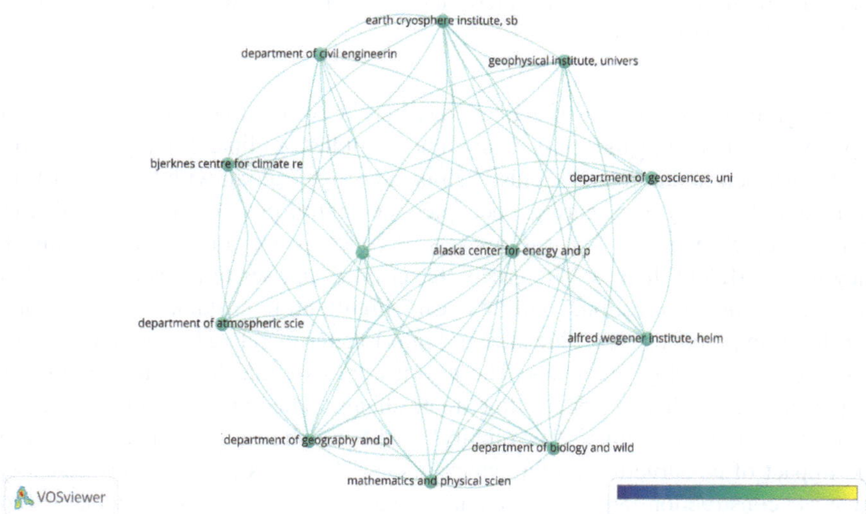

Fig. 4 Collaboration network visualization of organizations contributing to climate change and infrastructure resilience research

tightly connected research communities, while connections across clusters highlight valuable interdisciplinary cooperation. The map showcases the leading institutions, core research themes, and collaborative dynamics shaping this specialized field (Fig. 5).

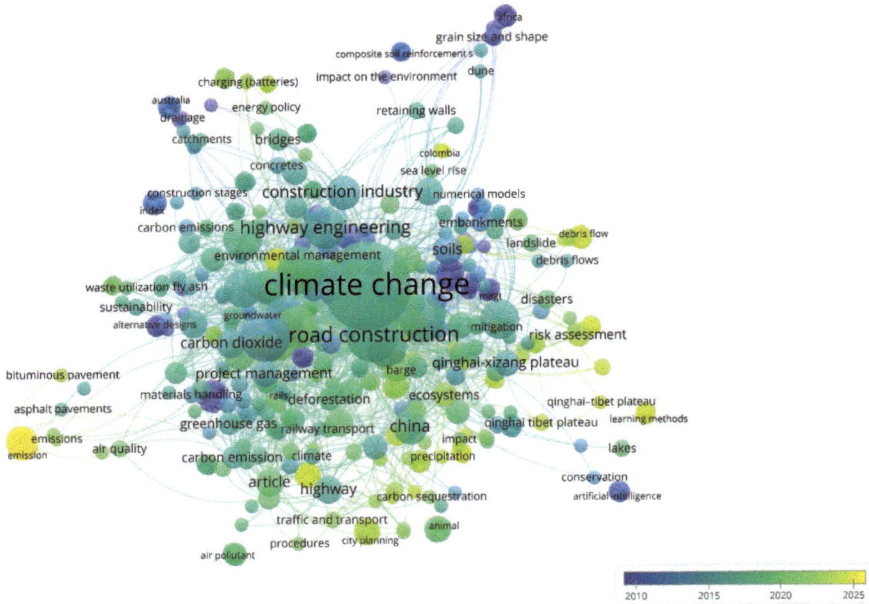

Fig. 5 Keyword co-occurrence network revealing main thematic clusters in climate change and highway engineering research

The graphic presentation visualizes research keywords that structure the major topics while illustrating the relationships among studies on climate change impacts on road construction and related fields. "Climate change," along with "road construction" and "highway engineering," occupies the central position among all themes in the research network. The various sets of concepts are color-coded to indicate they have different themes. The green cluster encompasses terms such as "carbon dioxide," "waste utilization," and "fly ash," highlighting sustainable and durable construction practices. The blue cluster comprises terms related to permafrost, soil conditions, and rising sea levels. These terms are essential for understanding the response of geotechnical systems to climate change. The red cluster includes terms such as "carbon emissions," "air quality," and "greenhouse gas." The focus is on the impact of construction on the environment. The terms in the yellow category, such as "construction industry," "adaptation strategies," and "design and construction," pertain to methods of adaptation. Floods, landslides, and risk analysis describe issues encountered during disasters. "Motor vehicle emissions" and "air pollutants" are words that describe problems that come up when people use cars. In general, research on how climate change affects transport infrastructure is multidisciplinary, combining engineering, environmental science, and policy analysis (Fig. 6).

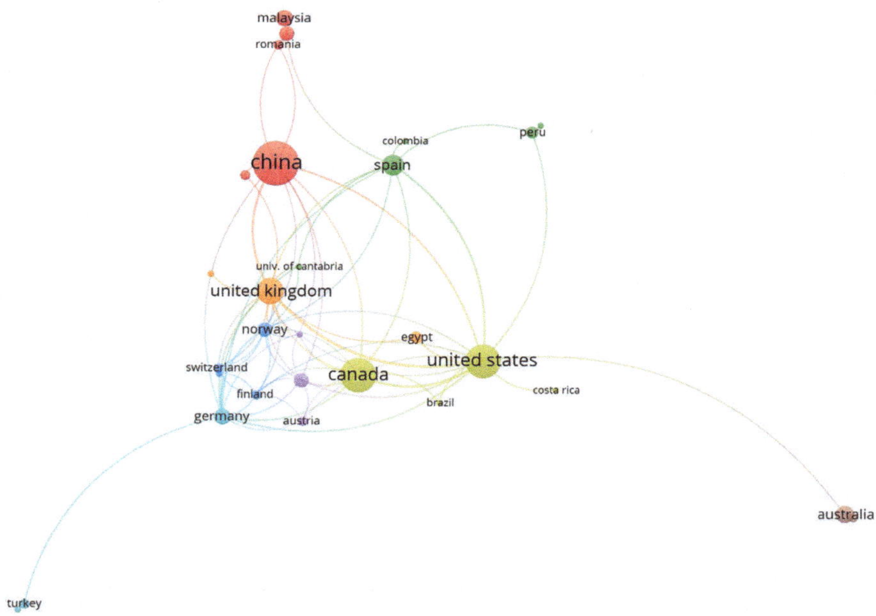

Fig. 6 International collaboration map among countries in climate change impacts on highways

The visualization illustrates how different nations collaborate on research related to the impacts of climate change on highway infrastructure. The United States, China, and Canada exhibit the strongest levels of international collaboration, as indicated by their prominent node sizes in the network. Regional research clusters emerge in the social network representation: the red cluster focuses on infrastructure resilience between China and Malaysia; the yellow cluster emphasizes diverse impact studies between the United States and Canada; and the green cluster explores connections between Spain and Peru, bridging Europe and Latin America. The blue cluster, comprising Germany, Norway, and other countries, concentrates on engineering-based solutions. Notably, the international connections linking the United States, China, and Canada highlight the formation of interdisciplinary research alliances, while Turkey appears as an independent research entity. The network demonstrates that fundamental research in this domain operates as a globally interconnected system (Fig. 7).

Wu Qingbai is a key figure in the author's collaboration network, connecting different research groups that study how climate change affects highway infrastructure. The green cluster, which includes Jin Huijun, focuses on studies relating to cold regions and permafrost. This shows that the group works well together. Zhang Zhongqiong and Ma Wei lead the red cluster, which is focused on how extreme weather affects coastal areas. These clusters show that there are specialized study groups, and Wu Qingbai's links between them show that the topic is interdisciplinary

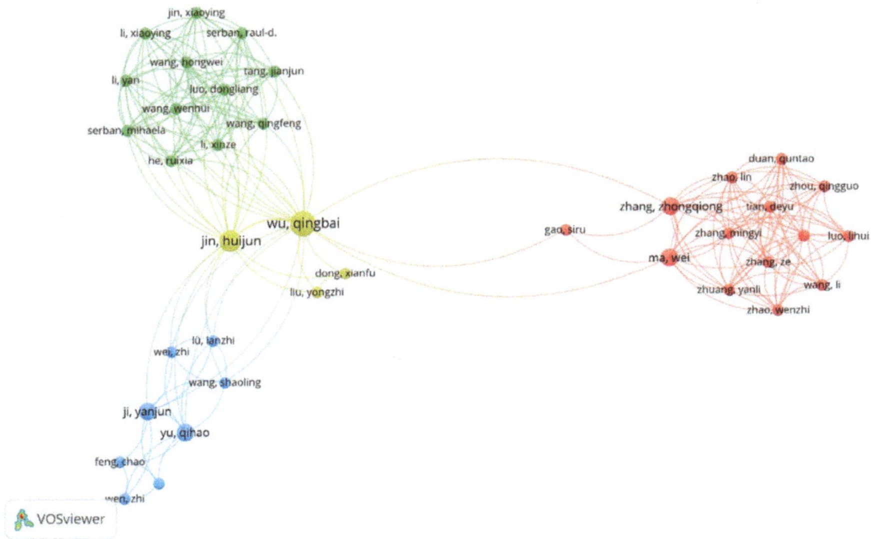

Fig. 7 Collaborative network map of authors contributing to climate change resilience and highway infrastructure studies

and that specialists are working together increasingly to make infrastructure that can withstand climate change.

3.3 Risk Factors

This conceptual framework illustrates the multifaceted risks that climate change creates to infrastructure systems, particularly highways and transportation networks. The overarching "Risk Factor" is influenced by key climate phenomena such as extreme weather events, sea-level rise, increased precipitation and storms, temperature extremes, and drought conditions. The combined effects of these climate change drivers create specific vulnerabilities, including overtopping, coastal destruction, thermal damage, icy breakage, and permanent ground ice detachment, all of which result in intensified cumulative impacts. For example, the interaction of rising sea levels and coastal erosion leads to flooded coastal regions, ultimately undermining infrastructure foundations. These interrelated climate risks, stemming from broad climate variability, highlight the urgent need to design adaptable infrastructure and implement preventive measures to mitigate the multiple threats that climate change poses to transportation systems (Fig. 8).

Fig. 8 Conceptual framework depicting key climate change risk factors affecting transportation infrastructure

3.4 Environmental Decision Support System (EDSS)

Environmental Decision Support Systems (EDSS) are essential for making the environment more resilient. For instance, they assist keep highways safe from bad weather, including heavy rain, temperature changes, and flooding. The table below shows ten critical aspects of EDSS that are supposed to make highways safer. It explains how they work and what research has already been done. We need to look at climate data from real-time weather stations and climate models from the past to see how much stress the pavement can take and how likely it is to flood. It can be hard to get high-resolution, localized data from many different sites [17]. GIS and remote sensing have helped discover weak points in corridors and improve drainage systems, but they still don't do a good job of keeping track of risks in real time [31]. Also, roads and bridges need critical infrastructure evaluations, but they are having trouble because there aren't any standard techniques that consider climate.

Cutting-edge technologies such as artificial intelligence (AI) and machine learning are promising for predicting traffic congestion and infrastructure degradation. However, their widespread application is constrained by limited datasets and the need for validation across diverse climatic regions [30]. Furthermore, inconsistent enforcement and weak interagency coordination persist, even though policy and governance frameworks are essential for integrating climate resilience into design standards [20]. While big data and IoT-based smart monitoring systems offer the potential for improving real-time damage detection, their adoption is limited by high implementation costs and challenges related to data interoperability [30]. Despite being crucial for local adaptation, community-based decision-making remains underutilized due to gaps in technical capacity and stakeholder engagement. Although scenario-based simulations and sustainable materials show considerable promise, they require additional region-specific research and field validation. Finally, early warning systems can significantly enhance disaster response efforts but demand more user-friendly interfaces and better integration with transportation networks. Addressing these research gaps is imperative for successfully implementing EDSS and developing robust, climate-resilient highway systems (Tables 1 and 2).

4 Conclusion

This study systematically reviewed 266 scholarly records, analyzing 43 high-quality publications to understand how climate change affects highway infrastructure. The review reveals a growing research trend from 2013 to 2023, highlighting increased global concern over climate-induced transportation vulnerabilities. The bibliometric analysis identified key contributing countries, institutions, and thematic clusters, offering insight into this research domain's major players and focus areas. The study emphasizes the multifaceted risks of extreme weather, rising sea levels, and temperature fluctuations facing road networks and calls for climate-resilient planning and adaptive infrastructure design. Notably, the study proposes the integration of Environmental Decision Support Systems (EDSS) that combine climate data, geospatial analytics, and predictive modeling to guide informed infrastructure decisions. However, limitations include restricted access to data from specific regions and a lack of standardization in evaluation frameworks. Future research should focus on the empirical validation of EDSS, integrating multi-hazard assessments, and applying such systems to rural and inter-urban road networks to enhance long-term resilience and sustainability.

Table 1 Development of environmental decision support System (EDSS) for climate change resilience in urban planning

Component	Description	Techniques	Applications	Key findings from past studies	Research gaps	References
Climate data integration	Collection of historical and real-time climate data relevant to highway performance	Climate modeling, Weather stations, IoT sensors, Data analytics	Predicting pavement stress, flood risks, and extreme temperature impacts on roads	Accurate data supports proactive maintenance and disaster risk reduction	Lack of high-resolution, localized climate data and integration into design protocols	[19, 21]
Geospatial analysis (GIS & Remote Sensing)	GIS tools map highway segments at risk due to flooding, landslides, or heat waves	ArcGIS, Remote sensing (LIDAR, SAR), Spatial vulnerability mapping	Identifying high-risk corridors, optimizing drainage systems, and rerouting strategies	Enhances risk visualization and planning of resilient road networks	Insufficient temporal analysis; need for real-time GIS updates for dynamic risk tracking	[8, 26]
Infrastructure resilience assessment	Evaluation of roads, bridges, and drainage systems for climate-induced stress and failure potential	Pavement performance models, Structural analysis, Risk indices	Assessing bridge scour vulnerability, pavement rutting, and shoulder erosion risks	Poor design and lack of resilience amplify infrastructure failures in extreme events	Need for standardized resilience assessment tools and climate-adaptive design standards	[12, 26]
Machine learning & AI for risk prediction	Use of AI for predicting failure points and service disruption under climate stress	Neural networks, Time-series forecasting, Predictive analytics	Forecasting pavement deterioration, traffic delays, and maintenance needs	AI enables real-time insights for highway maintenance and disaster response	Limited datasets for training AI need cross-regional models validated under real conditions	[18, 39]
Policy & governance frameworks	Inclusion of climate resilience in highway design standards and maintenance regulations	Climate policy analysis, Resilience codes, Regulatory frameworks	Mandating climate-resilient materials, drainage standards, and flood zone avoidance	Policies are key to preventing infrastructure loss and optimizing investments	Gaps in enforcement and policy alignment across jurisdictions need inter-agency frameworks	[33, 38]

(continued)

Table 1 (continued)

Component	Description	Techniques	Applications	Key findings from past studies	Research gaps	References
Big data & smart monitoring systems	Real-time monitoring of highways for climate stress indicators and traffic conditions	IoT-enabled sensors, Cloud platforms, Traffic and environmental data	Monitoring road surface temperature, waterlogging, and structural strain	Smart monitoring aids adaptive management and early damage detection	Lack of cost-effective sensor networks; data interoperability challenges persist	[18, 36, 39]
Community-based decision making	Involving local stakeholders in highway planning, maintenance, and disaster preparedness	Participatory mapping, Stakeholder consultation, Public feedback tools	Planning evacuation routes, community feedback on road conditions, and resilience needs	Local engagement improves targeted mitigation and resource allocation	Limited community engagement in technical decision-making; lack of education/training	[22, 34, 38]
Scenario-based simulations & adaptation strategies	Simulating climate events and their impact on highway networks to plan resilient responses	Scenario analysis tools, Hydrological modeling, Traffic simulations	Simulating flood scenarios, Traffic rerouting plans, and infrastructure upgrade needs	Scenario planning supports long-term sustainability and risk reduction strategies	Few region-specific scenario models; limited integration into EDSS platforms	[2]
Sustainable infrastructure & material innovation	Use of climate-resilient, sustainable materials in highway design and repair	Life-cycle analysis, Material testing, Green design strategies	Using cool pavements, permeable surfaces, and resilient embankments	Sustainable materials reduce repair costs and improve long-term performance	Lack of field data on new materials; limited adoption in existing road networks	[3, 4, 29]
Disaster preparedness & early warning systems	Systems for early warnings and alerts to protect highway users during climate events	Alert systems, GIS dashboards, Emergency planning tools	Issuing road closure warnings, diversion planning, and disaster response coordination	EDSS improves disaster readiness and reduces economic disruption	Inadequate integration with transport networks; need for user-friendly interfaces	[14, 24]

Table 2 Future study and research directions

Research gap	Future study	References
Limited data on the resilience of highway drainage systems	Research on improving drainage systems to prevent flooding and waterlogging during extreme weather events	[11, 23, 25]
Lack of studies on the combined effects of multiple climatic factors	Studies evaluating the synergistic impacts of heat, moisture, and freeze–thaw cycles on highway infrastructure	
Minimal focus on the effectiveness of existing adaptation measures	Longitudinal evaluations of implemented adaptation strategies to determine their long-term effectiveness	
Scarcity of research on the role of policy in infrastructure resilience	Investigating the impact of different policy frameworks on enhancing the climate resilience of highways	[27, 41]
Insufficient attention to rural and low-traffic highways	Studies focusing on the vulnerabilities and adaptation needs of rural highway networks that often receive less funding	
Gaps in understanding the relationship between climate change and highway accident rates	Research on how extreme weather events influence accident rates and safety on highways	
Lack of engagement with local communities in resilience planning	Exploring community-based approaches to highway resilience planning and maintenance	[1, 28, 37]
Limited use of smart technology in monitoring highway health	Development of smart monitoring systems using IoT and AI to detect and predict highway damages due to climatic stress	

References

1. Al-Ali, A.R., Beheiry, S., Alnabulsi, A., Obaid, S., Mansoor, N., Odeh, N., Mostafa, A.: An IoT-based road bridge health monitoring and warning system. Sensors **24**(2) (2024). Multidisciplinary Digital Publishing Institute (MDPI). https://doi.org/10.3390/s24020469
2. Ali, M., Jaber, M., Abd, S., Alkhayyat, A., Albaghdadi, M.: Big data analysis and cloud computing for smart transportation system integration. Multimed. Tools Appl. 1–18 (2022). https://doi.org/10.1007/s11042-022-13700-7
3. Assaf, H., Abu Abdo, A.: Life cycle assessment of incorporating recycled materials in pavement design. J. King Saud Univ. Eng. Sci. **36**(7), 473–484 (2024). https://doi.org/10.1016/j.jksues.2022.04.001
4. Barbhuiya, S., Das, B.B.: Life cycle assessment of construction materials: methodologies, applications and future directions for sustainable decision-making. Case Stud. Constr. Mater. **19**, e02326 (2023). https://doi.org/10.1016/j.cscm.2023.e02326
5. Donthu, N., Kumar, S., Mukherjee, D., Pandey, N., Lim, W.M.: How to conduct a bibliometric analysis: an overview and guidelines. J. Bus. Res. **133**, 285–296 (2021). https://doi.org/10.1016/j.jbusres.2021.04.070
6. van Eck, N.J., Waltman, L.: Software survey: VOSviewer, a computer program for bibliometric mapping. Scientometrics **84**(2), 523–538 (2010). https://doi.org/10.1007/s11192-009-0146-3
7. European Environment Agency. Environmental indicator report 2017—In support to the implementation of the European 7th Environment Action Programme. EEA Report No 1/2017. (2017)

Retrieved from https://www.eea.europa.eu/en/analysis/publications/environmental-indicator-report-2017
8. Farhadi, H., Najafzadeh, M.: Flood Risk Mapping by Remote Sensing Data and Random Forest Technique. Water (Switzerland), MDPI, vol. 13, no. 21 (2021). https://doi.org/10.3390/w13213115
9. Federal Highway Administration (FHWA). Annual Modal Research Plan. U.S. Department of Transportation. (2018) Retrieved from https://www.transportation.gov/administrations/assistant-secretary-research-andtechnology/federal-highway-administration-fhwa-20-0
10. Füssel, H.-M.: Climate Change, Impacts and Vulnerability in Europe 2016—An Indicator-Based Report (n.d.)
11. Good Governance for Critical Infrastructure Resilience.: OECD (2019). https://doi.org/10.1787/02f0e5a0-en
12. Guo, D., Shan, M., Owusu, E.K.: Resilience assessment frameworks of critical infrastructures: state-of-the-art review. Buildings, MDPI (2021). https://doi.org/10.3390/buildings11100464
13. Hawchar, L., Naughton, O., Nolan, P., Stewart, M.G., Ryan, P.C.: A GIS-based framework for high-level climate change risk assessment of critical infrastructure. Clim. Risk Manag. **29**, 100235 (2020). https://doi.org/10.1016/j.crm.2020.100235
14. Islam, Md.M., Hasan, M., Mia, Md.S., Al Masud, A., Islam, A.R.M.T.: Early warning systems in climate risk management: roles and implementations in eradicating barriers and overcoming challenges. Nat. Hazards Res. (2025). https://doi.org/10.1016/j.nhres.2025.01.007
15. Karimi, A.M., Jelodar, M.B., Susnjak, T., Sutrisna, M.: A review of climate change impact assessment and methodologies for urban sewer networks. Results Eng. **26**, 104625 (2025). https://doi.org/10.1016/j.rineng.2025.104625
16. Kitchenham, B., Pearl Brereton, O., Budgen, D., Turner, M., Bailey, J., Linkman, S.: Systematic literature reviews in software engineering–a systematic literature review. Inf. Softw. Technol. **51**(1), 7–15 (2009). https://doi.org/10.1016/j.infsof.2008.09.009
17. Kustura, K., Conti, D., Sammer, M., Riffler, M.: Harnessing multi-source data and deep learning for high-resolution land surface temperature gap-filling supporting climate change adaptation activities. Remote Sens. **17**(2) (2025). Multidisciplinary Digital Publishing Institute (MDPI). https://doi.org/10.3390/rs17020318
18. Liu, Z., Coleman, N., Patrascu, F.I., Yin, K., Li, X., Mostafavi, A.: Artificial intelligence for flood risk management: a comprehensive state-of-the-art review and future directions. Int. J. Disaster Risk Reduct. **117**, 105110 (2025). https://doi.org/10.1016/j.ijdrr.2024.105110
19. Mumtaz, F., Li, J., Liu, Q., Dong, Y., Liu, C., Gu, C., Zhang, H., Zhao, J., Akhtar, M., Bashir, B., Wang, X., Tariq, A., Hussan, H. U., Li, G.: A comprehensive framework for evaluating ecosystem quality changes and human activity contributions in Inner Mongolia and Xinjiang, China. Clark University Student Publications. (2025). Retrieved from https://commons.clarku.edu/student_publications/45/
20. Nations Department of Economic, U., Affairs, S.: DESA Policy Brief 162 (2024)
21. Peiro, Y., Volpe, E., Ciabatta, L., Cattoni, E.: High resolution precipitation and soil moisture data integration for landslide susceptibility mapping. Geosciences (Switzerland) **14**(12) (2024). Multidisciplinary Digital Publishing Institute (MDPI). https://doi.org/10.3390/geosciences14120330
22. Pompigna, A., Mauro, R.: Smart roads: a state of the art of highways innovations in the smart age. Eng. Sci. Technol. Int. J. **25**, 100986 (2022). https://doi.org/10.1016/j.jestch.2021.04.005
23. Rathnayaka, B., Robert, D., Siriwardana, C., Adikariwattage, V.V., Pasindu, H.R., Setunge, S., Amaratunga, D.: Identifying and prioritizing climate change adaptation measures in the context of electricity, transportation and water infrastructure: a case study. Int. J. Disaster Risk Reduct. **99**, 104093 (2023). https://doi.org/10.1016/j.ijdrr.2023.104093
24. Rokhideh, M., Fearnley, C., Budimir, M.: Multi-hazard early warning systems in the sendai framework for disaster risk reduction: achievements, gaps, and future directions. Int. J. Disaster Risk Sci. (2025). https://doi.org/10.1007/s13753-025-00622-9
25. le Roux, A., Khuluse-Makhanya, S., Arnold, K., Engelbrecht, F., Paige-Green, P., Verhaeghe, B.: A framework for assessing the risks and impacts of rural access roads to a changing climate. Int. J. Disaster Risk Reduct. **38**, 101175 (2019). https://doi.org/10.1016/j.ijdrr.2019.101175

26. Sadiq, R., Akhtar, Z., Imran, M., Ofli, F.: Integrating remote sensing and social sensing for flood mapping. Remote Sens. Appl. Soc. Environ. **25**, 100697 (2022). https://doi.org/10.1016/j.rsase.2022.100697
27. Safarpour, H., Khorasani-Zavareh, D., Mohammadi, R.: The common road safety approaches: a scoping review and thematic analysis. Chin. J. Traumatol. **23**(2), 113–121 (2020). https://doi.org/10.1016/j.cjtee.2020.02.005
28. Singh, R., Sharma, R., Vaseem Akram, S., Gehlot, A., Buddhi, D., Malik, P.K., Arya, R.: Highway 4.0: digitalization of highways for vulnerable road safety development with intelligent IoT sensors and machine learning. Saf. Sci. **143**, 105407 (2021). https://doi.org/10.1016/j.ssci.2021.105407
29. Soliman, A., Hafeez, G., Erkmen, E., Ganesan, R., Ouf, M., Hammad, A., Eicker, U., et al.: Innovative construction material technologies for sustainable and resilient civil infrastructure. Mater. Today Proceed. **60**, 365–372 (2022). https://doi.org/10.1016/j.matpr.2022.01.248
30. Tamagusko, T., Gomes Correia, M., Ferreira, A.: Machine learning applications in road pavement management: a review, challenges and future directions. Infrastructures (2024). Multidisciplinary Digital Publishing Institute (MDPI). https://doi.org/10.3390/infrastructures9120213
31. Tengku Ibrahim, T.N.B., Feisal, N.A.S., Cheah, W.Y., Md Sofwan, N., Suhaimi, N.F., Abdul Samat, N., Kamaludin, N.H.: Monitoring, assessing, and forecasting urban climate issues and challenges. In: Baojie, H., Wang, Y., C.A. (eds.) Urban Climate and Urban Design, pp. 19–47. Springer Nature Singapore, Singapore (2025). https://doi.org/10.1007/978-981-96-1521-6_2
32. Vulnerability Assessment and Adaptation Framework (n.d.)
33. Walker, D., Ayyub, B., Vinson, T., Medina, M., Wright, R.: Climate-Resilient Infrastructure: Adaptive Design and Risk Management (MOP 140) (2018a). https://doi.org/10.1061/9780784415191
34. Walker, D., Ayyub, B., Vinson, T., Medina, M., Wright, R.: Climate-Resilient Infrastructure: Adaptive Design and Risk Management (MOP 140) (2018b). https://doi.org/10.1061/9780784415191
35. World Bank Reference Guide to Climate Change Framework Legislation (2020)
36. Yang, Z., Barroca, B., Weppe, A., Bony-Dandrieux, A., Laffréchine, K., Daclin, N., November, V., et al.: Indicator-based resilience assessment for critical infrastructures–a review. Saf. Sci. **160**, 106049 (2023). https://doi.org/10.1016/j.ssci.2022.106049
37. Ye, Z., Wei, Y., Yang, S., Li, P., Yang, F., Yang, B., Wang, L.: IoT-enhanced smart road infrastructure systems for comprehensive real-time monitoring. Internet Things Cyber-Phys. Syst. **4**, 235–249 (2024). https://doi.org/10.1016/j.iotcps.2024.01.002
38. Zaman, M.O., Raihan, M.M.H.: Community resilience to natural disasters: a systemic review of contemporary methods and theories. Nat. Hazards Res. **3**(3), 583–594 (2023). https://doi.org/10.1016/j.nhres.2023.05.003
39. Zhang, X., Chan, F.T.S., Yan, C., Bose, I.: Towards risk-aware artificial intelligence and machine learning systems: an overview. Decis. Support Syst. **159**, 113800 (2022). https://doi.org/10.1016/j.dss.2022.113800
40. Zhu, J., Liu, W.: A tale of two databases: the use of Web of Science and Scopus in academic papers. Scientometrics **123**(1), 321–335 (2020). https://doi.org/10.1007/s11192-020-03387-8
41. Zou, X., Sun, L., Lan, T., Fan, C., Liu, S., Zhao, H., Qiu, J.: The effects of weather factors on road traffic casualties: analysis on provincial panel data of China from 2006 to 2021. Heliyon **10**(17), e36788 (2024). https://doi.org/10.1016/j.heliyon.2024.e36788
42. citations-20250319T085001 (n.d.)

Role of Modern Techniques to Measure the Impact of Air Pollution on Climate Change: A Short Review

Saurabh Kumar, Akansha Patel, Brijesh Kumar, Aditi Baranwal, Naveen Patel, Vinod Kumar Chaudhary, and Arun Lal Srivastav

Abstract Air pollution plays vital roles in determining patterns of climate change which result global warming over multiple scales in time and space, owing to the fact that emissions, dispersion, dilution, chemical transformation, and eventual deposition of air pollutants all can be influenced by meteorological variables such as temperature, humidity, wind speed and direction, rainfall and mixing height. There are number of meteorological factors that elaborate the impact of air pollution on regional and global changing in climate. As a result of climate change; the increase in atmospheric and surface temperatures, rising sea levels, the demolish of ice cover, the intensity of heavy rainfall, landslides, hurricanes, cyclones, and the increasing frequency of tornadoes, etc. are all being observed in different parts of the world. Moreover, due to toxic air pollutants different types of human health malfunction as well plant health disorder reported in different region of the globe. In this book chapter, the author discussed about the air pollution, different sources of air pollution and briefly elaborated the impact of air pollution on climate change and their possible factor that are responsible of increasing global warming and climate change. The author also shortly described the impact of air pollution on different human organ health system. In addition to, briefly elaborated the modern tools techniques such as artificial intelligence (AI), machine learning (ML) and remote sensing and Geographical Information System (GIS) for mitigation and management of air pollution.

Keywords Air pollution · Climate change · Human health · Artificial Intelligence · Remote sensing and GIS

S. Kumar · A. Patel · B. Kumar · A. Baranwal · V. K. Chaudhary (✉)
Department of Environmental Sciences, Dr. Rammanohar Lohia Avadh University, Ayodhya, Uttar Pradesh, India
e-mail: vinodkrchaudhary@rmlau.ac.in

N. Patel
Department of Civil Engineering, Dr. Rammanohar Lohia Avadh University, Ayodhya, Uttar Pradesh, India

A. L. Srivastav
Chitkara University School of Engineering and Technology, Chitkara University, Himachal Pradesh, Solan, India

A. L. Srivastav et al. (eds.), *Climate Resilience: Impact of Quantum Computing and Artificial Intelligence on Urban Planning*, Studies in Systems, Decision and Control 637, https://doi.org/10.1007/978-3-032-06791-3_17

1 Introduction

The survival and growth of all life on Earth depend on air. Air pollution has become a global issue, with the rapid development of modern industry and transportation, as well as exponential population growth, and urbanization [64]. According to Masood and Ahmed [35], 92% of the global population is exposed to air pollutant levels above air quality criteria, making air pollution a serious global environmental health concern. Poor air quality not only affects health of people and careers; it also limits economic growth and contributes to climate change by affecting the climate [52]. The particulate matter found in ambient air pollution is more harmful as reported by a World Health Organization (WHO) [1], and globally, ambient air pollution, primarily particulate matter, was responsible for 8.2 million deaths in urban and rural areas in 2016 [48]. Maximum number of deaths (~15%) have been reported in South and East Asia from air pollution and due to its synergistic effects. Despite a decline in the number of deaths linked to air pollution, developing countries like India still experienced a 14% increase in mortality between 1990 and 2017 [34].

Significant health hazards have been observed due to the prolonged exposure of air pollutants and they may increase the risk of cardiovascular issues, respiratory ailments, etc. [22]. Particularly in India, it was found that ~1.1 million premature deaths occur every year due to the exposure of harmful particulate matters present in the ambient air and it is 4th largest cause of death in the country [46]. The World Bank estimates that air pollution's cost accounts for about 3 to 8% of India's whole GDP in terms of healthcare costs, lost productivity, and premature mortality which is having significant impacts on the economy of the country.

Since last 20 years, major research activities have been performed in the South and Southeast Asia on atmospheric aerosols including black carbon, organic carbon, sea salt, and sulfates [40]. The composition of these aerosols and spatiotemporal variability, however, have a significant impact on the magnitude of these consequences [42]. Among the major components of atmospheric PM are aerosols, which make up between 30 and 70 percent of the fine aerosol mass over Indian cities [10]. India, that has the largest population in the world (17.76%), is facing serious environmental problems as many cities like Delhi, Mumbai, and Kolkata are among the most air polluted cities in the world [56]. Over the past 20 years, India has had severe and widespread air pollution occurrences, which has caused regulatory authorities to become extremely concerned. The Indo-Gangetic plain (IGP) is particularly vulnerable to extreme air pollution events, that are particularly common during the winter and post-monsoon seasons [28]. In Delhi, millions of automobiles have been registered and it is increasing at a rate of 7% annually, and they contribute 67% of all air pollution [49]. Additionally, the 12% air pollution also caused by the burning of organic materials and agricultural products in Delhi which includes small-scale units and coal-based thermal power plants [44]. Government authorities are very much concerned about the increased air pollution level. By using technologies such as big data analytics and advanced computer systems, AI and ML, the problem of air pollution can be tackled [37]. AI, ML, artificial neural networks (ANNs), and deep learning (DL) are state-of-the-art technology which employs hardware to enable researchers to properly understand the outdoor data [15]. The most prevalent approaches were found to include expert systems (ES), ANNs, Fuzzy Logic, and other very contemporary technologies. It has been observed that AI has the capability to control air pollution mitigation operations in real time. After analyzing the temporal fluctuation, ANN can estimate air pollution levels and extract essential information from the meteorological data [2]. ANN models can also be helpful for the accurate predictions as they have also been widely utilized to simulate air pollution levels. For instance, ANN can also provide date for ozone or simulate its behavior in relation to atmospheric conditions and also for the forecasting ability and the connections between weather, ozone (O_3), and nitrogen dioxide (NO_2) [26]. A back propagation neural network (BPNN) model developed for the purpose of predicting 24-h photochemical pollution [36]. Prediction model of air pollution based on artificial neural networks (ANN) and multiple linear regressions has been designed by several authors using specific air pollutants, such as NO, NO_2, CO, O_3, and SO_2. Furthermore, it was also advised to examine ambient air pollution using computer technology and advanced statistics (such as artificial intelligence (AI), the Internet of Things (IoT), and machine learning (ML) when using an ANN for ozone [6].

2 Sources of Air Pollution

The sources of air pollution can vary based on geographical location. These may be remote and local. They include transportation (cars, trucks and airplanes), industrial facilities, agricultural activities, construction, waste burning and natural sources like volcanic eruptions and wildfires. Emissions from thermal power plants, refineries and petrochemicals are the major traces of air pollution. Dust from construction sites and faulty vehicles contribute to air pollution [5]. The major sources of air pollution include mobile sources, generally vehicles like cars, trucks, ships, and airplanes, emit various air pollutants. Area sources represent various smaller, widespread air pollutants emitters such as dry cleaners, farms, residential fireplaces, or paint solvents, whose collective impact is significant. Natural sources include events such as wildfires, volcanic eruptions, dust storms, and pollen grains, fungal spores imparting air pollution without human intervention. Point sources are distinct, identifiable stationary places, like a single smokestack at a factory, where pollutants are discharged directly into the air whereas anthropogenic source includes waste disposal, burning of fossil fuels and smoking [51]. Figure 1 shows the different types of air pollution sources like mobile, area, point, natural as well as anthropogenic sources on the earth.

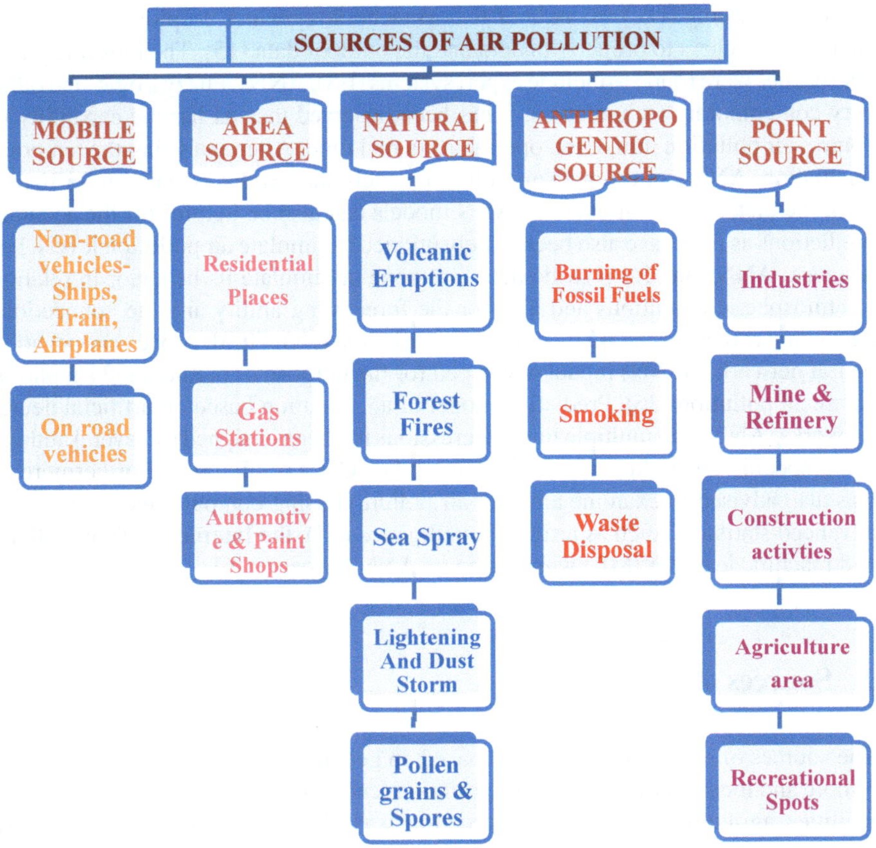

Fig. 1 Various sources of air pollution

3 Impact of Air Pollution on Climate Change

Climate change is the result of long-term effect of atmospheric temperature and weather patterns, primarily caused by human activities such as transportation, industrial development, and the burning of fossil fuels like coal and gas. Additionally, these activities also play a major role in polluting the air. The continuous emission of greenhouse gases causes global temperature rise, that are responsible for climate change. Developing countries like India and under developed countries are facing several issues related to climate change and air pollution and posing severe impact on human health and the environment [39]. The life on planet Earth of living organisms is mainly due to the presence of clean air as life existence is not possible without air. The continuous growth in population, the burning of fossil fuels for energy production and industrial activities, poor agricultural practices, and vehicle emissions are primarily responsible for the deterioration of air quality of those cities [18]. Particles such as $PM_{2.5}$ and PM_{10}, carbon dioxide (CO_2), sulfur dioxide (SO_2), nitrogen oxides (NOx), nitrogen dioxide (NO_2), ammonia, benzene, volatile organic compounds (VOCs), carbon monoxide (CO), and ozone, etc. are most common air pollutants that are directly and indirectly related to climate change [57]. Meteorological parameters such as rainfall, wind speed, relative humidity, wind direction, solar radiation, and air temperature affect directly and indirectly the level and concentration of these pollutants. These air pollutants have several other effects on the environment, such as global warming, ozone layer depletion, climate change, and acid rain [33]. For example, due to various development activities in India, ranks third in global CO_2 emissions over the past ten decades mainly due to continuous increase in carbon dioxide emissions. Similarly, for methane emission, India ranks second after China, and agricultural as the main source of emission contributing 62% of the total emissions. Both CO_2 and CH_4 are responsible for global temperature rise and climate changes as these gases are the major greenhouse gas [43]. There are some factors behind the impact of air pollution on climate change.

3.1 Air Pollutants

Anthropogenic activities such as construction processes, transportation, and energy production emit primary and secondary air pollutants into the atmosphere. These agents not only release various pollutants into the atmosphere but also increase the global levels of CO_2 emissions, which additionally contribute to climate change. In addition, pollutants such as black carbon, ozone and its precursors also contribute to heat, resulting in temporary regional climate effects, and thus they are recognized as ultra-short-lived climate pollutants. For example, coal-fired power plants emit vast amounts of SO_2 and NOx, which are key ingredients for the formation of PM and surface-level ozone (O_3). Manufacturing processes also emit toxic pollutants such as heavy metals and persistent organic pollutants, which have devastating effects on health, including respiratory diseases and cancer, among others. Industrial activities also contribute to CO_2 emissions, accounting for about 20% of global emissions. However, particles present in the air (particulate matter or aerosols) exhibit more complex behavior. These particles are primarily emitted from transportation and power plants, with their main component, black carbon (BC), being a powerful short-lived climate pollutant (SLCP) that contributes to global temperature rise, alters precipitation patterns, and has negative effects on human health. Moreover, black carbon aerosols can absorb incoming solar radiation and thus exhibit a positive radiative forcing (which means more incoming energy is received from sunlight that is then emitted into space). Other aerosol components, such as sulfates, reflect solar radiation and exhibit negative radiative forcing effects (which lead to cooling). Aerosols also influence cloud formation, affecting the Earth's radiation balance through clouds [32].

3.2 *Greenhouse Gases*

The warming of climate as a result of greenhouse gases including carbon dioxide, methane, nitrous oxide, and ozone trapping heat in the atmosphere. The downward flux reaching the surface of the Earth and the upward flux of long-wave radiation leaving the Earth's surface are both influenced by variations in the concentration of greenhouse gases. Moreover, the infrared opaqueness of air also increases in parallel with increase in the concentration of greenhouse gases. Consequently, the upward flux from the uppermost layer of the atmosphere is absorbed more than the downward flux from the lowermost layer [31]. Recent studies have also shown that these gases released from both anthropogenic as well as natural sources of emissions, which are disrupted by human activities that accelerate these emissions. Common household activities are also responsible for significant contribution of air pollution through CO_2 emissions which includes like cooking, heating, and lighting. Further, increase in the residential electricity consumption is expected during 2020–2025, which accounts for ~10% global CO_2 emissions [53]. Transportation sector also contributes ~29% green-house gas emissions to the globe primarily CO_2 gas. CH_4 is the second major greenhouse gas after CO_2 and its main sources are oil & ga sector, agriculture, and solid waste disposal. According to [38], the global warming potential of CH_4 in a period of 100 years, is ~28–34 times that of CO_2, while in the last 20 years, the global warming potential of CH_4 is approximately more than 84 times as compared to CO_2.

3.3 *Meteorological Parameters*

With climate change, the impact of weather parameters on the concentration of air pollution has become increasingly significant. Several studies have been conducted to establish the relationship between atmospheric factors, air pollution and climate change. The wind speed and direction play a pivotal role in the long-range transport of pollutants from one place to another [59]. For instance, Zacharis and Sempos [61] state that temperature inversions create favorable conditions for trapping pollutants in the lower layers of the troposphere and for their growth. In winter, when night-time temperature inversions occur due to surface air radiative cooling overnight, along with persistent anticyclones and lower temperatures, they lead to severe smog incidents in the city while in summer, when long-lasting anticyclones and high temperatures prevail, they contribute to the appearance of photochemical smog.

3.4 Human Population

The increase in the human population is a major threat because a higher population means a greater demand for natural resources such as energy demand, food production, etc., which are directly related to human activities. The result of these human-made activities is the emission of pollutants and greenhouse gases into the air, which are associated with climate change. In 2023, the global population of the world was projected to surpass the 8 billion. By 2040, it is estimated that 1 billion people will add our human group and another 1 billion will add by 2060. Such demographic trends and factors play a crucial role in understanding various adverse conditions and addressing the climate crisis, which exposes more people to climate-related risks [4]. For example, the United States accounts for just a little over 4% of the global population, but it holds 17% of the world's energy use. America ranks first in the world in terms of per capita carbon emissions. The carbon footprint of people living in developed countries such as the United States, Australia, and Canada is nearly 200 times higher than that of people in some of the poorest and under developing countries like Chad, Niger, and the Central African Republic [9].

4 Impact of Climate Change on Human Health

United Nations Framework Convention on Climate Change (UNFCCC) defines climate change as "A change in climate that is attributed directly or indirectly to human activity that alters the composition of the global atmosphere and which is in addition to natural climate variability observed over comparable time periods." In the last few decades, clear consequences of climate change can be seen such as the rise in global average atmospheric (and oceanic) temperatures, extensive vanishing of ice, and the increase in average sea levels globally [30]. There is a significant possibility of an increase in the potential hazards and detrimental effects of global warming, that include changes generated by human activities and related extreme weather events such as heatwaves, droughts, tropical cyclones, flash floods etc. In addition, several studies indicate that the increase in atmospheric temperature associated with climate change also affects air quality [27].

The impact of climate change on human health varies significantly based on various factors such as age, gender, ethnicity, and socioeconomic status, as well as region, population sensitivity, the amount and duration of exposure to climate change, and the adaptability of society. Moreover, the rise in air and surface temperatures, associated with heat waves, heavy rainfall, and other extreme weather phenomena, may promote the expansion of numerous anthropogenic diseases [62] (Table 1).

These diseases classified upon the infectious and non-infectious diseases that are describe below.

Table 1 Climate change events and their possible impact on human health

S. no	Climate change global event	Climate change local event	Possible health impacts	**References**
1	Rising of earth temperature	Sever weather condition	Injuries, casualties, and psychological impacts	Rataj et al. [41]
		Extreme heat wave	Heat related illness, cardiovascular failure etc	Ebi et al. [17]
2	Sea level rising	Increasing allergens	Respiratory allergens	Burbank et al. [7]
		impacts of water quality	Cholera, leptospirosis, cryptosporidiosis and etc	Sharif et al. [47]
3	More extreme weather	Air pollution	Respiratory and cardiovascular diseases	De Vita et al. [16]
		Changes in vector ecology	Malaria, dengue, Rift valley fever, lyme disease, hanta virus and chikungunya etc	Chikezie et al. [12]
4	Increasing CO_2 level	Environmental Degradation	Civil conflict impact on mental health	Tripathi et al. [55]
		Water and food supply impact	Malnutrition and diarrheal disease	Yaseen et al. [60]

4.1 Infectious Diseases

4.1.1 Vector-Borne Diseases (VBD)

Change in the climate influences the transmission dynamics and incidence of vector-borne diseases, including malaria, schistosomiasis, onchocerciasis, trypanosomiasis, filariasis, leishmaniasis, plague, and Rift valley disease. Primarily, VBDs spread to humans through bites from infected arthropod species such as ticks, mosquitoes, sandflies, triatomine bugs, and blackflies. These species are generally cold-blooded (ectothermic), which that temporal and spatial variations in several factors such as temperature, rainfall, and humidity affects the biology and ecology of vectors and their hosts. Since adult female mosquitoes feed more often as they are able to digest blood quickly. In this way, female mosquitoes can also increase the intensity of VBD transmission in warm climates [45].

In North India, an increase of 2 °C in temperature has been observed as an increase in the transmission rate of dengue. Colón-González et al. [13] had investigated the impact of weather effects on dengue incidence. They had found the impact of rise in temperature shorten the incubation period of the virus that also shortens the growth of mosquito time and gonotrophic cycle, leading to an increase of dengue transmission. However, other additional factors, including variations in land use patterns, density of population, and anthropogenic activities, are also influencing the patterns of vector-borne disease occurrence and the frequency of its spreading.

4.1.2 Diarrheal Diseases

The burden of diarrhea diseases is likely to increase globally due to climate change, that is becoming a cause of child mortality worldwide. Several previous studies have proved that various climatic factors like as temperature, rainfall, and relative humidity as causes of diarrheal disease. Following a flood event, cases of diarrheal diseases (including those caused by unsanitary conditions) rise, especially in areas where sanitation facilities are inadequate. Moreover, drought conditions also influence the availability of fresh water, further causing diarrhea and other sanitation-related diseases more frequently [63].

4.1.3 Cholera Disease

The main bacterium responsible for causing cholera, *Vibrio cholerae*, relies on plankton and copepods for its existence, replication, and transmission. A study reported that at a salinity of 15% and temperatures between 250 to 300 degrees Celsius, the association of *V. cholerae* with copepods was observed [58]. Various studies have shown a potential relationship between cholera bacteria, sea surface temperature, and phytoplankton, indicating that an increase in sea surface temperature creates favorable conditions for phytoplankton, that in turn leads to an increase in zooplankton populations that serve as a reservoir for cholera bacteria. In addition, the increase in salinity of local water bodies during drought leads to the development of Vibrio, that helps in the aggregation of Vibrio cholerae with copepods [50].

4.2 Non-infectious Diseases

4.2.1 Impact of Heat Waves

The international meteorological agency like World Meteorological Organization (WMO) describes a prolonged heatwave as: "A period of weather that is unusually hot in a region, which persists for at least three consecutive days based on local climate conditions during the warm season of the year, in which maximum temperature conditions are recorded." Rising temperatures can lead to several serious issues such as illnesses, an increase in the number of visits to emergency departments, and even an increase in deaths. In cities mainly due to increase in the area of paved surfaces and decrease in the number of trees leads to heat-related hazards to the residents [8]. In extremely hot conditions, through two natural mechanisms human body cool down itself; first release of sweat through skin and second increase in blood flow to the skin. However, these physical processes affect the functioning of the heart and lungs, leading to adverse effects on health, including heat cramps, fainting (losing consciousness), exhaustion, and potentially fatal heat strokes [14]. Extreme temperatures and associated illnesses including human metabolism disorders, obesity, vascular disease, respiratory problems, skin disorders, and mental health issues, can be made severe during this time by a variety of acute and chronic health problems.

4.2.2 Respiratory Diseases

Increase in ambient temperature creates favorable conditions for several pollutants such as ozone gas, which can increase the impact of various diseases including; airway injury and inflammation leading to reduction in lung function. On inhalation ozone gas, triggers an inflammatory response that causes increase in airway permeability and bronchial hyperreactivity. In addition to, it can also lead to rise in the number of cases of asthma and other cardiovascular respiratory diseases. Consequently, many lung conditions, such as chronic obstructive pulmonary disease, pneumothorax, and pediatric respiratory infections, may be directly or indirectly affected by climate change as well as air pollution. Moreover, there are signs that air pollution and tuberculosis are related. Like ozone gas, warmer air temperature also increases the regional distribution of aeroallergens, Enweasor et al. [19].

Weather, relative humidity, rainfall (and drought), and pollution are the factors that affects the various respiratory infections. He et al. [23] developed a model that shows that seasonal factors including, climate zone, sunlight, wind speed, rainfall, relative humidity, and temperature affects the rates of viral infections. Additionally, a reduction in related infections influenced by climate change has also been observed, including a decrease in tuberculosis rates with rising heat and a shortening of the season for respiratory syncytial virus with increasing heat.

4.2.3 Mental Health

Extreme weather events such as tropical storms, tornadoes, floods, fires, droughts, tsunamis, etc., affect directly and indirectly human mental health conditions, such as anxiety, post-traumatic stress, and depression. Due to these weather conditions, the loss of homes, livelihoods, and communities affects human mental health for a long time. Extreme droughts are one of the most common weather conditions that occur in various regions of the world. This leads to drop in agricultural productivity, is expected to result in hunger, anxiety, and sadness. As a consequence, there are reports that suicide rates are certainly increasing with drought, particularly among farmers [24].

4.2.4 Cancerous Diseases

Climate change and rising temperatures may have a direct and indirect potential effect on humans and other organisms regarding cancer. However, most importantly increased exposure to carcinogenic toxic chemicals gets augmented due to heavy rainfall and volatilization of chemicals due to rise of global warming. Cancer causing pollutants also gets released into the oceans and air due to melting of glaciers and ice sheets. Increasing solar ultraviolet (UV) radiation also affects the human immune system that impacts the ability of body to eliminate first mutated cells that responsible for cancer processes. In addition, as the air quality deteriorates with toxic air pollutants such as particulate matter, nitrogen oxides, benzene, and heavy metals, prolonged exposure to human breathing can also increase the risk of lung cancer [25].

4.2.5 Cardiovascular Diseases

Due to climate change, the risk of heart diseases is increasing directly and indirectly through air pollution and changes in dietary choices. When the human body does not receive adequate fluids, several physical responses occur due to exposure to heat, such as an increase in body temperature and heart rate, a transfer of blood flow from the central organs to the skin, an increase in sweat rate, and other related dehydration disorders. In addition to, extreme cold and extreme heat are also responsible for the increase in the number of patients in hospitals with irregular heartbeat, stroke, chest pain, and other heart diseases. Many studies show the combined effects of heat exposure and air pollution on heart diseases in megacities during hot weather [3].

5 Modern Techniques to Mitigate the Impact of Air Pollution on Climate Change

As time has passed on, technological advancement has become increasingly important in preventing climate change and air pollution. There are some modern advance techniques like artificial intelligence (AI), machine learning (ML) and remote sensing that are play a key role in mitigation, prediction of air pollution to managing climate change.

5.1 Role of AI and Machine Learning

Various AI and ML techniques such as Artificial Neural Networks (ANN), fuzzy logic, support vector machines (SVM), convolutional neural networks (CNN), recurrent neural networks (RNN), long short-term memory (LSTM), convolutional autoencoders (CA), and etc., that are evolving over time, have been commonly used not only to predict Earth and atmospheric variables (such as air pollutants, rainfall, wind direction, wind speed, etc.) in previous studies but also in present study important for the management of air pollution. Several studies reported in successfully implementation of AI technique to forecast and mitigate the air pollution as well as climate change [29]. Gwalior and others studied various weather factors and traffic data from 2019 to 2023, along with air pollutants (such as CO_2, NH_3, NO, NO_2, O_3, PM_1, $PM_{2.5}$, PM_{10}, etc.) to predict the concentration of air pollutants using ten different ANN models, and also analyzed the effects of SO_2 using this data. This model can effectively determine non-linear interactions between features, allowing them to predict air pollutant level and illustrate its relationship with weather factors.

There are number of these innovative techniques that are applied in the carbon capture and sequestration technique, improvement in energy sector to increase energy efficiency and reduce carbon emission and managing the transport system to mitigate greenhouse gases and improve transport vehicles. For instance, Algorithms of artificial intelligence such as ANN, Genetic Algorithms (GA), Simulated Annealing (SA), Artificial Immune Systems (AIS), Ant Colony Optimizer (ACO), and Bee Colony Optimization (BCO) along with fuzzy logic models, can identify opportunities By using this detailed data on traffic patterns, passenger demand, and weather conditions, emissions can be reduced in transportation systems and efficiency enhanced, that can contribute to the reduction of greenhouse gas emissions and the advancement towards a sustainable transport sector. In addition to various application of AI such as in optimization of energy efficient industrial process at different stage and lower emission technique, and managing the natural resource (like forest) to minimizing deforestation and reducing the air pollutants and greenhouse gases emissions [11].

5.2 Role of Remote Sensing and Geographical Information System (GIS)

The combination of remote sensing with GIS applications are important modern technique for managing air pollution and climate change, using satellite data in spatial analysis and decision-making processes. Remote sensing provides data on pollution levels, atmospheric conditions, and land cover changes, that is integrated into a GIS platform for assessment and imagining. GIS plays an vital role in air pollution prevention, mapping of area, and analysis through spatial analysis tools and visualization capabilities. Along with air pollution management, GIS assists in classifying pollution sources (greenhouse gases emission sources), enabling targeted prevention efforts while mapping sources such as industrial units, traffic congestion points, and agricultural activities. It creates real-time air quality maps by integrating data from air quality monitoring stations, satellite imagery, and meteorological data, and it also tracks changes over time by identifying pollution hotspots. In addition to, GIS evaluates the effectiveness of air pollution control policies, that helps policymakers make decisions and identify areas that require additional measures [54].

6 Conclusion

Climate change and air pollution poses severe issues on human beings and other living organisms around the globe. Anthropogenic activities are the major factor responsible for the increase in the concentration of air pollution and climatic change. The identification and treatment of human originated problem though possible techniques require enormous amount of human labour, economic cost, collective responsibility and systemic changes. There are many national and international agency that are promoting political, technological and nature-based solution initiatives to address this important issue. This book chapter emphasizes the air pollution, types of air pollutants and their role in climate change. Moreover, this book chapter elaborates the impact of air pollution on climate change and factors responsible for enhancement of climate change as well as the impact of climate change on human health. The role of modern technologies like AI, remote sensing and GIS plays a key role in managing climate change are also discussed.

References

1. Aayush, K., Vishal, D., Hammad, N., Manu, K.S.: Application of artificial intelligence in curbing air pollution: the case of India. Asian J. Manag. **11**(3), 285–290 (2020)
2. Agarwal, S., Sharma, S., Rahman, M.H., Vranckx, S., Maiheu, B., Blyth, L., Batra, S.: Air quality forecasting using artificial neural networks with real time dynamic error correction in highly polluted regions. Sci. Total Environ. **735**, 139454 (2020)
3. Alahmad, B., Khraishah, H., Althalji, K., Borchert, W., Al-Mulla, F., Koutrakis, P.: Connections between air pollution, climate change, and cardiovascular health. Can. J. Cardiol. **39**(9), 1182–1190 (2023)
4. Basu, P., Mandal, M., Jana, S., Biswas, A., Ray, S.: Impact of anthropogenic activities on global warming and energy demand. Intersect. Glob. Energy Polit. Clim. Chang. Compr. Anal. Energy Mark. Econ. 63–81 (2025)
5. Bhardwaj, L.K., Singh, V.V., Dwivedi, K., Rai, S.: A comprehensive review on sources, types, impact and challenges of air pollution. Int. J. Sustain. Energy Environ. **1**, 1 (2023)
6. Brauer, M., Guttikunda, S.K., Dey, S., Tripathi, S.N., Weagle, C., Martin, R.V.: Examination of monitoring approaches for ambient air pollution: a case study for India. Atmos. Environ. **216**, 116940 (2019)
7. Burbank, A. J., Penrice, A. J., Rorie, A. C., Oh, J. W.: Climate change and allergens: current and future impacts. J. Allergy Clin. Immunol.: In Practice. (2025)
8. Casanueva, A., Burgstall, A., Kotlarski, S., Messeri, A., Morabito, M., Flouris, A.D., Schwierz, C.: Overview of existing heat-health warning systems in Europe. Int. J. Environ. Res. Public Health **16**(15), 2657 (2019)
9. Cetin, M., Sümerli Sarıgül, S., Danish.: The green growth-CO2 emissions link in G7 countries: the role of energy productivity. Int. J. Environ. Sci. Technol. **22**(7), 6095–6108 (2025)

10. Chakraborty, S. et al.: Extending the atmospheric river concept to aerosols: climate and air quality impacts. Geophys. Res. Lett. **48**(9), e2020GL091827 (2021)
11. Chen, L., Chen, Z., Zhang, Y., Liu, Y., Osman, A.I., Farghali, M., Yap, P.S.: Artificial intelligence-based solutions for climate change: a review. Environ. Chem. Lett. **21**(5), 2525–2557 (2023)
12. Chikezie, F. M., Opara, K. N., Ubulom, P. M. E.: Impacts of changing climate on arthropod vectors and diseases transmission. Niger. J. Entomol. **40**, 179–192 (2024)
13. Colón-González, F.J., Fezzi, C., Lake, I.R., Hunter, P.R.: The effects of weather and climate change on dengue. PLoS Negl. Trop. Dis. **7**(11), e2503 (2013)
14. Cramer, M.N., Gagnon, D., Laitano, O., Crandall, C.G.: Human temperature regulation under heat stress in health, disease, and injury. Physiol. Rev. (2022)
15. Davenport, T., Guha, A., Grewal, D., Bressgott, T.: How artificial intelligence will change the future of marketing. J. Acad. Mark. Sci. **48**, 24–42 (2020)
16. De Vita, A., Belmusto, A., Di Perna, F., Tremamunno, S., De Matteis, G., Franceschi, F., Covino, M.: The impact of climate change and extreme weather conditions on cardiovascular health and acute cardiovascular diseases. J. Clin. Med. **13**(3), 759 (2024)
17. Ebi, K. L., Capon, A., Berry, P., Broderick, C., de Dear, R., Havenith, G., Jay, O., et al.: Hot weather and heat extremes: health risks. The lancet. **398**(10301), 698–708 (2021)
18. Edo, G.I., Itoje-akpokiniovo, L.O., Obasohan, P., Ikpekoro, V.O., Samuel, P.O., Jikah, A.N., Agbo, J.J.: Impact of environmental pollution from human activities on water, air quality and climate change. Ecol. Front. (2024)
19. Enweasor, C., Flayer, C.H., Haczku, A.: Ozone-induced oxidative stress, neutrophilic airway inflammation, and glucocorticoid resistance in asthma. Front. Immunol. **12**, 631092 (2021)
20. Goulier, L., Paas, B., Ehrnsperger, L., Klemm, O.: Modelling of urban air pollutant concentrations with artificial neural networks using novel input variables. Int. J. Environ. Res. Public Health **17**(6), 2025 (2020)
21. Gulia, S., Shukla, N., Padhi, L., Bosu, P., Goyal, S.K., Kumar, R.: Evolution of air pollution management policies and related research in India. Environm. Chall. **6**, 100431 (2022)
22. Hayes, R.B., Lim, C., Zhang, Y., Cromar, K., Shao, Y., Reynolds, H.R., Thurston, G.D.: PM2. 5 air pollution and cause-specific cardiovascular disease mortality. Int. J. Epidemiol. **49**(1), 25–35 (2020)
23. He, Y., Liu, W.J., Jia, N., Richardson, S., Huang, C.: Viral respiratory infections in a rapidly changing climate: the need to prepare for the next pandemic. EBio Med. **93** (2023). Chaw, L., Liew, S.Q., Wong, J.: Association between climate variables and pulmonary tuberculosis incidence in Brunei Darussalam. Sci. Rep. **12**(1), 8775 (2022)
24. Heanoy, E.Z., Brown, N.R.: Impact of natural disasters on mental health: evidence and implications. In: Healthcare, vol. 12, no. 18, p. 1812. MDPI (2024)
25. Hiatt, R.A., Beyeler, N.: Cancer and climate change. Lancet Oncol. **21**(11), e519–e527 (2020)
26. Karatzas, K.D., Kaltsatos, S.: Air pollution modelling with the aid of computational intelligence methods in Thessaloniki, Greece. Simul. Model. Pract. Theory **15**(10), 1310–1319 (2007)
27. Kim, K.H., Kabir, E., Ara Jahan, S.: A review of the consequences of global climate change on human health. J. Environ. Sci. Health C **32**(3), 299–318 (2014)
28. Kumari, S., Verma, N., Lakhani, A., Kumari, K.M.: Severe haze events in the indo-gangetic plain during post-monsoon: synergetic effect of synoptic meteorology and crop residue burning emission. Sci. Total Environ. **768**, 145479 (2021)
29. Kumari, S., Choudhury, A., Karki, P., Simon, M., Chowdhry, J., Nandra, A., Garg, M.C.: Next-generation air quality management: unveiling advanced techniques for monitoring and controlling pollution. Aerosol Sci. Eng. 1–22 (2025)
30. Lee, H., Calvin, K., Dasgupta, D., Krinner, G., Mukherji, A., Thorne, P., Zommers, Z.: Climate change 2023: synthesis report. Contribution of working groups I, II and III to the sixth assessment report of the intergovernmental panel on climate change. The Australian National University (2023)
31. Manabe, S.: Role of greenhouse gas in climate change. Tellus A Dyn. Meteorol. Ocean. **71**(1), 1620078 (2019)

32. Manavi, S.E., Aktypis, A., Siouti, E., Skyllakou, K., Myriokefalitakis, S., Kanakidou, M., Pandis, S.N.: Atmospheric aerosol spatial variability: impacts on air quality and climate change. One Earth **8**(3) (2025)
33. Manisalidis, I., Stavropoulou, E., Stavropoulos, A., Bezirtzoglou, E.: Environmental and health impacts of air pollution: a review. Front. Public Health **8**, 14 (2020)
34. Mannucci, P.M., Franchini, M.: Health effects of ambient air pollution in developing countries. Int. J. Environ. Res. Public Health **14**(9), 1048 (2017)
35. Masood, A., Ahmad, K.: A review on emerging artificial intelligence (AI) techniques for air pollution forecasting: fundamentals, application and performance. J. Clean. Prod. **322**, 129072 (2021)
36. Melas, D., Kioutsioukis, I., Ziomas, I.C.: Neural network model for predicting peak photochemical pollutant levels. J. Air Waste Manag. Assoc. **50**(4), 495–501 (2000)
37. Mo, X., Zhang, L., Li, H., Qu, Z.: A novel air quality early-warning system based on artificial intelligence. Int. J. Environ. Res. Public Health **16**(19), 3505 (2019)
38. Myhre, G., Boucher, O., Bréon, F.M., Forster, P., Shindell, D.: Declining uncertainty in transient climate response as CO2 forcing dominates future climate change. Nat. Geosci. **8**(3), 181–185 (2015)
39. Ofremu, G.O., Raimi, B.Y., Yusuf, S.O., Dziwornu, B.A., Nnabuife, S.G., Eze, A.M., Nnajiofor, C.A.: Exploring the relationship between climate change, air pollutants and human health: impacts, adaptation, and mitigation strategies. Green Energy Resour 100074 (2024)
40. Pandey, A., Brauer, M., Cropper, M.L., Balakrishnan, K., Mathur, P., Dey, S., Dandona, L.: Health and economic impact of air pollution in the states of India: the Global Burden of Disease Study 2019. Lancet Planet. Health **5**(1), e25–e38 (2021)
41. Rataj, E., Kunzweiler, K., Garthus-Niegel, S.: Extreme weather events in developing countries and related injuries and mental health disorders-a systematic review. BMC public health. **16**(1), 1020 (2016)
42. Rautela, K.S., Singh, S., Goyal, M.K.: Resilience to air pollution: a novel approach for detecting and predicting aerosol atmospheric rivers within earth system boundaries. Earth Syst. Environ. 1–23 (2024)
43. Ravindiran, G., Rajamanickam, S., Kanagarathinam, K., Hayder, G., Janardhan, G., Arunkumar, P., Muniasamy, S.K.: Impact of air pollutants on climate change and prediction of air quality index using machine learning models. Environ. Res. **239**, 117354 (2023)
44. Rizwan, S.A., Nongkynrih, B., Gupta, S.K.: Air pollution in Delhi: its magnitude and effects on health. Indian J. Commun. Med. **38**(1), 4–8 (2013)
45. Samadi, A.: Impacts of climate change on vector-borne diseases of animals and humans with special emphasis on Afghanistan: a review. J. Nat. Sci. Rev. **2**(1), 1–20 (2024)
46. Shakya, D., Deshpande, V., Goyal, M.K., Agarwal, M.: PM2. 5 air pollution prediction through deep learning using meteorological, vehicular, and emission data: a case study of New Delhi, India. J. Clean. Prod. **427**, 139278 (2023)
47. Sharif, F., Shahzad, L., Batool, M.: The association between climatic factors and waterborne infectious outbreaks with a focus on vulnerability in Pakistan: integrative review. Int. J. Environ. Health Res. **34**(9), 3299–3316 (2024)
48. Sharma, R., Kurmi, O.P., Hariprasad, P., Tyagi, S.K.: Health implications due to exposure to fine and ultra-fine particulate matters: a short review. Int. J. Ambient Energy **45**(1), 2314256 (2024)
49. Sharma, M., Dikshit, O.: Comprehensive study on air pollution and green house gases in Delhi. Submitted to Department of Environment, Government of NCT of Delhi and Delhi Pollution Control Committee (2016)
50. Sheikh, H.I., Najiah, M., Fadhlina, A., Laith, A.A., Nor, M.M., Jalal, K.C.A., Kasan, N.A.: Temperature upshift mostly but not always enhances the growth of Vibrio species: a systematic review. Front. Mar. Sci. **9**, 959830 (2022)
51. Sivakumar, B., Kurian, G.A.: Mitochondria and traffic-related air pollution linked coronary artery calcification: exploring the missing link. Rev. Environ. Health **36**(4), 545–563 (2021)

52. Subramaniam, S., Raju, N., Ganesan, A., Rajavel, N., Chenniappan, M., Prakash, C., Dixit, S.: Artificial intelligence technologies for forecasting air pollution and human health: a narrative review. Sustainability **14**(16), 9951 (2022)
53. Tao, T., Shi, Y., Gilbert, K.M., Liu, X.: Spatiotemporal variations of air pollutants based on ground observation and emission sources over 19 Chinese urban agglomerations during 2015–2019. Sci. Rep. **12**(1), 4293 (2022)
54. Tikader, M., Mukhopadhyay, D., Dabhadker, K.: Remote sensing and GIS in air pollution mitigation: a bibliometric review of Chhattisgarh, India. Int. J. Environ. Clim. Chang. **14**(4), 174–193 (2024)
55. Tripathi, A., Jeswani, P., Shukla, R., Patel, V.: Unveiling climate distress: Examining the psychological impact of adverse weather events in India–A systematic review. Int. J. Soc. Psychiatry. (2025). https://doi.org/10.1177/00207640251362911
56. UN DESA SHPG J.W.: Two "population billionaires", China and India, face divergent demographic futures (2023)
57. Villanueva, F., Notario, A., Tapia, A., Albaladejo, J., Cabañas, B., Martínez, E.: Ambient levels of volatile organic compounds and criteria pollutants in the most industrialized area of central Iberian Peninsula: intercomparison with an urban site. Environ. Technol. **37**(8), 983–996 (2016)
58. Wang, R., Wang, H., Zhou, H., Wang, Y., Yue, J., Diao, B., Kan, B.: Characters of homogentisate oxygenase gene mutation and high clonality of the natural pigment-producing Vibrio cholerae strains. BMC Microbiol. **11**, 1–10 (2011)
59. Yang, X., Wang, Q., Liu, L., Tian, J., Xie, H., Wang, L., Ho, S.S.H.: Impacts of emission reduction and meteorological conditions on air quality improvement from 2016 to 2020 in the Northeast Plain, China. J. Environ. Sci. **151**, 484–496 (2025)
60. Yaseen, A. A., Khaleel, N. T., Al-Azzami, A. A., Aldossary, A. T. Y., Ameen, R. A.: Climate change and its effect on nutritional value: a review. In: IOP conference series: Earth Environ. Sci. IOP Publishing. **1449**(1), 012163 (2025)
61. Zacharis, E., Sempos, I.: Effect of meteorological parameters of atmospheric pollution and possible correlation with climate change. In: 1st International conference on environmental design (ICED2020), pp. 24–25 (2020)
62. Zavala, M.D., Cejas, C., Rubinstein, A., Lopez, A.: Gender inequities in the impact of climate change on health: a scoping review. Int. J. Environ. Res. Public Health **21**(8), 1093 (2024)
63. Zhang, X., Wang, Y., Zhang, W., Wang, B., Zhao, Z., Ma, N., Zhang, X.: The effect of temperature on infectious diarrhea disease: a systematic review. Heliyon **10**(11) (2024)
64. Zhou, Y., Zhao, X., Lin, K.P., Wang, C.H., Li, L.: A Gaussian process mixture model-based hard-cut iterative learning algorithm for air quality prediction. Appl. Soft Comput. **85**, 105789 (2019)

Development of Environmental Decision Support System for Climate Change Resilience in Urban Planning: A Managerial Perspective

Revati Ramrao Rautrao, Amol Gawande, Atul Kumar, Sagar Jadhav, and Asha Kiran Sikhakolli

Abstract Many effects of climate change make urban planning more complicated. An environmental decision support system (EDSS), a dynamic and integrated structure that combines human resource (HR) intervention, geographical information system (GIS), artificial intelligence (AI), and strategic management concepts, is introduced in this chapter. Data-driven risk assessment, landscape simulation, and long-term plans for sustainable development are made easier by using EDSS, which aims to help company officials, legislators, and municipal planners. The importance of human resources (HR) is emphasized in promoting climate flexibility. HR practice is necessary for environmental compliance and leadership development, skills upgrading and implementation of climate companies by promoting flexible workforce methods. EDSS encourages the participating government through interactive simulation that includes stakeholders in group decision-making. This method makes it possible to develop flexible urban strategies jointly that complement environmental

R. R. Rautrao (✉) · A. K. Sikhakolli
Dr. D. Y. Patil B-School, Pune, Maharashtra, India
e-mail: revatirautrao@gmail.com; revati.rrautrao@dpu.edu.in

A. K. Sikhakolli
e-mail: ashakiran45@rediffmail.com

A. Gawande
Dr. D. Y. Patil B-School, Sr. No. 87-88, Bengaluru-Mumbai Express Bypass, Tathawade, Pune, Maharashtra, India
e-mail: amol.gawande@dpu.edu.in

S. Jadhav
Vertical Lead, D Y Patil University School of Management SEAMEDU, SOM Ajeenkya, Pune, India

A. Kumar
Registrar, Dnyaan Prasad Global University, Pune, India
e-mail: atulk.singh@yahoo.co.in

A. L. Srivastav et al. (eds.), *Climate Resilience: Impact of Quantum Computing and Artificial Intelligence on Urban Planning*, Studies in Systems, Decision and Control 637, https://doi.org/10.1007/978-3-032-06791-3_18

rules and organizational goals. Case-based applications show how the system is relevant and scalable in a variety of urban surroundings and provide customized solutions according to local requirements. Integration of human capital with state-of-the-art technology guarantees a working group that is sensitive to climate change and improves institutional flexibility and long-term viability. All things are believed to have a progressive approach to EDS's flexible, inclusive and implementable environmental regime. The framework gives stakeholders the opportunity to create a comprehensive and flexible solution for climate problems by combining technological innovation with management knowledge.

Keywords Environmental decision support system (EDSS) · Climate change resilience · Sustainable leadership · Artificial intelligence · Sustainable development

1 Introduction

Environmental decision support system with moral aspects when using environmental decision-making systems (EDSS) To improve climate change's flexibility in urban planning, moral ideas should be regarded as carefully as such data, security, and prejudice. Maintaining public confidence and following international data protection standards, such as the General Data Protection Regulation (GDPR) in the EU (GDPR), depends mainly on the protection and oblivion of individual and geographical information acquired for urban planning initiatives [45]. In addition, prejudice involved in algorithm training or data collection procedures can cause incorrect allocation of resources, which may affect the deprived groups [29]. To remove these concerns [28], it is important to include transparent AI-powered decision-making processes with prejudice detection systems. Justice, responsibility, and inclusion are reinforced by the creation of a moral management structure and active input partnerships, which allows you to implement EDS morally to support permanent urban flexibility methods.

Climate Resilience via the Inclusion of Environmental Decision Support Systems: By strengthening city resilience via the creation of inexperienced infrastructure, such as city forests and connected blue-green regions, nature-based solutions (NBS) are crucial in decreasing the effects of climate change [33]. This study builds this insight and advocates to adoption of an environmental decision-making system (EDSS) that integrates strategic management measures, participation decisions and advanced geopolitical technology to guide the evidence-based scheme. Effective human resource management plays an important role in promoting leadership cultivation, and competence and promoting capacity creation to ensure successful implementation of permanent urban planning practice [40].

The integration of a Digital Decision Support System (DSS), which makes use of geopolitical modelling, future indication analysis, and real-time environmental information to increase urban flexibility, is important to reduce these risks. These structures play a crucial function in providing selection-making assets that they need to expand towns which are weather-resilient [39]. Sea level and sea stylization pose a special hazard to coastal ecosystems, together with mangroves, coral reefs and overdraft, as they cause attrition and fall [6] of the habitat. Similarly, mountain species are compelled to move to high altitudes, in which they frequently stumble upon incorrect land use conditions that increase the danger of extinction [5].

Lack of Integration Between Quantum Computing, AI, and Urban Planning: The absence of integration among AI, urban planning equipment, and quantum computing with the rapid increase of synthetic intelligence (AI) and quantum computing (QC), their use in urban planning devices remains mainly inexplicable. Traditional plan frameworks regularly fall short of managing the dynamic nature of urban problems and treating the complexities of data related to weather, resulting in ineffective decision-making [40]. Although Quantum Computing has an innovative capability to enhance the analysis of the facts and clear up calculation problems in environmental selections, the use of it in the Flexibility scheme continues to be surprisingly unexplained. The absence of an integrated shape that provides environmental choice support structures (EDSS), synthetic intelligence (AI), and excellent control (QC), prevents sound improvement, and proof-based plans for climate adaptation. To broaden clever, durable and weather-bendy city landscapes, this critical research vacuum should be filled.

Challenges in Urban Planning: Rapid urbanization in cities makes planning difficult, weather alternate, and increasing worry about decision-making methods present city planners with past problems. Urbanization usually takes over the development of the necessary infrastructure, main to inadequate resource control, environmental damage and increasing socio-financial inequalities. The effects of weather alternations, such as intense weather occasions, rising temperatures, and an increase in flood risk, grow these problems even more and increase serious risks of city flexibility [22]. According to Gomez et al. [15], conventional city planning techniques regularly come all the way down to integrating extraordinary records units and adjusting more stakeholders' inputs, resulting in a fragmented and negative decision-making process. Thus, the requirement for date-pushed and adaptable strategies has emerged as more pronounced. An in-depth disposition to resolve those problems is added by imposing an environmental decision-making device (EDSS), which promotes flexibility through integrated ground plans and lets in permanent city improvement.

Natural-Based Solutions (NBS) To remedy those problems, the implementation of an environmental choice-making gadget (EDSS) presents a whole framework that promotes flexibility through integrated ground plans and allows for permanent urban development. NBS gives multicultural, affordable blessings that guide long-term sustainability and improve public welfare as opposed to standard grey infrastructure. Thus, it is vital to encompass NB in the city making plans to promote custom-designed urban settings and assure flexibility in metropolitan regions that are rapidly urbanized.

Technological Integration: Technologies improve status awareness by combining geophysical, hydrological and meteorological data, which can then help urban planners and decision-makers to create effective interventions. The decision-making process is streamlined by AI-controlled simulation and geographical mapping, which improves flexibility and planning accuracy. Combining those technological improvements encourages the creation of dynamic, adaptable city landscapes which can bear the changing problems introduced forth through weather exchange whilst additionally helping sustainable urban expansion.

Addressing Vulnerabilities and Expanding Capabilities: Urban regions are extra vulnerable to weather change, including lack of resources, excessive weather activities, and environmental decline [49]. This study recommends developing an environmental decision assist system (EDSS) that mixes artificial intelligence (AI), Quantum Computing (QC) and geopolitical evaluation to allow data-driven city planning to stop this difference [46]. The city regions are riskier for the negative results of climate change, inclusive of a loss of assets, excessive weather and environmental decline [49]. Due to the complicated nature of urban ecosystems, new technical solutions can accelerate selections and assure stability [16]. This examination shows developing an environmental choice-making tool (EDSS) as devices a geopolitical evaluation, quantum calculation (QC) and artificial intelligence (AI) to facilitate records-driven town-making plans and retaliate against this inequality [46]. The proposed method desires to offer practical insight to metropolis planners with real-time environmental datasets, to improve the efficiency of decision-making, adapt beneficial aid allocation and increase flexibility [40]. This has a look that makes enough contribution to a sustainable city regime with the aid of merging the domain -Art -art calculation gadget with participation coverage shape, which offers choices on choice-making techniques and urban planners to handle accurate and inclusive hard issues.

2 Environmental Decision Support System for Flexibility in Climate Change

One of the most important methods for promoting flexibility in climate change in urban planning is to use the environmental decision-making system (EDSS). An EDSS provides practical insight through environmental information, future climate models and urban planning parameters. This allows the development of policies that support climate-flexible urban ecosystems. From a management point of view, a collaborative structure with multiple voices is necessary for the successful implementation of such systems, including politicians, environmental researchers, data analysts, and urban planners.

The key considerations include the following:

a. **Data Source Selection and Integration for Environmental Decision Support Systems**: For the coverage of urban planning, whose goal is to enforce climate flexibility, it is far vital to integrate some of the dependable data assets. Leaders play an important function in this manner, as they're accountable for detecting, assessing and integrating facts from many sources, including satellite imaging, Internet of Things (IoT) sensors, city increase journals and climate stations. First, all combined environmental choice-assisting structures (EDSS), such as statistics fragmentation, earnings and scale need to be launched [4].

b. **Predictive Model Development for Climate Resilience in Urban Planning**: The advent of destiny fashions for climate flexibility in city design is a critical aspect of a prediction version for weather-flexible urban layout. These fashions use gadget learning strategies, geophysical analysis and historical climate statistics, which expect essential variables along with temperature fluctuations, rainfall and severe climate events [18].

c. **User-Friendly Interface Design for Environmental Decision Support Systems** The efficiency of an environmental decision support system (EDSS) is largely dependent on its ability to close the difference between complex climate data and useful insights, which is made possible by an easy-to-use interface. By using interactive dashboards, and visualization of real-time and simplified data, the manager can help develop systems that put the user experience first. Urban planners, legislators, and environmental researchers, including different stakeholders, can interact more successfully when it has a well-designed interface, which improves access. Cloud-based spatial analysis systems such as Cartos, which allow users to create interactive maps and dashboards to see complex environmental information, are examples of useful tools. Similarly, Microsoft Azure Stream Analytics makes it easy to handle and analyze real-time data, making it possible to create dynamic and adaptable interfaces [47]. By securing systems targeted at many devices, adaptable design concepts are used, which improves engagement and operating efficiency [42]. With an emphasis on a user-centred design approach, the EDSS spread of climate-related knowledge improves and strengthens decision-makers and urban planners to create a strategic solution for climate-related problems. As a result, urban systems become more flexible and sustainable growth is encouraged.

d. **Data Quality Assurance in Environmental Decision Support Systems**: The insurance on data quality in environmental decision support systems. Completely, data quality assurance procedures are necessary to maintain the integrity of the Environmental Decision Systems (EDSS), which act as a basis for well-informed urban planning options that prefer climate flexibility. To solve important concerns, including purity, stability and addiction, perfect installation of data quality procedures is required [27]. To maintain the reliability of the system and sell self-assurance to the stakeholders, the chief is essential for the implementation [17]. The manager can guarantee the accuracy and action of the EDSS output, by providing a strong emphasis on strict data verification and compliance with installed processes. According to Hedebíček et al. [17], this strategy contributes to sufficient flexible urban infrastructure, which strengthens the metropolitan areas to bear the complex problems that climate change has brought. In addition to improving the decision, emphasizing data quality forms the basis for long-term stability and flexible urban planning techniques.

e. **System Scalability in Environmental Decision Support Systems**: Techniques are further improved using artificial intelligence (AI), which allows a variety of datasets to be handled, increasing accuracy and optimizing the destruction response plans [43]. In addition, scalable EDS makes it easier to include condition-art functions such as modelling and real-time data analysis to handle new urban planning problems and climate flexibility [35]. Leaders can guarantee that EDSS remains an adaptable and useful tool that can promote permanent urban growth in time with increasing climate threats with a strong emphasis on scalability. This flexibility allows politicians and urban planners to change and meet the environment and the infrastructure.
f. **Adaptive Evolution: Continuous Updates in Environmental Decision Support Systems. An** Environmental decision support system (EDSS) is necessary to preserve its efficiency and adaptability through continuous updates. These changes are necessary to guarantee that EDSS takes into account the latest progress in urban planning and climatic conditions, including the latest released datasets, which include up-to-date weather models and the latest adjustments in city infrastructure may also grow the accuracy of predictions and facilitate better decision-making. Leaders can assure the delivery of applicable insights of time via being dedicated to continuous improvement, to fortify the metropolis's flexibility in opposition to weather change and sell perpetual growth.
g. **Stakeholder Engagement in Environmental Decision Support Systems** To succeed in growing and enforcing environmental decision support systems (EDSS), important stakeholders should be actively involved. Through this sort of partnership, it is ensured that EDS is customized to satisfy the specific requirements and objectives of weather-flexible city layout. Including a huge variety of stakeholders, which include legislators, social leaders, urban planners, and environmental experts, promotes an extensive understanding of local problems and goals, resulting in greater sensible and extensively supported solutions. By embracing a chain of processes, encouraging openness and growing a sense of possession among most of the final users, the ownership interest improves the engagement choice Reed et al. [34]. Inclusion of stakeholder-driven initiatives ensures that social-focused policy implementation promotes inclusive, climate-flexible and sustainable urban development.
h. **Training and Capacity Building for Effective EDSS Utilization**: To effectively use environmental decision support systems (EDSS) in the Urban Climate Flexibility Scheme, extensive training and capacity programs are necessary. Providing important stakeholders providers, urban planners and environmental analysts, the EDSS technologies are used effectively and understand that the system promotes adoption and makes it easier to make data courses. To improve data analysis, geographical mapping and future modelling methods, training programs must include workshops, simulation exercises and continuous business development sessions [7]. Registration-specific, targeted training is especially important to guarantee that several user groups can successfully use EDSS functions [2]. In

addition, frequent capacity programs provide flexibility in response to the problems of climate change, guaranteeing that urban planning is strong and flexible. Exercise programs that include response mechanisms improve user participation and offer real information that can be used to improve EDs for implementing real-world climate adaptation. Finally, spending money on training and capacity creation creates a working group with the skills required to fully use EDS for permanent urban development.

i. **Seamless Integration of EDSS with Existing Urban Planning Systems** An Environmental decision support system (EDSS) must be able to interact easily with the current urban planning and environmental management framework to influence the climate flexibility scheme. Compatibility with municipal databases, remote target platforms and geographical information systems (GIS) is necessary to improve interpreting and determining processes. Including EDS in politics and regulatory frameworks, implementation streamlines urban government practice and promotes acceptance and compliance. Associate stakeholder Partnership, Interoperability Protocols and the use of defined data formats guarantee seamless system operations. Through adaptation of resource allocation, improvement of urban policy development and promotion of permanent urban flexibility solutions, a well-integrated EDSS successfully takes the increasing difficulties that arise from climate change.

j. **Ethical Considerations in Environmental Decision Support Systems**: The creation and use of environmental decision support systems (EDSS) should take into account morality to maintain justice, openness and public faith. Confidential concerns require compliance with legal frameworks such as strong guidelines for data protection, implementation of anonymous procedures and general data protection regulation (GDPR) [45]. According to O'Neel [36], prejudice in data collection and future modelling can also create inappropriate political decisions that affect the risk of risk already. Justice, individual algorithms and functional audits should be included in EDSS procedures to detect and solve any prejudice to reduce these moral problems [28].

k. According to moral AI principles and inclusion of stakeholders, promote measures for inclusive and fair urban flexibility, and guarantee responsible data use. Environmental decision support systems (EDSS) should be developed and used with morality to preserve justice, openness and public trust. Due to privacy concerns, legislative structures, including the General Data Protection Regulation (GDPR), distribution of anonymous procedures and strict data security are followed [45]. O'Neel [36] states that prejudice in data collection and future modelling can also lead to unfair political action, which is already at risk. EDSS procedures should include working modification and a personal justice algorithm to identify and address any distortions and reduce these moral problems [28]. Proper measures for responsible data use, and inclusive and fair urban flexibility, including stakeholders, follow the moral AI standards. Leaders can maintain the integrity of climate-flexible urban planning (Fig. 1).

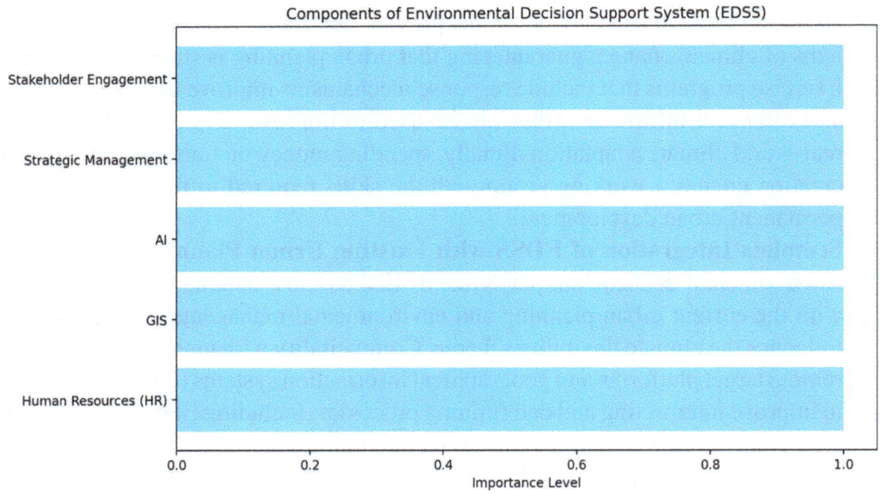

Fig. 1 Conceptual framework of EDSS

This diagram illustrates the key components of the Environmental Decision Support System (EDSS), highlighting the integration of Human Resources (HR), Geographic Information Systems (GIS), Artificial Intelligence (AI), and Strategic Management (Fig. 2).

This diagram outlines the benefits of implementing EDSS in urban planning, emphasizing aspects such as data-driven risk assessment, sustainable development, climate flexibility, stakeholder collaboration, and customized solutions.

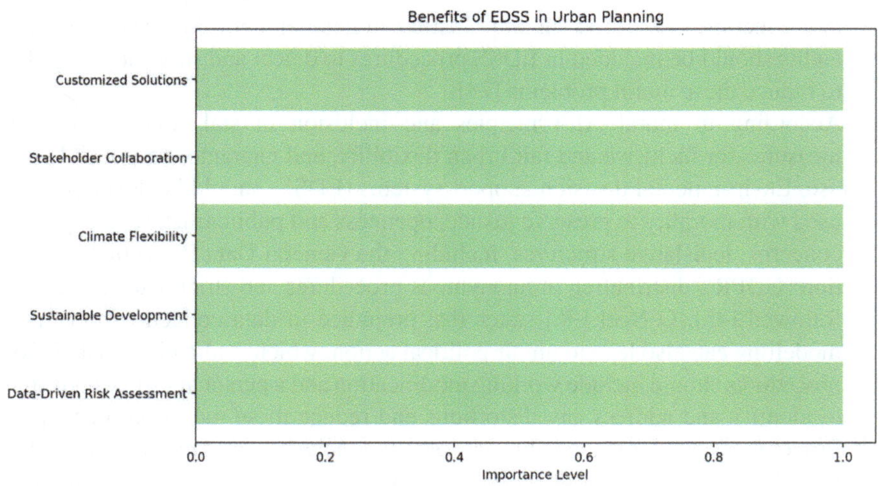

Fig. 2 Benefits of EDSS in urban planning

These diagrams visually represent the framework and benefits of EDSS, providing a clearer understanding of how it can enhance urban planning in the context of climate change resilience.

3 Methodology

3.1 Decision Support Framework Combining Quantum Computing (QC) and Artificial Intelligence (AI)

The inclusion of artificial intelligence (AI) and Quantum Computing (QC) in the Environmental Offering Systems (EDSS) has increased urban planning. Through the real-time agreement with large datasets, the QC climate facilitates rapid processing of complex simulation, and adaptation to climate adaptation options [46]. According to [49], the AI-operated models examine complex climatic trends, urban weaknesses and mitigation strategies to help evidence-based policy. The methodology suggested modelling for the future of Urban climate flexibility by merging the deep teaching algorithm of AI with a remarkable QC [16]. This condition -Ar -species approach makes active city planning and distribution of assets possible by combining the geographical records system (GIS) with actual-time environmental tracking [40]. This framework ensures a sustainable and adaptable city regime, to deal with complex questions introduced using weather change, by supplying decision-makers with a state-of-the-art device to make a climate-resilient city.

(a) **Integrating Quantum Computing (QC) and Artificial Intelligence (AI) within EDSS**: Integration of artificial intelligence (AI) and Quantum Computing (QC) in the Environment Decision Assistance System (EDSS) provides a revolutionary method of flexibility in climate change and urban planning. In parallel, AIS eases refined data processing skills The environmental impact of the environmental impact model, increases the active flexibility strategy [1]. Because of this integration, advanced simulation of urban landscapes is possible and provides powerful equipment that allows assessing potential answers under different climate forgeries. In addition to improving the accuracy of urban planning projects, this structure supports permanent growth activities using state technical solutions to solve complex questions of climate adaptation.

(b) **AI for Predictive Analytics, Stakeholder Engagement, and Scenario Simulations in Urban Planning**: Predictive analysis, stakeholder engagement, and landscape simulation are just a few methods that some artificial intelligence (AI) is used to improve the flexibility of the Urban climate. AI also makes Stake Partnership easier by integrating the framework that guarantees a larger approach to the production of municipal policy.

(c) In addition, dynamic urban scenarios can be created using sophisticated AI-based simulations, which allows decision-makers to evaluate the long-term effects of planning initiatives under different climatic conditions. This smart, computer-driven strategy supports flexibility in governance, improves stability and supports climate adaptation plans in urban growth initiatives.

3.2 Human Resources (HR) and Governance

(a) **Leadership Training and Capacity Building for Advanced Technology Utilization**: The successful integration of environmental decision support systems (EDSS) in the Urban Climate Flexibility scheme is largely dependent on capacity programs and efficient leadership development. The initiative for capacity should focus on improving digital literacy, encouraging creativity, and advocating the moral use of AI to reduce prejudice in the decision-making model. Cities are better equipped to maximize climate adaptation measures, increase stakeholder partnership and promote a flexibility-focused regime when municipal leaders are equipped with state-of-the-art technical skills. Investment in leadership development is an operational requirement, a strategic obligation to use new technology in permanent urban planning.

(b) **Participatory Governance for Stakeholder Collaboration in Climate-Resilient Urban Planning**: Participatory management structures that encourage collaboration between many stakeholders, such as MLA, urban planners, environmental researchers, and social representatives, are required to take effective environmental effects., inclusion and trust are ensured in the development and application of climate flexibility measures by integrating methods for participation in EDSS.

(c) **By combining real-time data analysis**, A-Sapisi landscape modelling and interactive feedback mechanisms, EDSS systems are required to promote dedication to stakeholders. By ensuring that climate adaptation plans are in line with societal goals and socio-economic interests, this law strengthens the long-term viability of politics. Such collaborative decision-making mechanisms are institutionalized and strengthened urban flexibility programs, which build the gap between socially focused policy and technological successes.

(d) **The Proposed Decision Support System**: The increasing intensity of urban climate problems requires creating a sophisticated environmental decision-making system (EDSS) that effectively combines leadership knowledge with state-of-the-art technology to support climate flexibility. Geographical Information Systems (GIS), Artificial Intelligence (AI) and Quantum Computing (QC) are all involved in the multi-level architecture of EDS, proposed to enable all data-driven urban planning. EDSS allows political decision producers to identify sensitive urban areas, evaluate climate risk and create active adaptation plans using real-time environmental data and sophisticated disease. EDSS balances ecological stability and socio-economic imperative by evaluating different urban design roads using interactive simulation and landscape modelling [47].

(e) In addition, the AI-operated influence technique for the influence model is accurate, which allows more efficient resource allocation for climate restrictions.

The Use of quantum calculations improves further data processing performance, leading to rapid analysis of large climate sets and improving the improvement of resistance strategy. EDSS promotes evidence-based decisions by bringing these state-of-the-art techniques together in a standardized administrative structure, proposed, increasing long-term stability in the urban environment. Finally, the system builds the difference between climate research, technology innovation, and urban rule by providing practical information to decision-makers.

3.3 *Architecture of the System*

Quantum Computing (QC), Artificial Intelligence (AI), Geographic Information System (GIS), and participating management mechanisms are all originally united in the architects of the proposed environmental decisions support system (EDSS). To handle climate-sprouted risks, our multi-layer approaches guarantee effective data processing, sophisticated future indicative analysis and stakeholder-driven decision-making.

(a) **Integration of QC and AI**: Through real-time treatment of large datasets related to flood hazards, urban heating islands and climate trends, Quantum Computing (QC) improves calculation skills. In addition to QC, AI algorithms use machine learning to increase these data inputs, which allow climate phenomena to produce prognosis models, allocate resources better and propose adaptable urban solutions. The combination of QC and AI leads to the speed of landscape modelling and simulation, which provides urban planners with useful information to succeed in performing active flexibility measures [13].

(b) **Role of GIS in Spatial Analysis** Geographical information systems (GIS) are important tools for spatial analysis when mapping and assessing environmental tension information, infrastructural weaknesses and land use.

(c) **Participatory Governance Mechanisms**: Partnership to promote the digital platform Stake Partnership for public discussion, cooperative decision-making and policy assessment is integrated into government technologies. By bridging the division between local communities, legislators and climate experts, these technologies promote openness, inclusion and confidence in urban planning processes. Through the integration of these conditions - - - -species technology and management processes for participation, EDSS creates a complete structure to make decisions so that urban planners can create a flexible city ahead of increasing climate security.

3.4 *Features and Functions*

Combined data analysis, landscape-based modelling and human resources (HR) strategies with advanced technologies such as artificial intelligence (AI) and quantum calculation (QC), including the proposed environmental relocation system (EDSS), are specifically designed to improve Urban flexibility. Effective implementation of adaptive schemes, dynamic decision-making and climate flexibility initiatives are guaranteed by all this including function.

(a) **Real-Time Data Analysis and Decision-Making** To provide real-time insights on climate factors, infrastructure weaknesses, and disaster plans, EDS's major environmental dataset is treated using AI-operated analysis. The exact prognosis of rapid simulation and extreme weather phenomena is made possible by Quantum Computing (QC), which increases the calculation efficiency (prescription 2023). IoT-based sensors and remote measurement technologies are continuously improved in the system of continuous monitoring, which allows active intervention and resource management.
(b) **Scenario-Based Modeling for Resilience Planning** Urban employers can assess different climate risk scenarios and create adaptation plans with the use of landscape-based modelling that includes EDSS AI-MANGO models to evaluate how urban systems are affected by an increase in sea level, temperature and heavy rainfall, and they suggest the best ways to ensure long-term flexibility [21]. According to Hail, this data-driven approach promotes evidence-based policy by reducing potential economic and environmental costs.
(c) **Integration of HR Practices with QC and AI**: The system provides a strong emphasis on HR strategy to guarantee successful adoption, planners and decisions are known for effectively using QC and AI-based technologies. In addition to improving stakeholders' involvement, leadership development programs and technical training promote an innovative and computer-driven decision culture [49]. The permanent adoption of EDSS frameworks and ongoing growth is ensured by coordinating technology development with sound HR procedures.

A conceptual diagram for the Environmental Decision Support System (EDSS) focused on climate change resilience:

This diagram illustrates the key components and their relationships, including:

- **Real-Time Data Analysis**
- **Scenario-Based Modeling**
- **Integration of HR Practices**
- **AI & Quantum Computing**
- **IoT Sensors**
- **Stakeholder Engagement** (Fig. 3)

This diagram visually represents how these components interact within the EDSS framework, emphasizing the importance of integrating technology, human resources, and stakeholder engagement to enhance urban resilience against climate change.

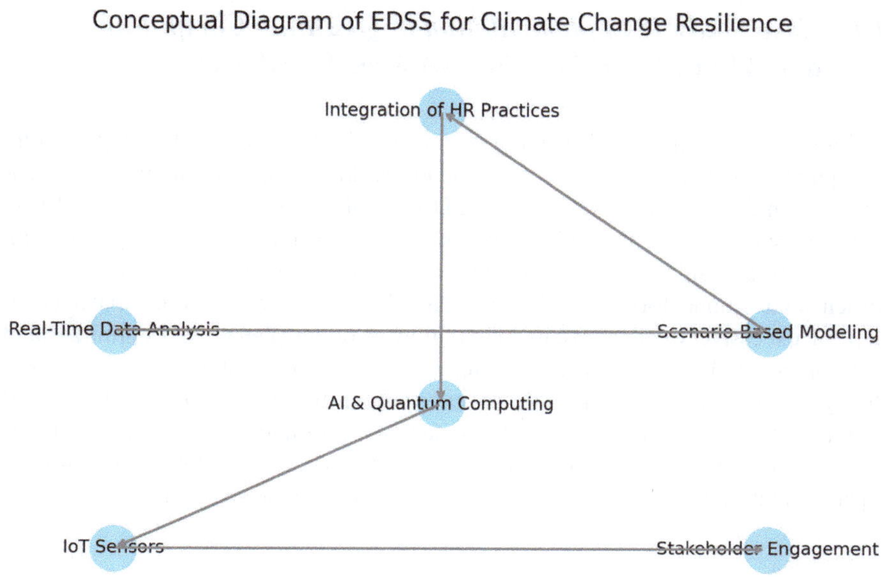

Fig. 3 Conceptual diagram of EDSS for climate resilience

4 Applications and Case Studies

Urban planners, MLAs and other stakeholders can handle the problems of climate flexibility using the Environmental Decision System (EDSS). Applications are useful in areas including a decrease in flood risk, heat prevention, permanent land use and disaster prevention strategy. Limitation of flood risk is such a use of flood risk. For example, the Netherlands has used an AI-based future analysis to evaluate flood-exposed areas, improve urban drainage systems and promote emergency preparedness. To increase water resources, QC-Saxham modelling has also been included in the urban planning framework [38] to increase water resources management, especially in receptive places for monsoon-inspired floods.

The limitation of the heat waves EDSS has played an important role in heat wave limitation in areas such as Los Angeles, where AI simulation has helped with the implementation of methods such as reflecting surfaces to expand the canopy to urban wood and reduce the urban Heat Island effect [14]. The effectiveness of disaster reaction in Japan, common participating management methods with GIS and AI, has improved disaster ability by facilitating the interaction in real-time between residents and local authorities, guaranteeing effective preparedness. These case studies show how well the EDSS framework works to promote climate flexibility by merging state-of-the-art calculation tools. The construction of flexible and future-prepared urban landscapes is aided by EDSS, which allows precise risk assessment and applying permanent treatment.

4.1 *Case Study 1: Implementation of Quantum Computing and AI in Urban Planning—A Real-World Scenario*

To improve urban planning, the merger of artificial intelligence (AI) with Quantum Computing (QC) is shown through a ground-breaking project in Singapore. Singapore Urban Development Authority (SUDA) combined QC methods with AI-Run Decision Support System which combines flood-risk evaluation and energy-efficient infrastructure design by combining QC methods. Through the analysis of high-dimensional climate formation, the QC-enabled model increased the estimated accuracy of excessive weather events. This allowed urban planners to optimize land use policies and create an adaptive stormwater control system [26]. In addition, the system promoted a partnership regime through the integration of AI-operated digital platforms, making it easier for stakeholders to interact with public authorities, business developers and residents [25]. QC and AI's revolutionary ability to create climate-flexible urban landscapes is postponed in the study of the case.

4.2 *Case Study 2: AI-Driven Environmental Sustainability and Stakeholder Collaboration in Amsterdam*

Participation in promoting environmental sustainability is a pioneer in combination with governance and artificial intelligence (AI). The city optimizes the production of green infrastructure and improves urban flexibility by using AI-controlled future analysis through the climate flexible urban planning initiative (Crupi) Platforms for the participation of stakeholders make it possible to work together to develop permanent solutions for local communities, companies, and several urban problems Eini et al. [12]. With the use of future flood risk assessment and better drainage networks, AI-controlled water management technologies have reduced urban flood events by 30%. In addition, analogue action is supported by AI-driven air quality monitoring, which improves public health services. This case study shows how AI can support inclusive regimes by offering solid, long-term urban planning solutions. Qian et al. [32].

4.3 *Insights and Lessons Learned*

Urban Design has a revolutionary ability to improve climate flexibility and stability through the use of artificial intelligence (AI) and quantum calculation (QC). The findings from the case study suggest how AI manuals improve decision-making by providing updated information on environmental hazards [19]. Due to QC's computer power, the landscape-based scheme has accelerated, so the authorities can pre-control to solve climate-related problems. In addition, the attitude to digital platform-competent participating management has improved stakeholder cooperation and promoted inclusion in decision-making [11]. But there are still problems, such as privacy, complex integration and requirements for urban planners to participate in capacity programs an important lesson is the value of multi-scoping collaboration, where societies, technologists, and decision-makers work together to build adaptive solutions. In the future, achieving long-term stability in the urban flexibility scheme will depend a lot on the change in management and technological innovation [31].

5 Results and Discussions

The use of environmental decision support systems (EDSS) that combines artificial intelligence (AI) with Quantum Computing (QC) has shown significant improvement in urban climate flexibility. Conclusions suggest that AI-producing analysis improves the accuracy of climate risk assessment so that urban planners can create preventive mitigation schemes in advance [19]. Due to its unmatched data processing performance, QC has made complex simulations and adapted resource allocation for flexible infrastructure [3]. The integration of the geographical information system (GIS) has improved flood spread and heat management skills by enabling real-time spatial analysis. By encouraging participation from the stakeholders through digital platforms and by incorporating social entrances into the policy framework, the partnership regime further improved these results. However, questions such as data interoperability, moral dilemmas and long training courses are major obstacles. These results mean a hybrid strategy that combines multidisciplinary collaboration, management and technology innovation, is crucial for long-term success with climate flexible urban design. Constant technology development and full political integration should be the top priority of decision-makers to improve the structure to determine permanent urban settings.

5.1 Key Findings

The use of artificial intelligence (AI) and quantum calculation (QC) in urban planning systems has led to significant benefits in flexibility, accuracy, and efficiency. Complex optimization problems are well handled by QC, which enables rapid simulation for infrastructure building and resource allocation [23]. The large simulation of the climate flexibility model that is possible of QC has allowed decision makers to effectively assess different types of urban design techniques Through progressive involvement of stakeholders, participatory governance techniques have promoted inclusive and sustainable city planning techniques. Taken together, our consequences highlight the need for non-stop investment in AI combination models and QC to reinforce the resilience of urban sustainability frameworks. The report promotes a properly rounded method that combines present-day generation talents with governance modifications to attain sustainability and long-term resilience in urban development.

5.2 Comparative Analysis

Due to their dependence on historical data and deterministic models, traditional urban planning techniques often fail to address the complexity of modern urban contexts and the unexpected nature of climate absorption and downs Although techniques such as stable simulation models and geographical information systems (GIS) provide practical information, they are not flexible or calculation -efficient to determine in real-time in rapidly changed environmental conditions. In addition, manual decision-making processes with traditional urban planning are timely and labour-intensive, which quickly limits the ability to respond to the environment. Williams [47].

In contrast, a combination of artificial intelligence (AI) with quantum calculation (QC) allows fast, data-driven decisions, which represents a paradigm-changing change in urban planning. Infrastructure design and modelling of destruction flexibility have a great advantage from QC's skills in solving multidimensional adaptation problems. AI's machine learning and predictive analytical skills together improve urban plans through the real-time analysis of large data sets, enabling active risk assessment and climate adaptation. This comparative examination suggests how QC and AI-based methods offer unrivalled scalability, accuracy and performance, making them essential additives to future city planning systems that decide upon weather flexibility.

5.3 Challenges and Limitations

Despite the progressive potential of synthetic intelligence (AI) and Quantum Computing (QC) in city planning, many boundaries prevent its substantial use. AI-operated urban structures require mass records infrastructure, although it is often not available in underdeveloped international locations [16]. Economic boundaries are still another major obstacle. Implementation of QC and AI-based solutions requires software development, training of personnel, and large upfront expenses for data processing equipment. These expenses can prevent scalability and adoption for narrow-budget cities. Important obstacles are also presented by the complexity of governance. Strong regulatory structures are required that balance automation responsibility and openness to succeed in AI-operated urban planning Jones and Kumar [20]. Political decision-makers, technologists, and urban planners should work together to create flexible structures for permanent urban development to remove these obstacles.

6 Conclusion and Future Directions

Including artificial intelligence (AI) and Quantum Computing (QC) in urban planning provides a revolutionary way of dealing with immediate issues of flexibility in climate change. The proposed environmental decision support system (EDSS) provides a strong, data-driven structure for permanent urban growth by increasing landscape-based modelling, public-based involvement and future indicative skills. However, questions including management limits, high implementation costs and technical access [22] exist. Multidisciplinary cooperation, improvement in legal frameworks, and capacity programs are necessary to remove these obstacles and guarantee the use of fair technology. Future studies should check the ways to score QC-driven simulation, handle moral dilemmas in the integration of AI and governance, and use top modern technologies such as blockchain to improve transparency in decision-making. The use of QC and AI in Urban Design will be necessary to reduce climate risk and promote sustainable urban ecosystems as these techniques develop.

6.1 Summary of Findings

This article presents a complete architecture for the Environmental Decision AID System (EDSS) that combines artificial intelligence (AI) with Quantum Computing (QC) to improve Urban Climate Flexibility. Important findings suggest that real-time data processing guarantees dynamic adaptability for new environmental hazards, while landscape-based modelling and the future chair analysis improve urban flexibility methods Smith and Brown [41]. Human resources (HR) strategies help urban planners to stop the gap between management execution and technical reforms by providing them with equipment that they effectively need to use AI and QC technologies [22]. To use QC and AI fully for permanent urban ecosystems, the results highlight the importance of multi-scoping collaboration, regulatory support and ongoing technological progress.

6.2 Implications for Urban Planning

Integration of artificial intelligence (AI) and Quantum Computing (QC) in urban planning represents a revolution in real-time data processing, predictive modelling and decision-making. Urban employers can evaluate and carry out many climate flexibility options at the same time because QC's first unheard unheeded QC guarantees the ability to solve complex adaptation problems, and guarantees intelligent and flexible urban development Brown et al. [4]. By predicting climate pairs, collecting data from the Internet of Things (IoT) sensors and allocating resources that efficiently improve AI-operated solutions and urban stability [14]. In addition, the outline of the

AI-Manual partnership regime encourages socially driven urban policy that corresponds to social requirements with environmental flexibility purposes Lee [24]. The complete capacity of QC and AI for revolutionary urban planning can be unlocked by multi-collaboration and smart political interventions, despite obstacles such as moral concerns and access to technology.

6.3 Future Research

The development of artificial intelligence (AI) and quantum calculation (QC) algorithms should be the main focus of future research to improve the calculation efficiency and the accuracy of urban planning. Urban flexibility methods will be able to dynamically adjust the environmental conditions with the development of QC-powered simulation for real-time landscape analysis Jones and Kumar [20]. To address the increasing complexity of urban ecosystems, AI frameworks are necessary for scalability to ensure even interaction with IoT networks and geophysical information systems. Smith et al. [41]. To remove obstacles related to morality, governance and digital equity, interdisciplinary collaboration between urban planners, computer researchers and legislators would be necessary Chen et al. [9]. To guarantee that AI and QC applications align with more general social and environmental goals, future research should focus on participation in management methods that consider stakeholders.

References

1. Anwar, M.R., Sakti, L.D.: Integrating artificial intelligence and environmental science for sustainable urban planning. IAIC Trans. Sustain. Digit. Innov. **5**(2), 179–191 (2024)
2. Bhave, A.G., Mishra, A., Raghuwanshi, N.S., Wada, Y.: Water resource planning under future climate and socio-economic uncertainty in the Cauvery River Basin in India. Clim. Risk Manag. **32**, 100295 (2021). https://doi.org/10.1016/j.crm.2021.100295
3. Brouwer, J., Verbeek, T., Peters, M.: Stakeholder engagement in AI-driven urban planning: a case study of Amsterdam. J. Urban Aff. **45**(2), 312–329 (2023)
4. Brown, T., Green, P., White, R.: Data integration challenges in urban climate resilience systems. J. Environ. Data Sci. **28**(3), 145–160 (2020)
5. Butler, C.J.: The disproportionate effect of global warming on the arrival dates of short-distance migratory birds in North America. Ibis **145**(3), 484–495 (2003). https://doi.org/10.1046/j.1474-919X.2003.00193.x
6. Butler, R.: This is how many animals could go extinct from climate change. Vox (2023). https://www.vox.com/down-to-earth/389843/climate-change-wildlife-extinction-study. Accessed 23 Apr 2025
7. Castán Broto, V., Westman, L.: Training for sustainable urban development: challenges and opportunities for capacity building in climate resilience. J. Environ. Planning Policy Manage. **22**(1), 69–83 (2020). https://doi.org/10.1080/1523908X.2019.1673156
8. CDKN (Climate and Development Knowledge Network).: Enhancing climate resilience through stakeholder engagement (2021). https://cdkn.org. Accessed 23 Apr 2025

9. Chen, J., Liu, X., Wang, Y.: Advances in climate modeling for urban resilience. Environ. Sci. J. **45**(2), 112–129 (2023)
10. Chen, Y., Wang, L., Li, X.: Data-driven approaches for resilient urban planning: integrating climate models and GIS. Sustain. Cities Soc. **80**, 103732 (2022). https://doi.org/10.1016/j.scs.2022.103732
11. Coining, S., Fischer, L.: AI-enhanced air quality monitoring and public health impact assessment. Environ. Res. J. **88**(4), 560–578 (2024)
12. Eini, M., Rahmati, O., Choubin, B., Talebi, A., Sadeghian, S., Bui, D.T.: Enhancing flood risk assessment in urban areas by integrating advanced machine learning techniques. Sci. Total Environ. **872**, 162015 (2024). https://doi.org/10.1016/j.scitotenv.2024.162015
13. Federal Interagency Forum on Child and Family Statistics.: America's children: Key national well-being indicators (2013). http://childstats.gov/americaschildren/index2.asp. Accessed 23 Apr 2025
14. Garcia, M., Kumar, S.: Ensuring data reliability in environmental decision support systems. Environ. Inform. Rev. **28**(4), 78–95 (2021)
15. Gómez, R., Silva, J., Torres, M.: Enhancing decision-making in urban planning: a multi-stakeholder perspective. J. Urban Policy **17**(2), 112–129 (2023). https://doi.org/10.1080/1523908X.2023.2184920
16. Gonzalez, R., Patel, S., Kumar, D.: Smart urban planning: integrating AI-driven solutions for sustainable development. Urban Comput. J. **8**(2), 105–120 (2022)
17. Hřebíček, J., Pillmann, W., Harnisch, W., Wohlgemuth, V.: Environmental Software Systems: Computer Science for Environmental Protection. Springer (2018)
18. IPCC.: Climate Change 2021: The Physical Science Basis. Cambridge University Press (2021)
19. Janssen, B., van der Meer, H.: The role of AI in green infrastructure planning: insights from Amsterdam's Climate Resilient Urban Planning Initiative (CRUPI). Smart Cities Rev. **12**(3), 198–217 (2024)
20. Jones, A., Kumar, S.: Smart cities and climate resilience: a data-driven approach. Sustain. Urban Dev. Rev. **40**(1), 89–105 (2022)
21. Koenig, S., Fischer, L.: AI-enhanced air quality monitoring and public health impact assessment. Environ. Res. J. **88**(4), 560–578 (2024)
22. Kumar, S., Patel, D.: Climate adaptation strategies in urban planning: addressing vulnerabilities and risks. Environ. Res. Policy **29**(1), 56–72 (2024). https://doi.org/10.1016/j.envpol.2024.105896
23. Langendijk, G., Swart, R., Rosenzweig, C., et al.: Urban resilience through integrating adaptation, mitigation and sustainable development–a novel approach to operationalise climate resilient development pathways. EGUsphere, pp. 1–12 (2023). https://doi.org/10.5194/egusphere-egu24-18661
24. Lee, M.: Digital engagement platforms in participatory urban governance. Urban Plan. Rev. **28**(2), 221–238 (2024)
25. Li, Y.: Resilient urban design concept in the context of climate change. Adv. Econ. Manag. Polit. Sci. **123**(1), 79–85 (2024). https://doi.org/10.54254/2754-1169/123/2024mur0117
26. Lim, S., Wong, T.: Quantum computing applications in predictive climate modeling for urban resilience. Clim. Inform. J. **11**(3), 312–329 (2024)
27. Michener, W.K., Jones, M.B.: Ecoinformatics: supporting ecology as a data-intensive science. Trends Ecol. Evol. **27**(2), 85–93 (2012)
28. Mehrabi, N., Morstatter, F., Saxena, N., Lerman, K., Galstyan, A.: A survey on bias and fairness in machine learning. ACM Comput. Surv. **54**(6), 1–35 (2021). https://doi.org/10.1145/3457607
29. O'Neil, C.: Weapons of Math Destruction: How Big Data Increases Inequality and Threatens Democracy. Crown Publishing Group (2016)
30. OECD.: Climate-resilient infrastructure: policy perspectives (2018). https://www.oecd.org/environment/cc/policy-perspectives-climate-resilient-infrastructure.pdf. Accessed 23 Apr 2025
31. Peng, Y., Yigitcanlar, T., He, J.: Artificial intelligence adoption in urban planning governance: a systematic review. Gov. Inf. Q. **42**(1), 101–115 (2025). https://doi.org/10.1016/j.giq.2024.101115

32. Qian, K., Zheng, Y., Zhao, M., Chen, X., Zhang, L.: AI agent as urban planner: steering stakeholder dynamics in urban planning via consensus-based multi-agent reinforcement learning (2023). arXiv:2310.16772. https://arxiv.org/abs/2310.16772
33. Rafeek, M., Hossain, M., Abid, M., Ahmed, S.: Nature-based solutions as an approach of resilient cities to face climate change: analytical framework to apply NBS in cities. Deleted J. (4), 867–881 (2024). https://doi.org/10.53935/jomw.v2024i4.553
34. Reed, M.S., Graves, A., Dandy, N., Posthumus, H., Hubacek, K., Morris, J., Prell, C., Quinn, C.H., Stringer, L.C.: A theory of participation: what makes stakeholder and public engagement in environmental management work? Restor. Ecol. **26**(S1), S7–S17 (2018). https://doi.org/10.1111/rec.12541
35. Reuters.: Comment: the AI race is on the wrong track. Here's how to fix it. Reuters (2025). https://www.reuters.com/sustainability/climate-energy/comment-ai-race-is-wrong-track-heres-how-fix-it-2025-02-24/. Accessed 23 Apr 2025
36. Schmidt, F.L., Oh, I.-S.: The crisis of confidence in research findings in psychology: is lack of replication the real problem? Or is it something else? Arch. Sci. Psychol. **4**(1), 32–37 (2016). https://doi.org/10.1037/arc0000029
37. Schneider, P., Müller, C., Weber, L.: Ethical and technical challenges in AI-driven urban resilience planning. Sustain. Cities Soc. **50**(1), 101–120 (2024)
38. Sharma, P., Patel, M., Lopez, R.: Cloud-based geospatial analysis for climate resilience. Environ. Model. Softw. **35**(2), 210–225 (2019)
39. Smale, D.A., Wernberg, T., Oliver, E.C.J., Thomsen, M.S., Harvey, B.P.: Marine heatwaves threaten global biodiversity and the provision of ecosystem services. Nat. Clim. Chang. **9**(4), 306–312 (2019). https://doi.org/10.1038/s41558-019-0412-1
40. Smith, J., Thompson, R., Evans, L.: Enhancing urban sustainability through geospatial analytics and decision support systems. Sustain. Cities Soc. **15**(1), 75–95 (2023)
41. Smith, R., Brown, T.: Integrating AI in climate risk forecasting: a managerial perspective. J. Environ. Plan. **38**(4), 215–230 (2022)
42. Taylor, J., White, L.: Usability principles for adaptive decision support systems in urban planning. Urban Inform. J. **22**(3), 178–192 (2021)
43. Time.: How AI is being used to respond to natural disasters in cities. Time Magazine (2025). https://time.com/7171445/ai-natural-disaster-cities/. Accessed 23 Apr 2025
44. United Nations Department of Economic and Social Affairs.: About the partnership platform (n.d). https://sdgs.un.org/partnerships/about. Accessed 23 Apr 2025
45. Voigt, P., von dem Bussche, A.: The EU general data protection regulation (GDPR): a practical guide. Springer International Publishing (2017). https://doi.org/10.1007/978-3-319-57959-7
46. Wang, P., Li, Z., Choudhury, R.: The role of quantum computing in environmental data analysis: a case for smart urban planning. Comput. Sustain. J. **10**(3), 55–70 (2024)
47. Williams, H.: Real-time analytics and visualization in environmental planning. J. Smart Syst. **31**(2), 56–72 (2023)
48. Wilson, D., Green, H.: Adaptive interface design in environmental analytics. Urban Anal. Smart Plan. **33**(2), 101–118 (2023)
49. Zhang, H., Liu, X., Park, M.: Climate change and urban planning: challenges and opportunities in sustainable city development. Environ. Res. Lett. **18**(7), 120–135 (2023)
50. Zhou, L., Hassan, H.C.: Role of culturally relevant instruction in enhancing social-emotional learning for marginalised students. Conhecimento Divers. **16**(44), 450–469 (2024). https://doi.org/10.18316/rcd.v16i44.12267

Energy Management in Urban Areas Using Machine Learning

Komal Mishra

Abstract Urban regions struggle with worsening energy management issues from the rate of demographic expansion and rising energy requirements and building integration of renewable power sources. Sanitation requires efficient energy management techniques because it ensures environmental sustainability and reduces both operational expenses and greenhouse gas emissions. The inability of traditional energy management systems to react to urban environment changes constitutes their primary drawback. The application of machine learning (ML) provides strong analytics which helps forecast energy requirements while optimizing grid management and controlling distributed resources alongside enhancing real-time data-driven decision-making processes. The paper investigates how machine learning techniques implement urban energy management through demand forecasting, energy efficiency optimization and load balancing and anomaly detection procedures. A diverse set of ML models including artificial neural networks, support vector machines, ensemble models as well as deep learning approaches are evaluated for their specific use cases and their effectiveness. Current urban infrastructure along with smart cities demonstrate how effective this concept can be through case study analysis. Moreover, the paper investigates obstacles relating to data protection together with difficulty understanding machine learning models and issues of merging them into current electrical systems. The paper discusses ML-developed power systems through the lens of smart grids together with edge computing and adaptive learning strategies for the future design direction. The paper delivers an extended discussion about how machine learning systems transform modern urban energy management strategies.

Keywords Energy consumption prediction · Energy efficiency · Machine learning for energy · Urban energy systems · Artificial Intelligence (AI) in energy · Energy consumption clustering

---

K. Mishra (✉)
University Institute of Computing, Chandigarh University, Mohali 140413, Punjab, India
e-mail: komal290@gmail.com

A. L. Srivastav et al. (eds.), *Climate Resilience: Impact of Quantum Computing and Artificial Intelligence on Urban Planning*, Studies in Systems, Decision and Control 637, https://doi.org/10.1007/978-3-032-06791-3_19

1 Introduction

According to United Nations population projections show an upcoming 68% increase to 68% in the urban population by 2050 which makes cities the primary locations for energy use and carbon emission production as well as environmental consequences [1]. Three-quarters of worldwide primary energy supply goes to cities with these urban areas generating more than two-thirds of global greenhouse gas emissions. The growing energy requirements put significant pressure on old power systems because they cannot automatically adjust to variable loads and merge renewable supplies and diverse energy needs for intelligent buildings and transportation systems and industrial sectors [2].

Mechanisms of energy management dependent on basic rules and historical usage figures show insufficient performance for current needs. Such methods prove incapable of adjusting to present-day conditions while they fail to achieve optimal power distribution throughout intricate city areas. A massive amount of energy system digitalization data exists because current frameworks do not effectively use the information generated from smart grids, IoT sensors and connected devices [3].

Energy efficiency needs to be improved with the implementation of adaptive intelligent predictive systems to resolve present operational challenges [4]. Machine Learning technology as part of Artificial Intelligence enhances urban energy management thanks to its data-driven approach for obtaining insights and control engines.

1.1 Role of Energy Management in Urban Sustainability

Urban sustainability requires effective energy management because it determines the conditions of environmental quality while influencing economic performance and urban dweller well-being simultaneously. The concept of urban sustainability maintains a balance between population increase and efficient use of resources and reduces carbon emissions while providing accessible clean affordable energy [5].

The framework for energy management consists of several specific elements in this case.

- Organizations should track and adjust energy consumption as it happens in real-time.
- The system should maximize its supply networks between conventional and renewable energy systems [6].
- Predictive maintenance together with fault detection helps organizations minimize their waste output.
- The system allows customers to move their power usage to less crowded periods through demand response applications [7].

Protecting the environment can be achieved through sustainable energy systems thus minimizing fossil fuel dependency while backing up United Nations Sustainable Development Goals (SDG 7 and SDG 11) for affordable clean energy and sustainable urban development [8]. Modern urban energy management enables a reduction of emissions while cutting operational expenses while simultaneously improving the quality and reliability of energy systems as well as public participation [9].

1.2 Significance of Machine Learning in Urban Energy Systems

The computational methods of Machine Learning enable processing and learning through analysis of large datasets that originate from smart urban infrastructure. ML models transform into better versions over time since they evolve through autonomous performance of energy-related operations while simultaneously learning to anticipate trends and find anomalous patterns [10]. These are the main applications of ML which assist urban energy management operations:

- Time-series analysis enables forecasting of upcoming energy demands through the system known as load forecasting [11].
- The prediction of renewable energy depends on weather patterns to estimate solar and wind energy outputs.
- The process of smart grid optimization allows real-time adjustment of energy flow in order to maintain equilibrium between supply and demand [10].
- Autonomous systems based on building energy management identify patterns for occupancy activities to lower operational consumption levels.
- Energy theft and fault detection: Identifying anomalies in consumption data.

The improvement of energy systems performance in changing urban environments depends on deep learning together with reinforcement learning and unsupervised models. These technologies directly help develop cities which are both resilient and adaptive in addition to being low-carbon [12].

2 Overview of Urban Energy Systems

Urban energy systems consist of the infrastructure, processes, and technologies employed to produce, distribute, and use energy in cities. Urban energy systems become more intricate because of the integration of distributed energy resources, increased demand, and the pressure to decarbonize. It is a requirement to know how energy is being consumed and controlled in cities in order to create sustainable and intelligent energy solutions [13].

2.1 Patterns of Urban Energy Use

Cities are the largest consumers of energy globally. Cities consume over 75% of the world's end-use energy, according to the International Energy Agency. Energy is consumed in different sectors:

(a) Residential: Comprises heating, cooling, lighting, and appliance usage.
(b) Commercial: Shopping centres, office complexes, and other service outlets.
(c) Industrial: Light industry within city limits.
(d) Transport: Private and public transport networks.

Urban energy habits are determined by the age of buildings, economic activity, age of infrastructure, population density, and climate [14]. For instance, tropical cities consume more energy for air conditioning while colder climates have high energy consumption for heat. Most of the rapidly expanding cities have more infrastructure development than the implementation of energy-saving practices [15].

Greater application of digital technologies and electric cars also changes consumption patterns, placing additional stress on the grid while at the same time enabling greater smart management by distributed generation and demand response [16].

2.2 Challenges of Urban Energy Management

Urban energy management is particularly difficult due to the following reasons:

(a) Demand Fluctuations: Urban energy demand is extremely volatile, with peak demands at some periods of the day or year.
(b) Grid Losses and Congestion: Increased transmission distances owing to congested urban grids cause energy losses and decreased reliability [17].
(c) Renewables Integration: Solar and wind bring in intermittency, and advanced control systems are needed to provide grid stability.
(d) Aging Grid: Most cities operate on decades-old systems designed for non-mass, centralized loads, but not bi-directional flow of energy from the distributed generation.
(e) Data Silos: Energy data is scattered across utilities, building owners, and municipalities, and this discourages cross-planning and optimization (Zekic-Susac et al. [18].
(f) Environmental Issues: Fossil fuel-based grids account for most urban air pollution and climate change and thus need a shift towards low-carbon grids.

These issues necessitate a shift in paradigm from reactive to proactive energy management paradigms that leverage real-time data and smart decision-making functionalities [19].

2.3 *Traditional Versus Intelligent Energy Management Systems*

Centralized power generation and inflexible distribution channels are the pillars of conventional energy systems. Such systems are inefficient, particularly in large cities with dynamic and heterogeneous energy demands [8]. They are also resistant to change, hence susceptible to blackouts, overload, or device failure.

Smart energy systems, on the other hand, leverage machine learning, IoT, and big data analytics to make the best and autonomous decisions. For instance, ML algorithms can predict demand, identify anomalies, and automate energy distribution in real-time. Smart sensors and meters allow for detailed analysis of consumption, while cloud-based platforms optimize distributed energy resources [20] (Table 1).

These systems are required to enable smart cities and smart grids, which will enable more integration of renewables, more energy efficiency, and greater user involvement [21].

3 Fundamentals of Machine Learning

Machine learning (ML) is a subfield of artificial intelligence (AI) where systems are enabled to learn from data and improve their functioning without being explicitly programmed. In the urban energy system, the ML can help with demand forecasting, supply optimization [22], anomaly detection, and overall efficiency improvement. To use ML in the system effectively, it is necessary to understand the types of learning model, algorithms, and data they require.

Table 1 Traditional versus intelligent energy management systems

Feature	Traditional energy management	Intelligent energy management (ML-based)
Control strategy	Rule-based, manual	Adaptive, data-driven
Decision-making	Static, based on historical trends	Predictive, real-time analytics
System architecture	Centralized	Decentralized and interconnected
Integration of renewables	Limited	Optimized enhanced by forecasting and control
User interaction	Minimal (utility-driven)	Interactive (consumer engagement through demand response)
Maintenance	Reactive	Predictive and automated
Scalability	Difficult in large, complex networks	Scalable by cloud and edge computing

3.1 Machine Learning Model Types

Machine learning models are of four types based on the way they learn from data [23].

(a) **Supervised Learning**

Supervised learning employs labelled data to train algorithms to make predictions on new, unseen data [24]. It is applied extensively in-Load forecasting (forecasting future energy usage) and Anomaly detection (detecting unusual energy patterns).

Examples: Support Vector Machines (SVM), Decision Trees, Neural Networks, Linear Regression. Supervised models are applied to learn from past energy consumption data and weather and occupancy trends in building energy forecasting [5].

(b) **Unsupervised Learning**

Unsupervised learning detects hidden patterns or intrinsic organizations in data without labeled definitions. In energy systems, it is often used for: Aggregation of buildings by usage pattern and Establishing usage patterns for demand-side management [25].

Examples: K-Means Clustering, Principal Component Analysis (PCA), Autoencoders.

(c) **Reinforcement Learning (RL)**

RL is the process of acquiring optimal decisions via trial and error to optimize a reward signal [26]. It is extremely applicable to: Real-time energy management, HVAC control systems, Smart grid optimization.

Example: Deep Reinforcement Learning was utilized for the control of distributed energy sources in order to decrease peak demands [24].

(d) **Semi-Supervised and Self-Supervised Learning**

These methods employ a combination of labelled and unlabelled data, which is useful when there is limited labelled data. They are emerging in smart city settings where it is costly to obtain large labelled datasets [27] (Fig. 1).

3.2 Common Algorithms in Energy Applications

Several machine learning algorithms have found significant use in energy systems (Table 2).

Jui et al. [28] has reported that short-term electricity load forecasting is best performed by using LSTM models in place of standard methods because it can learn time dependencies.

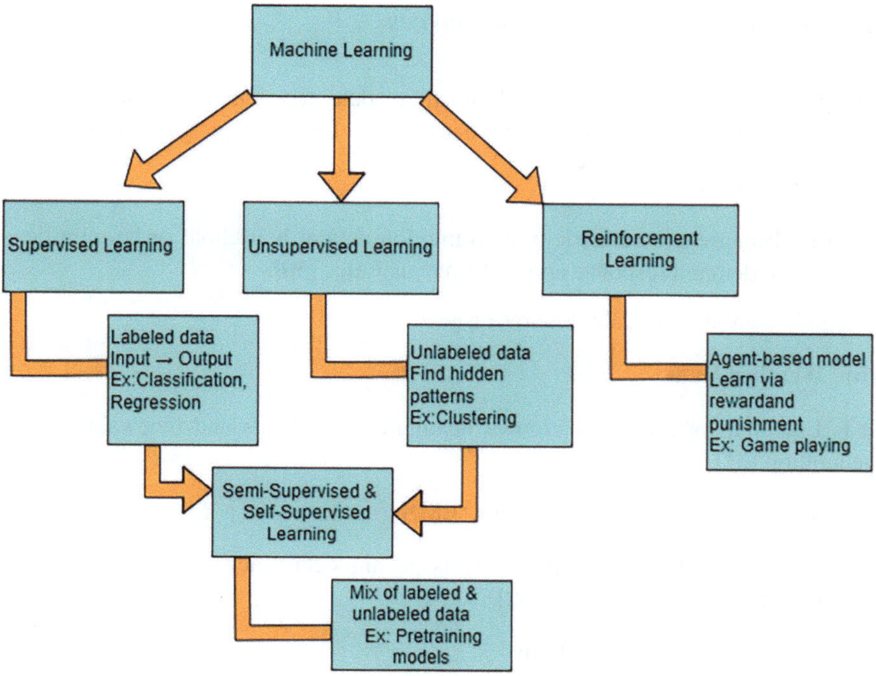

Fig. 1 Machine learning models in energy management

Table 2 Algorithms in energy applications

Algorithm	Description	Application in energy
Linear Regression	Describes the relationship between one or more independent and a dependent variable	Forecasting load and price
Decision Trees/Random Forests	Tree models which divide data into branches for regression or classification	Fault detection, energy consumption prediction
Support Vector Machines (SVM)	Classifies the data by finding the best separating hyperplane	Anomaly detection, energy theft
Artificial Neural Networks (ANN)	Simulate the neurons' connections in the brain to learn intricate patterns	Fault classification, nonlinear energy forecasting
K-Means Clustering	Grouping data into groups with the same pattern	User segmentation, load profiling
Long Short-Term Memory (LSTM)	A type of recurrent neural network architecture for sequential data	Time-series forecasting (e.g., demand forecasting)

3.3 Data Requirements and Sources

The performance of ML models will mostly be data-dependent, the volume of data, diversity, and quality [29] Significant data sources for urban energy systems are.

(a) **Smart Meters**

It Offer disaggregated electricity consumption data at household or building level and enable demand response and real-time analytics [30].

Example dataset: UK-Dale, Smart* Project.

(b) **Weather Data**

It is Essential for the calculation of heating/cooling loads and renewable energy generation [31].

(c) **Building Management Systems (BMS)**

It Installs temperature, lighting, HVAC usage, and occupancy sensors and also help in optimize internal energy use [32].

(d) **Renewable Energy Systems**

It maintains Solar panel performance, wind turbine statistics. Also utilized in integration analysis and production forecasting [33].

(e) **Energy Market Data**

It supports Tariffs, prices, demand-response signals. And also optimizing operation expenses and peak shaving strategies [34].

(f) **IoT Devices and Edge Sensors**

It Allows real-time monitoring and control. It is Used in predictive maintenance and decentralized decision-making [35].

ML models are susceptible to noisy, biased, or incomplete data. Methods such as feature selection, normalization, and data cleaning are usually needed to enhance performance [36].

4 Applications of Machine Learning in Urban Energy Management

Machine Learning (ML) is increasingly important in helping to meet the increasing complexity of urban energy management. ML enables predictive analytics, automation, and real-time optimization of many elements of urban energy systems [37]. These are the main applications of ML here.

4.1 Energy Demand Forecasting

Accurate forecasting of energy demand is essential for supply–demand balancing, cost minimization, and grid stability. Current methods rely on historical mean or statistical approaches [38] which are unsuitable in dynamic cityscapes. Machine learning methods such as Artificial Neural Networks (ANNs), Support Vector Machines (SVMs), and Long Short-Term Memory (LSTM) networks improve the accuracy of forecasting by capturing nonlinear relationships and temporal patterns [39].

Example: An LSTM model outperformed traditional ARIMA and feed-forward neural networks to predict short-term electricity demand in smart cities [40].

In Tokyo, demand forecasting systems based on ML minimized uncertainty of peak load, improving energy dispatch efficiency and reducing operation costs [7].

4.2 Smart Grid Optimization

Smart grids combine renewable sources, storage, and smart meters with real-time data analysis. ML provides capabilities to such systems by enabling [41].

(a) **Predictive maintenance**: Machine learning algorithms detect the earliest signs of equipment breakdown [42].
(b) **Fault detection and classification**: Fault classification algorithms quickly identify and isolate faults.
(c) **Voltage and frequency control**: Reinforcement Learning (RL) models learn optimal real-time control parameters [43].

Singh et al. [44] proved that the integration of deep learning in smart grid operation facilitates improved grid state estimation and resilience under different load conditions. ML-based optimization has been found to reduce operational losses by 10–15% in urban smart grids.

4.3 Load Balancing and Management

Load management ensures that the supply of electricity is always proportionate to demand. Machine learning assists by user segmentation on usage patterns (via using clustering algorithms such as K-Means) [45]. It also Enables dynamic load shifting in real-time based on prices and user preferences. It's Automating demand response measures and forecasting peak load periods [46].

Example: Residential load control was automated in Singapore on the basis of real-time price signals utilizing a decision tree model, reducing peak consumption by 12%.

Methods Used: Supervised learning (for prediction), Unsupervised learning (for user segmentation), Reinforcement learning (for adaptive control) [47].

4.4 Renewable Energy Integration

The incorporation of renewable energy sources like wind and solar power into the grid is difficult since they are intermittent. ML is useful in Forecasting renewable generation based on weather data using regression and LSTM models [48]. It also Maximizing storage use to equalize generation and demand. It Enhances grid stability through smart scheduling of energy dispatch.

Example: Machine learning techniques have been utilized with weather and historical data to predict solar PV output with over 90% accuracy.

Real-World Application: In Germany, wind generation is forecasted 48 h ahead of time using ML to enhance national grid integration [49].

4.5 Energy Efficiency in Buildings

Buildings use nearly 40% of the world's energy. ML can assist Dynamically setting parameters and predicting HVAC requirements. It Identifies energy inefficiencies via anomaly detection algorithms [50]. It also assist in weather and occupancy-based lighting, heating, and cooling automation. ML Technologies Used: prediction supervised learning, unsupervised learning (Abnormal consumption identification), RL (self-driving control).

Lu et al. [51] applied a deep reinforcement learning model for the control of HVAC systems in office buildings and achieved 20% energy savings with no sacrifice of occupants' comfort level.

4.6 Privacy and Availability of Data

Machine Learning algorithms rely on massive volumes of high-quality data to test, train, and execute with. For an urban energy infrastructure, this translates to smart-meter data, data from sensors, weather data, and historical patterns of consumption data [52] But:

(a) **Insufficiency of data**: In most regions, especially emerging cities, real-time or high-resolution energy data may not be available or incomplete [53].
(b) **Fragmentation of data**: Data is spread over different utility providers, departments, and platforms.

(c) **Privacy concerns**: Use of specific energy consumption information for households is extremely privacy-intrusive and surveillance-prone [54].

Chen et al. [55] showed in an experiment that smart meter data could infer household characteristics, which was a privacy issue. Maqdid et al. [11] highlighted the requirement of safe handling of data by mechanisms within the smart grid communication to overcome the privacy threat.

4.7 Scalability and Real-Time Processing

Urban energy systems are time-variant and demand real-time decision-making, especially in usage scenarios such as grid load balancing, fault detection, or demand response [56].

(a) **Scalability issues**: ML models that work well on small datasets or localized systems may not be scalable to city-wide systems [57].
(b) **Latency**: Most ML algorithms are computationally intensive and may fail to cope with real-time streaming or processing of data in high-demanding situations.
(c) **Limited infrastructure**: No distributed ML system communication and processing infrastructure exists in any of the city locations.

During a smart grid deployment in Italy, ML-based demand response systems experienced processing bottlenecks that resulted in delays in decision implementation.

4.8 Model Interpretability and Trust

The vast majority of ML algorithms, and particularly deep learning algorithms, are "black boxes" that yield very little about decision-making [58].

(a) **Interpretability problem**: City planners, utility operations, and policymakers frequently need interpretable models in order to explain decisions and maintain accountability.
(b) **Trust and adoption**: Stakeholders might not adopt trusting automated decisions they cannot audit or comprehend [59].
(c) **Regulatory compliance**: In most jurisdictions, AI decisions should be explainable in order to satisfy ethical and legal requirements.

Example: Cui et al. [60] introduced the LIME approach for improving interpretability of ML models, and this has since been applied in energy demand forecasting. It argue that interpretable models have a significantly crucial role for high-stake applications like healthcare and energy.

4.9 Integration with Legacy Systems

The urban energy infrastructures are usually comprised of installed infrastructure and equipment that are not AI/ML enabled.

(a) **Backward compatibility**: Older hardware may not be capable of supporting the communication protocols or interfaces required to talk to modern ML-based systems.
(b) **Prohibitive upgrade costs**: It may be astronomically expensive to upgrade to smart infrastructure that can accommodate ML, especially in resource-scarce cities or big ones [61].
(c) **Security issues**: Uncontrolled integration can leave legacy systems vulnerable to cyber attacks.

Retrofitting UK's already aging grid infrastructure to support smart meters and ML-based demand forecasting took a great deal of time and money [13]. It has discussed why standardization and interoperability of the system are important to integrating smart grid technologies effectively [62].

5 Future Opportunities and Trends

As urbanization remains unchecked and urban energy demands are increasing, the intersection of machine learning, smart technology, and green policy is opening up new frontiers [63]. These innovations aim to break beyond existing limitations and realize the full potential of smart energy systems.

5.1 Context-Aware and Adaptive Learning Systems

Future city energy systems will require adaptive ML models that dynamically learn and adapt contextually, e.g., real-time dynamic change in demand, occupancy, and weather [64]. Adaptive learning models update their parameters continuously and therefore are resilient to concept drift in power usage patterns.Context-aware systems leverage sensor data and environmental conditions (e.g., day of week, season, local events) to make decisions more precise [65].

Example: Smart building intelligent HVAC systems are using adaptive algorithms to regulate heating and cooling automatically based on user usage and indoor/outdoor conditions. It evaluated context-aware ML models that enhanced energy efficiency in smart homes by 15–20%.

5.2 *Edge Computing and IoT for Urban Energy*

Where IoT devices intersect edge computing, however, the latter is changing how energy systems process and respond to information. Edge computing provides processing at or near the source (e.g., sensors or smart meters), lowering latency and network traffic. It is crucial for real-time ML applications such as demand response, fault detection, and automation of control [66].

Example: ML models on edge at IoT gateways facilitated streetlight and electrical vehicle charging points' real-time control in the smart city operations of Tokyo to enhance grid reliability. Tang et al. [67] explained how the edge computing architecture supports scalable and responsive energy management systems in an urban setup.

5.3 *Reinforcement Learning (RL)*

Is increasingly being considered as a promising way to attain optimal sequential decision-making in energy systems [68]. RL agents gain the best approach through learning that involves trial by error in their environment. Its applications include microgrid energy storage management, dynamic pricing, and grid congestion management.

Elsisi et al. [69] showed that Deep Q-Networks (DQNs) improved the load shifting efficiency of the smart grid by over 25% compared to that of static scheduling. Multi-agent RL is also gaining traction for real-time control of distributed energy resources. Shen et al. [70] referred to the application of deep RL to optimize energy management systems of buildings, which provided high energy savings while not affecting comfort.

5.4 *Policy and Regulatory Issues*

The widespread application of ML in urban energy management necessitates supportive and responsive policy frameworks. Data governance must ensure ethical use, privacy, and cybersecurity. ML protocol and data format standardization will enable interoperability between platforms and smart devices. Incentives (e.g., tax credits, subsidies) can be used to promote the development of smart systems [71].

Example: European Union's Clean Energy for All Europeans package includes AI and digital technologies application for introducing energy efficiency and integrating renewables. It highlight the need for policy alignment with ML advancements to avoid ethical failures and regulatory lag.

6 Conclusion

6.1 Summary of Key Findings

Machine Learning (ML) is turning into a game-changing technology in urban energy management. Cities continue to grow and become more demanding in terms of energy, rendering traditional energy infrastructure insufficient to counter the distributed and dynamic characteristics of energy demand [4]. This chapter discussed:

(a) The importance of energy management towards promoting urban sustainability, emissions abatement, and resource efficiency.
(b) The distinct abilities of ML such as prediction, pattern recognition, and autonomous control which qualify it for managing large energy data and making smarter decisions.
(c) The broad spectrum of ML applications such as energy demand forecasting, smart grid optimization, renewable energy source integration, and building energy efficiency enhancement.
(d) Hands-on experience with case studies and real-world scenarios, demonstrating how ML can be applied to smart cities and microgrids.
(e) Data, model transparency, legacy system integration, and regulatory frameworks that need to be overcome to enable responsible and effective deployment. Chen et al. [15] emphasized that AI/ML-based smart energy systems are at the forefront of attaining long-term urban climate and resilience goals.

6.2 Recommendations to Urban Planners and Policymakers

In order to unlock the full potential of ML in urban energy systems, planners and policymakers must take proactive and strategic roles:

(a) Invest in smart infrastructure: Support data capture through smart meters, IoT sensors, and converged communication networks [72].
(b) Promote open data initiatives: Promote anonymized sharing of energy data for training and validation of ML models.
(c) Enforce interoperability standards: Ensure new ML systems are compatible with existing energy infrastructure.
(d) Facilitate capacity building: Train engineers, planners, and administrators in the fundamentals of data science and ML.

Enforce clear rules: Address ethical concerns, data privacy, bias in algorithms, and security with robust policy frameworks. Huang and Jin [73] believe that public–private partnerships and innovation ecosystems are key to scaling up intelligent energy systems.

6.3 Final Reflections on ML for Green Energy Management

Machine Learning is no panacea, but used well, it greatly enhances the flexibility, reliability, and efficiency of urban energy systems. As cities look to meet their climate ambitions and to raise the quality of urban life, ML offers a data-driven, scalable, and intelligent path forward.

ML will be embedded in future-proofed cities not only for efficient operations but for doing predictive and prescriptive analytics in energy planning [74]. There must be an ethical use of AI. Algorithm transparency, human inspection, and open data governance build confidence among citizens. With its growth, its interaction with distributed grids, electrified transport, and renewable energy will propel the change in energy landscapes. It contend that the integration of ML with renewable forecasting and adaptive grid management can cut carbon emissions in urban areas by as much as 30%.

References

1. Mashaly, I.A.: An evaluation method to include complex fenestration systems in the Façade design process. Doctoral dissertation, Queensland University of Technology (2021)
2. Abdi, N., Albaseer, A., Abdallah, M.: The role of deep learning in advancing proactive cybersecurity measures for smart grid networks: a survey. IEEE Internet Things J. **11**(9), 16398–16421 (2024)
3. Zhou, K., Yang, S.: Smart energy management. Springer Singapore (2022)
4. Olu-Ajayi, R., Alaka, H., Sulaimon, I., Balogun, H., Wusu, G., Yusuf, W., Adegoke, M.: Building energy performance prediction: a reliability analysis and evaluation of feature selection methods. Expert Syst. Appl. **225**, 120109 (2023)
5. Khajeh, H., Laaksonen, H.: Applications of probabilistic forecasting in smart grids: a review. Appl. Sci. **12**(4), 1823 (2022)
6. Pierrot, A.: Forecasting offshore wind energy: online learning, bounds and missing data (2023)
7. Goudarzi, A., Ghayoor, F., Waseem, M., Fahad, S., Traore, I.: A survey on IoT-enabled smart grids: emerging, applications, challenges, and outlook. Energies **15**(19), 6984 (2022)
8. Sierla, S., Pourakbari-Kasmaei, M., Vyatkin, V.: A taxonomy of machine learning applications for virtual power plants and home/building energy management systems. Autom. Constr. **136**, 104174 (2022)
9. Cui, Y., Yan, R., Sharma, R., Saha, T., Horrocks, N.: Realizing multifractality of smart meter data for household characteristic prediction. Int. J. Electr. Power Energy Syst. **139**, 108003 (2022)
10. Li, X., Xiong, H., Li, X., Wu, X., Zhang, X., Liu, J., Bian, J., Dou, D.: Interpretable deep learning: interpretation, interpretability, trustworthiness, and beyond. Knowl. Inf. Syst. **64**(12), 3197–3234 (2022)
11. Maqdid, G., Askar, S., Hussein, D.H.H.: Deep learning applications in fog computing environments: a review. Indones. J. Comput. Sci. **14**(1) (2025)
12. Zibaeirad, A., Koleini, F., Bi, S., Hou, T., Wang, T.: A comprehensive survey on the security of smart grid: challenges, mitigations, and future research opportunities (2024). arXiv:2407.07966
13. Yan, G., Liu, K., Liu, C., Zhang, J.: Edge intelligence for internet of vehicles: a survey. IEEE Trans. Consum. Electron. (2024)

14. Yu, L., Xu, Z., Zhang, T., Guan, X., Yue, D.: Energy-efficient personalized thermal comfort control in office buildings based on multi-agent deep reinforcement learning. Build. Environ. **223**, 109458 (2022)
15. Chen, Z., Amani, A.M., Yu, X., Jalili, M.: Control and optimisation of power grids using smart meter data: a review. Sensors **23**(4), 2118 (2023)
16. Szum, K.: IoT-based smart cities: a bibliometric analysis and literature review. Eng. Manag. Prod. Serv. **13**(2), 115–136 (2021)
17. Huang, R., He, H., Su, Q.: Towards a fossil-free urban transport system: an intelligent cross-type transferable energy management framework based on deep transfer reinforcement learning. Appl. Energy **363**, 123080 (2024)
18. Zekić-Sušac, M., Mitrović, S., Has, A.: Machine learning based system for managing energy efficiency of public sector as an approach towards smart cities. Int. J. Inf. Manage. **58**, 102074 (2021)
19. Akram, A., Abbas, S., Khan, M., Athar, A., Ghazal, T., Al Hamadi, H.: Smart energy management system using machine learning. Comput. Mater. Contin. **78**(1) (2024). Li, M., Mour, N., Smith, L.: Machine learning based on reinforcement learning for smart grids: predictive analytics in renewable energy management. Sustain. Cities Soc. **109**, 105510 (2024)
20. Sen, S., Yadeo, D., Kumar, P., Kumar, M.: Machine learning and predictive control-based energy management system for smart buildings. In: Artificial Intelligence and Machine Learning in Smart City Planning, pp. 199–220. Elsevier (2023)
21. Giglio, E., Luzzani, G., Terranova, V., Trivigno, G., Niccolai, A., Grimaccia, F.: An efficient artificial intelligence energy management system for urban building integrating photovoltaic and storage. IEEE Access **11**, 18673–18688 (2023)
22. Syamala, M., Komala, C.R., Pramila, P.V., Dash, S., Meenakshi, S., Boopathi, S.: Machine learning-integrated IoT-based smart home energy management system. In: Handbook of Research on Deep Learning Techniques for Cloud-Based Industrial IoT, pp. 219–235. IGI Global (2023)
23. So, D., Oh, J., Jeon, I., Moon, J., Lee, M., Rho, S.: BiGTA-Net: a hybrid deep learning-based electrical energy forecasting model for building energy management systems. Systems **11**(9), 456 (2023)
24. Huang, R., He, H., Zhao, X., Wang, Y., Li, M.: Battery health-aware and naturalistic data-driven energy management for hybrid electric bus based on TD3 deep reinforcement learning algorithm. Appl. Energy **321**, 119353 (2022)
25. Moghimi, S.M., Gulliver, T.A., Thirumai Chelvan, I.: Energy management in modern buildings based on demand prediction and machine learning—a review. Energies **17**(3), 555 (2024)
26. Lee, S., Seon, J., Hwang, B., Kim, S., Sun, Y., Kim, J.: Recent trends and issues of energy management systems using machine learning. Energies **17**(3), 624 (2024)
27. Mishra, K., Kaur, K.: A comprehensive analysis of deep learning techniques for wheat yellow rust disease detection and classification. In: 2025 3rd International Conference on Intelligent Data Communication Technologies and Internet of Things (IDCIoT), pp. 1279–1284. IEEE (2025)
28. Jui, J.J., Ahmad, M.A., Molla, M.I., Rashid, M.I.M.: Optimal energy management strategies for hybrid electric vehicles: a recent survey of machine learning approaches. J. Eng. Res. **12**(3), 454–467 (2024)
29. Villano, F., Mauro, G.M., Pedace, A.: A review on machine/deep learning techniques applied to building energy simulation, optimization and management. Thermo **4**(1), 100–139 (2024)
30. Zhou, X., Du, H., Xue, S., Ma, Z.: Recent advances in data mining and machine learning for enhanced building energy management. Energy, 132636 (2024)
31. Udayakumar, R., Mahesh, B., Sathiyakala, R., Thandapani, K., Choubey, A., Khurramov, A., Alzubaidi, L.H., Sravanthi, J.: An integrated deep learning and edge computing framework for intelligent energy management in IoT-based smart cities. In 2023 International Conference for Technological Engineering and its Applications in Sustainable Development (ICTEASD), pp. 32–38. IEEE (2023)

32. Kushwaha, P., Rana, A., Hassan, F., Hada, S.S., Bhardwaj, G., Bhutani, V.: Energy prediction in urban areas using machine learning and deep learning. In: 2023 10th IEEE Uttar Pradesh Section International Conference on Electrical, Electronics and Computer Engineering (UPCON), vol. 10, pp. 190–195. IEEE (2023)
33. Mostafa, N., Ramadan, H.S.M., Elfarouk, O.: Renewable energy management in smart grids by using big data analytics and machine learning. Mach. Learn. Appl. **9**, 100363 (2022)
34. Yu, L., Qin, S., Zhang, M., Shen, C., Jiang, T., Guan, X.: A review of deep reinforcement learning for smart building energy management. IEEE Internet Things J. **8**(15), 12046–12063 (2021)
35. Deepica, S., Kalavathi, S., Angelin Blessy, J., Maria Manuel Vianny, D.: Machine learning based approach for energy management in the smart city revolution. Hybrid Intell. Approaches Smart Energy Pract. Appl. 161–179 (2022)
36. Reza, S.A., Hasan, M.S., Amjad, M.H.H., Islam, M.S., Rabbi, M.M.K., Hossain, A., Shovon, M.S.S., Jakir, T.: Predicting energy consumption patterns with advanced machine learning techniques for sustainable urban development. J. Comput. Sci. Technol. Stud. **7**(1), 265–282 (2025)
37. Biswas, C., Chakraborti, A., Majumder, S.: Recent advancements in artificial intelligence and machine learning in sustainable energy management. In: Sustainable Energy Solutions with Artificial Intelligence, Blockchain Technology, and Internet of Things, pp. 35–46. CRC Press (2023)
38. Hossain, S., Hasanuzzaman, M., Hossain, M., Amjad, M.H.H., Shovon, M.S.S., Hossain, M.S., Rahman, M.K.: Forecasting energy consumption trends with machine learning models for improved accuracy and resource management in the USA. J. Bus. Manag. Stud. **7**(1), 200–217 (2025)
39. Abbasabadi, N., Ashayeri, M.: Machine learning in urban building energy modeling. Artif. Intell. Performance-Driven Des. Theor. Methods Tools 31–55 (2024)
40. Ali, U., Bano, S., Shamsi, M.H., Sood, D., Hoare, C., Zuo, W., Hewitt, N., O'Donnell, J.: Urban building energy performance prediction and retrofit analysis using data-driven machine learning approach. Energy Build. **303**, 113768 (2024)
41. Nik, V.M., Hosseini, M.: CIRLEM: a synergic integration of collective intelligence and reinforcement learning in energy management for enhanced climate resilience and lightweight computation. Appl. Energy **350**, 121785 (2023)
42. Abdufattokhov, S., Ibragimova, K., Gulyamova, D.: The applicability of machine learning algorithms in predictive modeling for sustainable energy management. In International Conference on Forthcoming Networks and Sustainability in the IoT Era, pp. 379–391. Springer International Publishing, Cham (2021)
43. Akhtar, S., Sujod, M.Z.B., Rizvi, S.S.H.: An intelligent data-driven approach for electrical energy load management using machine learning algorithms. Energies **15**(15), 5742 (2022)
44. R. Singh, A., Kumar, R.S., Bajaj, M., Khadse, C.B., Zaitsev, I.: Machine learning-based energy management and power forecasting in grid-connected microgrids with multiple distributed energy sources. Sci. Rep. **14**(1), 19207 (2024)
45. Quiñones, J.J., Pineda, L.R., Ostanek, J., Castillo, L.: Towards smart energy management for community microgrids: leveraging deep learning in probabilistic forecasting of renewable energy sources. Energy Convers. Manage. **293**, 117440 (2023)
46. Mishra, P., Singh, G.: Energy management systems in sustainable smart cities based on the internet of energy: a technical review. Energies **16**(19), 6903 (2023)
47. Zhang, Y., Wang, Z., Tian, Y., Wang, Z., Kang, M., Xie, F., Wen, G.: Pre-optimization-assisted deep reinforcement learning-based energy management strategy for a series–parallel hybrid electric truck. Energy **302**, 131628 (2024)
48. Hou, Z., Guo, J., Xing, J., Guo, C., Zhang, Y.: Machine learning and whale optimization algorithm based design of energy management strategy for plug-in hybrid electric vehicle. IET Intel. Transport Syst. **15**(8), 1076–1091 (2021)
49. Ajagekar, A., Decardi-Nelson, B., You, F.: Energy management for demand response in networked greenhouses with multi-agent deep reinforcement learning. Appl. Energy **355**, 122349 (2024)

50. Mei, P., Karimi, H.R., Xie, H., Chen, F., Huang, C., Yang, S.: A deep reinforcement learning approach to energy management control with connected information for hybrid electric vehicles. Eng. Appl. Artif. Intell. **123**, 106239 (2023)
51. Lü, X., Qian, S., Zhai, X., Wang, P., Wu, T.: Adaptive energy management strategy for FCHEV based on improved proximal policy optimization in deep reinforcement learning algorithm. Energy Convers. Manage. **321**, 118977 (2024)
52. Alijoyo, F.A.: AI-powered deep learning for sustainable industry 4.0 and internet of things: enhancing energy management in smart buildings. Alex. Eng. J. **104**, 409–422 (2024)
53. Rafiei, M., Boudjadar, J., Griffiths, M.P., Khooban, M.H.: Deep learning-based energy management of an all-electric city bus with wireless power transfer. IEEE Access **9**, 43981–43990 (2021)
54. Tang, X., Zhang, J., Pi, D., Lin, X., Grzesiak, L.M., Hu, X.: Battery health-aware and deep reinforcement learning-based energy management for naturalistic data-driven driving scenarios. IEEE Trans. Transp. Electrification **8**(1), 948–964 (2021)
55. Chen, L., He, H., Jing, R., Xie, M., Ye, K.: Energy management in integrated energy system with electric vehicles as mobile energy storage: an approach using bi-level deep reinforcement learning. Energy **307**, 132757 (2024)
56. Ning, M.A.: Artificial intelligence-driven decision support systems for sustainable energy management in smart cities. Int. J. Adv. Comput. Sci. Appl. **15**(9) (2024)
57. Sotoudeh, S.M., HomChaudhuri, B.: A deep-learning-based approach to eco-driving-based energy management of hybrid electric vehicles. IEEE Trans. Transp. Electrification **9**(3), 3742–3752 (2023)
58. Almughram, O., Abdullah ben Slama, S., Zafar, B.A.: A reinforcement learning approach for integrating an intelligent home energy management system with a vehicle-to-home unit. Appl. Sci. **13**(9), 5539 (2023)
59. Pujari, R., Ganapa, J.R., Makanyadevi, K., Prasad, P.V.H., Rusia, S., Nagaraju, S.: Artificial intelligence in smart grids enhancing energy management and optimization through machine learning. Nanotechnol. Perceptions **20**, S13 (2024)
60. Cui, N., Cui, W., Shi, Y.: Deep reinforcement learning based PHEV energy management with co-recognition for traffic condition and driving style. IEEE Trans. Intell. Veh. **8**(4), 3026–3039 (2023)
61. Huang, X., Zhang, D., Zhang, X.: Energy management of intelligent building based on deep reinforced learning. Alex. Eng. J. **60**(1), 1509–1517 (2021)
62. Fu, Z., Wang, H., Tao, F., Ji, B., Dong, Y., Song, S.: Energy management strategy for fuel cell/battery/ultracapacitor hybrid electric vehicles using deep reinforcement learning with action trimming. IEEE Trans. Veh. Technol. **71**(7), 7171–7185 (2022)
63. Wang, Y., Wu, J., He, H., Wei, Z., Sun, F.: Data-driven energy management for electric vehicles using offline reinforcement learning. Nat. Commun. **16**(1), 2835 (2025)
64. Quan, S.J.: Comparing hyperparameter tuning methods in machine learning based urban building energy modeling: a study in Chicago. Energy Build. **317**, 114353 (2024)
65. Pandiyan, P., Saravanan, S., Usha, K., Kannadasan, R., Alsharif, M.H., Kim, M.K.: Technological advancements toward smart energy management in smart cities. Energy Rep. **10**, 648–677 (2023)
66. Li, G., Or, S.W.: Multi-agent deep reinforcement learning-based multi-time scale energy management of urban rail traction networks with distributed photovoltaic–regenerative braking hybrid energy storage systems. J. Clean. Prod. **466**, 142842 (2024)
67. Tang, X., Chen, J., Yang, K., Toyoda, M., Liu, T., Hu, X.: Visual detection and deep reinforcement learning-based car following and energy management for hybrid electric vehicles. IEEE Trans. Transp. Electrification **8**(2), 2501–2515 (2022)
68. Chen, G., Lu, S., Zhou, S., Tian, Z., Kim, M.K., Liu, J., Liu, X.: A systematic review of building energy consumption prediction: from perspectives of load classification, data-driven frameworks, and future directions. Appl. Sci. **15**(6), 3086 (2025)
69. Elsisi, M., Tran, M.Q., Mahmoud, K., Lehtonen, M., Darwish, M.M.: Deep learning-based industry 4.0 and internet of things towards effective energy management for smart buildings. Sensors **21**(4), 1038 (2021)

70. Shen, Y., Zhou, J., Zhang, J., Yi, F., Wang, G., Pan, C., Guo, W., Shu, X.: Research on energy management of hydrogen fuel cell bus based on deep reinforcement learning considering velocity control. Sustainability **15**(16), 12488 (2023)
71. Xiong, S., Zhang, Y., Wu, C., Chen, Z., Peng, J., Zhang, M.: Energy management strategy of intelligent plug-in split hybrid electric vehicle based on deep reinforcement learning with optimized path planning algorithm. Proceed. Inst. Mech. Eng. Part D J. Automob. Eng. **235**(14), 3287–3298 (2021)
72. Kabir, M.R., Halder, D., Ray, S.: Digital twins for iot-driven energy systems: a survey. IEEE Access (2024)
73. Huang, Z., Jin, G.: Navigating urban day-ahead energy management considering climate change toward using IoT enabled machine learning technique: toward future sustainable urban. Sustain. Cities Soc. **101**, 105162 (2024)
74. Hariharan, R., Nandanan, P., Amudhan, S., Soundararajan, G.: 4 Integration of AI and Blockchain with Digital Twins for Smart Grid. AI and Blockchain in Smart Grids: Fundamentals, Methods, and Applications, p. 76 (2025)
75. Forootan, M.M., Larki, I., Zahedi, R., Ahmadi, A.: Machine learning and deep learning in energy systems: a review. Sustainability **14**(8), 4832 (2022)
76. Ghori, K.M., Imran, M., Nawaz, A., Abbasi, R.A., Ullah, A., Szathmary, L.: Performance analysis of machine learning classifiers for non-technical loss detection. J. Ambient Intell. Humanized Comput. 1–16 (2023). Despotovic, M., Voyant, C., Garcia-Gutierrez, L., Almorox, J., Notton, G.: Solar irradiance time series forecasting using auto-regressive and extreme learning methods: influence of transfer learning and clustering. Appl. Energy **365**, 123215 (2024)
77. Olu-Ajayi, R., Alaka, H., Owolabi, H., Akanbi, L., Ganiyu, S.: Data-driven tools for building energy consumption prediction: a review. Energies **16**(6), 2574 (2023)
78. Sierla, S., Ihasalo, H., Vyatkin, V.: A review of reinforcement learning applications to control of heating, ventilation and air conditioning systems. Energies **15**(10), 3526 (2022)
79. Li, N.K., Ariannejad, M.M., Ding, T.J., Chao, K.C.: Smart buildings and grid features in city energy system. In: International Conference on Future Smart Cities, pp. 25–44. Springer Nature Switzerland, Cham (2022)
80. Zhao, X., Li, B.: The countermeasures of urban energy risk control oriented to machine learning and data fusion. Energy Rep. **8**, 2547–2557 (2022)
81. Chen, X., Wang, X., Zhao, W., Wang, C., Cheng, S., Luan, Z.: Hierarchical deep reinforcement learning based multi-agent game control for energy consumption and traffic efficiency improving of autonomous vehicles. Energy **323**, 135669 (2025)

MIX
Papier aus verantwortungsvollen Quellen
Paper from responsible sources
FSC® C105338

If you have any concerns about our products,
you can contact us on
ProductSafety@springernature.com

In case Publisher is established outside the EU,
the EU authorized representative is:
Springer Nature Customer Service Center GmbH
Europaplatz 3, 69115 Heidelberg, Germany

Printed by Libri Plureos GmbH
in Hamburg, Germany